A Lifecycle Approach
to Knowledge Excellence in
the Biopharmaceutical Industry

BIOTECHNOLOGY AND BIOPROCESSING SERIES

Series Editor
Anurag Rathore

A Lifecycle Approach to Knowledge Excellence in the Biopharmaceutical Industry

Edited by
Nuala Calnan
Martin J. Lipa
Paige E. Kane
Jose C. Menezes

CRC Press
Taylor & Francis Group
Boca Raton London New York

CRC Press is an imprint of the
Taylor & Francis Group, an **informa** business

Contents

Section III Practices, Pillars and Enablers: Foundations for Successful KM

Section IV Practices and Case Studies in Enabling Knowledge Flow

Editors

Nuala Calnan, PhD, has over 20 years' experience in the pharmaceutical industry and is currently an adjunct research fellow with the Pharmaceutical Regulatory Science Team at Dublin Institute of Technology, Ireland, where she leads a number of patient-focused regulatory science research projects at masters and PhD level.

Calnan's focus is on the integration of *Knowledge Excellence, Operational Excellence, and Cultural Excellence* in delivering enhanced quality outcomes for the patient and has led a recent Irish Industry research study in this field examining the *Product Recall and Quality Defect* data at the Irish medicines regulator, Health Products Regulatory Authority. She is currently a member of the St. Gallen University, St. Gallen, Switzerland—led team who were awarded a one-year research grant by the Food and Drug Administration examining the role of quality metrics in determining risk-based inspection planning. She also works closely with industry in the areas of quality excellence and metrics, data integrity, and quality culture development.

Calnan co-leads the ISPE Quality Culture Team and the ISPE/PQLI Task Team on knowledge management.

Martin J. Lipa is an executive director in the Merck Manufacturing Division (MMD) and a leader of the MMD Knowledge Management Center of Excellence. For the last 8 years, he has worked to create knowledge management strategy for MMD and other functions across the Merck & Co., Inc. enterprise and to build capabilities and competency in knowledge management as an enabler of the company's strategy.

Lipa has over 20 years' experience in the biopharmaceutical industry. Before focusing on knowledge management, he started his career as a technical operations engineer and then progressed through roles in shop floor automation, computer systems validation, new GxP facility start-up, project management, and clinical supplies. He is also a certified Lean Six Sigma Black Belt and specializes in change management techniques.

Lipa is a recognized industry knowledge management thought leader and has helped organize multiple knowledge management conferences. He has also published industry-specific works and features as a regular speaker sharing his experience and learnings.

Paige E. Kane, CPIP, is a director in the Merck Manufacturing Division Knowledge Management Center of Excellence. In addition, she is a regulatory science researcher at Dublin Institute of Technology in Dublin, Ireland, where she is pursuing a PhD focusing in the realization of the ICH Q10 enabler—knowledge management.

For the last 10 years, she has been involved in developing and implementing knowledge management strategies for Wyeth and Pfizer Global Supply with a strong focus on people, collaboration, and processes. Kane has over 25 years' experience working in a regulated environment (including Genetics Institute, Wyeth, Pfizer, Monsanto, and the U.S. Government). She has been responsible for developing and implementing quality systems for GLP, GCP, and GMP, as well as computer system validation programs. In addition, she provided leadership and subject matter expertise for multiple biotechnology start-ups in the United States and Europe.

Kane is the co-chair of the ISPE/PQLI Task Team on Knowledge Management and serves locally as the co-chair of the Boston ISPE Student Development Committee.

Jose C. Menezes is the founder and director of one of the earliest pharmaceutical engineering programs in Europe and a pioneer on the use of PAT and QbD tools in bioprocessing. He is the recipient of the Presidential Awards for Excellence in university–industry collaborations. Dr. Menezes has been a professor at the University of Lisbon, Lisbon, Portugal, since 2005. He has published extensively on the subject of manufacturing sciences and technologies applied to pharmaceutical and biotechnology products. He holds BSc and MSc degrees in chemical engineering and a PhD in bioengineering from University of Lisbon. He worked for Ciba-Geigy and the nuclear industry in Switzerland, before starting an academic career. During his 20 years at ULisbon, he spent multiple sabbatical periods in industry at different pharmaceutical companies in Europe, where he built, trained, or managed several PAT and QbD teams. He is an active member of different societies and a senior member of AIChE, ISPE, PDA, and IFPAC among other organizations. Dr. Menezes co-founded 4Tune Engineering Ltd., Lisbon, Portugal, in 2004—an award-winning ISO 9001:2008 engineering services company—for developing and implementing worldwide state-of-the-art manufacturing sciences and technologies solutions to deliver quality risk management and excellence in corporate knowledge management over lifecycle.

Contributors

Barbara Allen
Global Quality Systems
Eli Lilly and Company
Kinsale, Cork, Ireland

Amr Arisha
Department of International
  Business
Dublin Institute of Technology
Dublin, Ireland

Jon Boyle
Chief Operations Officer
Knowledge Engagement, LLC
  (formerly NASA)
Burke, Virginia

Sandra Bush
Operational Excellence
Genentech (formerly with Amgen)
Hillsboro, Oregon

Vincent Capodanno
Merck Manufacturing Division –
  Global Science, Technology &
  Commercialization
Merck & Co., Inc.
Rahway, New Jersey

Anando Chowdhury
Office of the CEO and Leader,
  Strategic Initiatives
Merck & Co., Inc.
Kenilworth, New Jersey

Paul Czech
Research Center for Data-driven
  Business and Big Data Analytics
Know-Center GmbH
Graz, Austria

Dave deBronkart
Patient Advocate
e-Patient Dave, LLC
Nashua, New Hampshire

Adam Duckworth
MMD Knowledge Management
  Center of Excellence
Merck & Co., Inc.
Kenilworth Whitehouse Station,
  New Jersey

Angela Fessl
Research Center for Data-driven
  Business and Big Data Analytics
Know-Center GmbH
Graz, Austria

Francisca F. Gouveia
Technical Operations
4Tune Engineering GmbH
Lisbon, Portugal

Anne Greene
School of Chemical and
  Pharmaceutical Sciences
Dublin Institute of Technology
Dublin, Ireland

Kayhan Guceli
Pharmaceutical Quality Technical
  Systems
Genentech, Inc.
South San Francisco, California

Yukio Hiyama
Visiting (retired) Scientist
National Institute of Health Sciences
  (NIHS)
Tokyo, Japan

Ed Hoffman
Founder & CEO
Knowledge Engagement, LLC
  (formerly NASA)
Crofton, Maryland

Joseph Horvath
Global IT Systems–Quality
Takeda Pharmaceuticals
  International
Cambridge, Massachusetts

Cindy Hubert
Knowledge Management Advisory
  Services
APQC
Houston, Texas

Alton Johnson
Global Technical Services
Pfizer Global Supply
New York, New York

Beth Junker
Principal Consultant & Owner
BioProcess Advantage LLC
  (formerly Merck & Co., Inc.)
Westfield, New Jersey

Hansjürg Leuenberger
Swiss Agency for Therapeutic
  Products
Swissmedic
Bern, Switzerland

Phil Levett
Pharmaceutical Sciences
Pfizer
Sandwich, Kent, UK

Stefanie N. Lindstaedt
Research Center for Data-driven
  Business and Big Data Analytics
Graz University of Technology
Graz, Austria

Thomas Loughlin
Merck Manufacturing
  Division – Global Science,
  Technology & Commercialization
Merck & Co., Inc.
Rahway, New Jersey

David Lowndes
Technical Operations
Shire Pharmaceuticals
Wayne, Pennsylvania

Eda Ross Montgomery
Technical Operations
Shire Pharmaceuticals
Wayne, Pennsylvania

Ronan Murphy
Site Engineering
MSD
Swords, Dublin, Ireland

Matthew Neal
Product Management
LIQUENT, a PAREXEL
  (formerly Amgen)
Los Angeles, California

Bill Paulson
Editor-in-Chief
International Pharmaceutical
  Quality
Bethesda, Maryland

Christelle Pradines
Global Quality System & Processes
F. Hoffmann-La Roche AG
Basel, Switzerland

Mohamed A. F. Ragab
3S Group, College of Business
Dublin Institute of Technology
Dublin, Ireland

Doug Redden, DBA
IT Professional (on Sabbatical)
Allentown, Pennsylvania

David Reifsnyder
Global Biological Manufacturing
and Sciences
Genentech, Inc.
South San Francisco, California

Michael Renaudin
Swiss Agency for Therapeutic
Products
Swissmedic
Bern, Switzerland

Gabriele Ricci
Technical Operations
Shire Pharmaceuticals
Lexington, Massachusetts

Joseph Schaller
Merck Manufacturing Division
Chief of Staff
Merck & Co., Inc.
Whitehouse Station, New Jersey

Jodi Schuttig
MMD Knowledge Management
Center of Excellence
Merck & Co., Inc.
Whitehouse Station, New Jersey

Catherine Shen
Data Strategy & Acquisition
Bristol-Myers Squibb
West Windsor, New Jersey

Siân Slade
Worldwide Medical Contact Content
Insights
Bristol-Myers Squibb
Mulgrave, Victoria, Australia

Christopher Smalley
Pharmaceutical and Compounding
Pharmacy Consultant
Smalley Compliance, LLC
West Chester, Pennsylvania

Marco Strohmeier
Zero Deviation Team
Roche Pharma Biotech
Penzberg, Germany

Mani Sundararajan
Technical Operations
Shire Pharmaceuticals
Wayne, Pennsylvania

Michael P. Thien
Merck Manufacturing
Division – Global
Science, Technology &
Commercialization
Merck & Co., Inc.
Whitehouse Station, New Jersey

Lauren Trees
Knowledge Management Research
Services
APQC
Houston, Texas

Renee Vogt
MMD Knowledge Management
    Center of Excellence
Merck & Co., Inc.
West Point, Pennsylvania

Kate Waters
Pharmaceutical Quality Technical
    Systems
Genentech, Inc.
South San Francisco, California

P.K. Yegneswaran
Global Technical Operations, Merck
    Manufacturing Division
Merck & Co., Inc.
West Point, Pennsylvania

Introduction

This book addresses the rapidly emerging field of knowledge management within the biopharmaceutical and other life science industries. In particular, its central theme explores the role that knowledge management (KM) can play in ensuring the timely and reliable delivery of safe and effective products to patients around the globe. A key driver for the editors in the development of this book was to provide practical examples of how organizations are now turning their attention to the effective use of their knowledge assets as a path toward business excellence.

The origins of this project grew from a scientific symposium that took place in Dublin, Ireland, in March 2015 entitled *Knowledge Management: From Discovery to Patient* hosted by Regulatory Science Dublin Ireland. The KM Dublin 2015 symposium was the first of its kind to bring together international regulators, life science industry practitioners, academics, and KM thought leaders to discuss and explore the integration of knowledge management and risk management in the development, manufacture, surveillance, and regulation of biopharmaceutical and medical device-related health products. Many of the contributions shared within these pages came from the passionate and knowledgeable speakers at the Dublin symposium from within the industry and beyond. The keynote speakers Dr. Ed Hoffman, Chief Knowledge Officer with National Aeronautics and Space Administration (NASA), Crofton, Maryland; Dr. Michael Thien, Head of New Product Commercialization at Merck & Co., Whitehouse Station, New Jersey; and Cindy Hubert, Knowledge Management Thought Leader at the American Productivity and Quality Center, Houston, Texas, have each provided valuable and thought-provoking chapters, as have several other members of the biopharmaceutical industry and international regulatory community.

In addition, as knowledge practitioners, teachers, and researchers in this fast-changing field, our editorial stance has leaned strongly toward providing a practical handbook of case studies and good practice examples currently underway within organizations that have commenced their knowledge excellence journey. It is appropriate here to note this material applies to the greater biopharmaceutical industry—that is—the challenged faced in traditional pharmaceuticals are not unique from those in biologics, and it is well known many companies are evolving into biopharmaceuticals companies. While references within individual contributions may be to pharmaceuticals or to biopharmaceuticals, these are effectively interchangable for the context of this publication.

To this end, we have developed a holistic *House of Knowledge Excellence Framework* that depicts the foundations for successful KM by outlining the relationships between knowledge *enablers*, *pillars*, *practices*, and the strategic

EXTERNAL TRENDS & DRIVERS

FIGURE 1
The *House of Knowledge Excellence* Framework.

objectives of the business. The four pillars—people, process, technology, and governance—provide the strength of the framework. Martin J. Lipa and Paige E. Kane describe this framework in greater detail in Section III and assert that the power of this framework lies not only in explaining the function and role of each element of the "house," but in the top-to-bottom integration that clearly links the KM program to the overall business strategy.

The framework also provides an opportunity to define what we mean by *knowledge excellence* and how it exceeds the mere management of knowledge. Knowledge excellence is not simply the application of a series of knowledge solutions or the provision of sets of tools but rather about enabling and sustaining knowledge-focused business capabilities. The essence of the *House of Knowledge Excellence* Framework is about enabling knowledge to flow in order to achieve the desired business outcomes. This requires a deep understanding about "how" work gets done on a day-to-day basis and how best to influence the behaviors of the employees or knowledge workers within the organization. Employees must be encouraged and enabled to think and act differently in how they seek and share knowledge. They must also recognize the value and importance of achieving knowledge flow rather than knowledge hoarding. Indeed, many of the case studies and insights shared in the chapters that follow describe how organizations have addressed a specific business challenge or strategic objective by identifying and overcoming barriers to knowledge flow.

In terms of structure, we have compiled this handbook in four sections to facilitate the reader using the book as a reference resource at different points along their own knowledge excellence journey. There are materials included to support the establishment of cogent, sponsored KM strategies and the development of holistic KM programs rounded off by an extensive array of KM practices and case studies in Section IV.

Section I, *Making the Case for Knowledge Excellence in the Biopharmaceutical Industry*, opens with a key contribution from the technology and commercialization industry leaders at Merck who share their perspectives on why KM is a critical business enabler for them. They discuss how their focus on managing the flow of knowledge across the product lifecycle helps to deliver agility, speed, reliability, and competitiveness from product realization, through manufacturing excellence to supply chain execution. Then, to set the scene for the rest of the material to follow, the Austrian-based Know-Center outlines several important and fundamental knowledge management concepts and theories. This is followed by a unique Japanese perspective on KM authored by Dr. Yukio Hiyama, a leading international regulator of long-standing, and comes to us by kind permission of ISPE as a reprint of an article originally published in May 2014 in an e-journal on Knowledge Management for the pharmaceutical industry. Section I closes with recommendations from the American Productivity and Quality Center on best practices that could help to accelerate the implementation of KM within the pharmaceutical industry.

In Section II, we bring you *perspectives on knowledge*, which opens with the patient perspective provided by Dave deBronkart who tells of when he learned he had a rare and terminal cancer, he turned to a group of fellow patients online and found the information that helped to save his life. He passionately proposes that proactive collaboration between patients, the medicinal products scientific community, and their clinicians can lead to better outcomes for all. Two regulatory perspectives are then shared by the Swiss and Irish medicines regulators, respectively, who outline recent KM initiatives undertaken within their agencies in efforts to value the knowledge held within their knowledge-intensive organizations.

Section II also includes a significant non-pharmaceutical contribution from our colleagues at NASA, who share insights gained from their long tenure in developing their own KM program. In particular, they outline how they manage the flow of knowledge across increasingly complex organizational boundaries where 80% of every program or project undertaken at NASA is an international collaboration. Two academic perspectives are as follows: the first of which examines the central role that KM plays in the ICH Q10 Pharmaceutical Quality System guidance yet coins the phrase the *Orphan Enabler* due to the lack of practical implementation guidance. The second academic contribution introduces how effective *knowledge assessment* can empower organizations to better evaluate the stock of knowledge assets within an organization in order to create value for the business and introduces an exciting new model (and Smart APP) to manage individual knowledge, known as the *MinK* Framework. This section closes with the perspectives from the editor-in-chief at a leading industry journal, *International Pharmaceutical Quality* (*IPQ*), on the direction that KM may take following the publication of the much-anticipated ICH Q12 on Pharmaceutical Product Lifecycle Management.

As noted, Section III introduces the pillars and enablers as foundations for successful KM and the *House of Knowledge Excellence* Framework. Here we also present chapters from Pfizer Global Supply, Merck Manufacturing Division*, and Eli Lilly and Company in which each shares the specific drivers, direction, and success factors for their KM programs. This section closes with two complementary perspectives on the importance of developing a strong partnership between KM champions and their information technology counterparts as a catalyst that can accelerate and enhance the KM value proposition.

Finally, Section IV contains a collection of case studies from across the biopharmaceutical industry that profile a variety of KM efforts, some of which are full-blown KM programs, whereas others are grassroots initiatives with a limited area of focus. All of these case studies represent real and tangible KM efforts currently under way in our industry, often in the context of a pressing business problem, and share how those involved leveraged KM to address this business problem. To help the reader navigate these stories, and in acknowledgment that not all languages used may be consistent with how we have delineated the KM practices, pillars, and enablers, we have included a useful KM matrix as a reference guide to understand how the various companies have approached KM.

We offer our sincere gratitude to each and every one of the authors who generously shared their insightful contributions. We have learned much and continue to learn each day. For the editors, this book has been a labor of love born from a conviction to spread the word and encourage more widespread adoption of KM. It is our fervent hope you find this handbook a valuable companion in your journey toward knowledge excellence.

Nuala Calnan
Martin J. Lipa
Paige E. Kane
Jose C. Menezes

---

* All references to the term 'Merck' in this publication and any derivatives including 'Merck Manufacturing Division' or 'Merck Research Laboratories', refer solely to Merck & Co., Inc., Kenilworth, NJ, USA.

Section I

Making the Case for Knowledge Excellence in the Biopharmaceutical Industry

This opening section sets the stage for the entire book and puts forth the case that the time is now right for the biopharmaceutical industry to embrace the journey toward *knowledge excellence*. This section brings four major contributions. Chapter 1 is from one of the leaders in the biopharmaceutical industry and one of the pioneers in the adoption of knowledge management (KM) as a business excellence measure, Merck & Co., Inc. In Chapter 2, an academic group from the Know-Center in Austria introduces a perspective on what other industries are undertaking with respect to KM and the impact that these measures are having. Chapter 3 is written by a well-known former Japanese regulator, Dr. Hiyama, who played a key role in the development of the ICH Q8 through ICH Q10 guidance documents, and considers the perspective of the *learning organization* school-of-thought and considers the different types of knowledge (i.e., explicit and tacit) and their relationships and conversion. This section closes with Chapter 4 and some key insights from one of the leading think tanks on KM, the American Productivity and Quality Center (APQC). These four chapters create the foundation for a better understanding of KM as a lifecycle approach to excellence in the biopharmaceutical industry and provide a first overview of why, how, and when to use KM, which will be further illustrated by specific case studies later in the book.

1

Why Knowledge Management Is Good Business

P.K. Yegneswaran, Michael P. Thien, and Martin J. Lipa

CONTENTS

Unprecedented changes in the pharmaceutical industry are creating both challenges and opportunities. Addressing these changes requires enhanced agility, speed, reliability, and competitiveness across the entire pipeline from product realization, through manufacturing excellence and supply chain execution. Managing the flow of knowledge across the product lifecycle is a key enabler to overcoming these challenges and in realizing the opportunities. Here we glean insights from the technology and commercialization industry leaders at Merck (Merck & Co., Inc., Kenilworth, NJ, USA) as they share their perspectives on why knowledge management (KM) is a critical business enabler for them.

Editorial Team

The Challenges

The challenges facing the biopharmaceutical industry today are many. These include pricing pressures, vanishing of the *innovator's premium* due to simultaneous competitive product introductions, loss of exclusivity from the blockbuster era, low-cost entrants, supply chain security challenges, decreased research productivity, and more (PWC, 2016). At the same time, there is a steady stream of entrants into the traditional biopharmaceutical space from the fast growing health care technology companies that provide patient empowerment, data access, and measurement of outcomes (PWC, 2016). In addition, the supply chains responsible for clinical materials, new product introductions, and reliable supply through the product lifecycle continue to remain fragmented and complex, and in many ways designed for the era of blockbuster products including high capacities and long lead times. With the industry pressures outlined above, there is a renewed effort within the industry to enhance the efficiency, agility, and velocity across the supply chain, to lower the cost of goods, and to reduce inventory levels. The challenging business environment has also driven the industry to increased mergers, acquisitions, and outsourcing activities resulting in additional complexities in the supply network. In addition, the pressures toward localization of manufacturing in the emerging markets results in further fragmentation of the supply chain, resulting in multiple supply chains for the same product, thereby adding further complexities and barriers to efficient execution. This has also driven the need for agility to meet the rapidly changing requirements in the areas of product demands in different markets and the supply chain configuration. It is imperative for the industry to transform in order to enhance responsiveness and efficiency across an increasingly complex supply network while maintaining high quality and low cost for its products. A typical supply chain is shown in Figure 1.1 and includes chemical (active pharmaceutical ingredient) manufacturing, formulation, and packaging followed by distribution to the markets. Each of these steps is often performed at multiple locations across the internal and external (contract manufacturing organizations) nodes in the supply network. Critical knowledge about the performance, risks, and solutions is generated at all of these nodes in the supply chain and also across the lifecycle of the product starting in development.

As in many industries, a largely untapped resource is the knowledge possessed by an organization, and as importantly, the organization's ability to flow that knowledge to where and when it is needed in a robust, timely, and repeatable manner. The biopharmaceutical industry as a whole underutilizes the flow of knowledge, so it can be readily used to reliably deliver high-quality, low-cost products. In this chapter, we will explore the role of managing the free flow of knowledge in this complex environment to overcome these challenges in a manner that ultimately benefits the patients and stakeholders of the company including employees and shareholders.

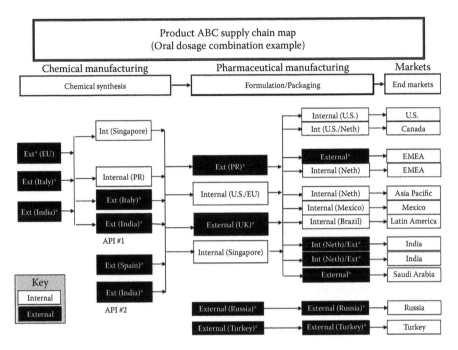

FIGURE 1.1
A typical pharmaceutical supply chain.

Managing Knowledge as Part of the Solution

Knowledge about products and processes is generated at each stage of a product's lifecycle, from development through supply. How this knowledge is captured, maintained, enhanced, shared, and leveraged by flowing freely through the network is essential to the continuous improvement of operational performance and the achievement of the goals of reliable and cost-competitive supply of high-quality products to the patients.

There are different types of knowledge generated during the development and supply of biopharmaceuticals including explicit and tacit knowledge. A majority (~80%) of knowledge is tacit (experiences, expertise, insights, and so on.) and is not easily captured in a written form (Calnan, 2014). The remaining knowledge is explicit—recorded in a form that can be straightforwardly used by others. Both types of knowledge are essential, and both must freely flow to where they are needed, when they are needed. To capture the benefits of both the explicit and tacit knowledge generated by the technical groups in late stage development and manufacturing at Merck & Co., Inc. a Knowledge Management (KM) Program Office was created with dedicated resources within the science and technology area in the Merck's

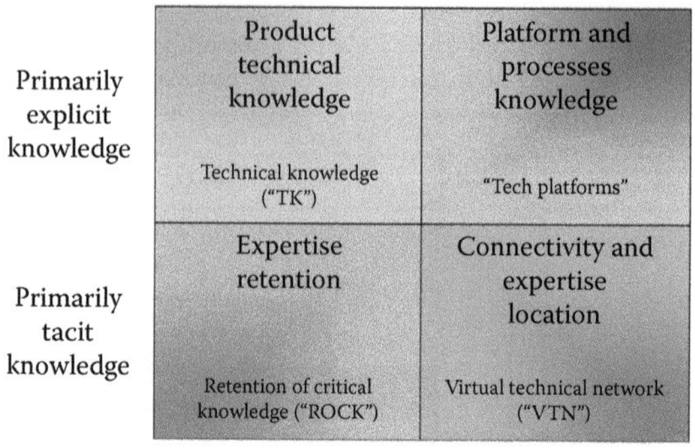

	Product technical knowledge	Platform and processes knowledge
Primarily explicit knowledge	Technical knowledge ("TK")	"Tech platforms"
Primarily tacit knowledge	Expertise retention Retention of critical knowledge ("ROCK")	Connectivity and expertise location Virtual technical network ("VTN")

FIGURE 1.2
The four core KM capabilities.

Manufacturing Division (MMD). A framework for knowledge-based business capabilities was created with the goal of ensuring flow and reuse of both tacit and explicit knowledge within this area (Lipa, Bruno, Thien, and Guenard, 2013). It was strongly believed that the realization of such business capabilities would create a much more efficient, effective, and agile use of knowledge while avoiding the wastes associated with knowledge search and knowledge recreation. The key elements of the framework (Figure 1.2) along with their intent and benefits are described below:

1. Product technical knowledge: Knowledge specific to products and the associated process and analytical methods. The technical knowledge (TK) platform provides a unified framework for storage, retrieval, and use of product knowledge across the entire lifecycle of the product, starting at the development. The TK platform includes standard templates with which to capture knowledge, a taxonomy to consistently organize and tag knowledge, an electronic repository to store knowledge, and a flexible search and filter mechanism (from multiple business perspectives) with which to find knowledge, stewardship roles to maintain the platform, and a governance structure to sustain and improve this capability. The TK content for a product enables rapid access to historical knowledge; known robustness issues and solutions; key steps in the evolution of the product, process, and analytical methods; and process performance during tech transfers and source changes, all enabling rapid response to challenges and opportunities to enable compliant, reliable, and cost-effective supply. In addition, given the distributed nature of the supply chains, the product TK allows rapid sharing of problem-solving and continuous improvement

across the different supply nodes including CMOs (contract manufacturing organizations), contributing to rapid deployment of this knowledge.

2. Process, platform, and technology knowledge: Knowledge about core manufacturing technology platforms. The technology platform (TP) is a framework for the capture, storage, maintenance, and use/reuse of general knowledge, both tacit and explicit, which applies to a given technology or technical platform. It includes application of analytical methods and PAT, process development, equipment, manufacturing science, and operations. The knowledge stewarding communities of practice (CoP) is a critical element of the framework and ensures capture of new knowledge relevant to the platform acquired internally and externally and translates lessons learned into best practices. The TP platform can be used as a training tool for a wide range of expertise levels and used to solve problems related to performance of the platform at a given manufacturing node. It allows for the rapid sharing of best practices to enhance equipment, process and turnaround performance across the network, and enables creation of standardized work and continuous improvement to improve product quality and the speed of manufacturing and release.

3. Connectivity and expertise location: Connections to tacit and experiential knowledge involving a wide variety of technical topic areas. The virtual technical network (VTN) is a platform that provides on demand networking capability for connecting with experience and expertise and the sharing of technical knowledge across the company. A series of *helping communities* serve as a platform for sharing and innovating while solving problems and sharing proven practices. Although some communities are narrow and deep in their scope, other workhorse communities, such as the *Biologics & Sterile Processing Community* or the *Processing of Powders Community*, have very large populations that connect many practitioners who face similar challenges on a daily basis. These communities provide a means of direct access to a large and diverse community, eliminating organizational, geographic, and hierarchical boundaries. Participants are no longer confined to their own personal networks: our surveys indicate that more than 50% of respondents to a query are not known to the person asking the question. It is critical that each community is served by one or more steward who sees to the health of the community. These stewards are critical to the success of the VTN. Our experience indicates that the VTN is used extensively to obtain a variety of inputs from subject matter experts (SMEs) and practitioners across the network, thereby driving toward a *solving a problem only once* mindset. In the process, several new connections are made across the network with each exchange resulting in new

opportunities to collaborate and innovate, leading to tangible business outcomes. As an added but important benefit, the VTN also provides talent development and growth opportunities for the community stewards and the subject matter experts (Guernard, Katz, Bruno, and Lipa, 2013).

4. Expertise retention: Unique and tacit technical knowledge held by an individual. The Retention of Critical Knowledge (ROCK) is a platform (adapted from pioneering work in this area by Royal Dutch Shell plc, the Hague, the Netherlands) that enables the capture of knowledge from people with unique expertise and experience. A structured interview process is used to transfer critical knowledge from individuals with valuable and unique knowledge, before the transition from their current situation (e.g., transfer and retirement). Such knowledge is generally tacit in nature and cannot be comprehensively codified into guidelines and standard operating procedures. Examples include deep knowledge of product, process, and analytical technologies that have significant empirical elements, approaches to organizational effectiveness and Lean manufacturing systems, approaches to investigations, all of which are difficult to recreate and critical to maintain a compliant, reliable, and cost effective supply of products. Using ROCK to capture and share such knowledge mitigates the risk of losing this capability. Our experiences using ROCK for capturing knowledge on legacy products and processes have been particularly beneficial. That said, it is our experience that the utility and scope of the captured information is directly related to the amount of preparative work invested in preinterview activities.

The Business Case

When embarking on specific projects to manage knowledge using the framework described earlier, our goal has been to prioritize the work based on what is most value added for the business. This was critical for prioritizing the work. We also then followed some simple operational principles: benchmark where possible; learn by doing; *think big, start small, but start*; do not invent if you can borrow. This meant the initial deployment of this approach in the Science & Technology area within MMD for managing technical knowledge followed by scaling of applications to other areas within the company using the same framework and platforms. In our experience, we have observed that many of the knowledge flow problems we are tackling are universal; while the content may differ from area to area, the KM capabilities we have developed are broadly applicable to different areas, thus enhancing the value proposition of meaningful KM focus. This also provides an opportunity to leverage experiences from other industries and institutions such as consumer goods, consulting, petrochemical, and aerospace that have developed mature KM capabilities over a longer time period.

The benefits of managing knowledge in a structured way are both tangible and intangible. These benefits affect the *top line* (minimizing disruption to supply, thus allowing maximum sales), the *bottom line* (promoting best practices and avoiding previous issues, thus decreasing the product cost), and the *pipeline* (providing use of previous knowledge on products and platforms to accelerate development and thus enabling shorter time to market). Some of the benefits are discussed in this section.

Patient Benefit

Manufacturing in the biopharmaceutical industry exists to ensure that safe and efficacious product reaches patients in a robust and reliable manner. To that end, our KM business capabilities have ensured that we are using the most updated knowledge when creating the technical translations of target product profiles. Similarly, risk assessments and control strategies are informed by previous product's use of similar technology platforms. From a supply perspective, the TK platform, the VTN, and our technology platform capabilities provide the means to accelerate investigations, leading to more comprehensive and faster closure of investigations. In addition, the VTN plays an added role of helping to rapidly locate specific or unusual reagents or equipment, better ensuring continuity of supply. These benefits can make all the differences when it comes to continuity of supply, ultimately benefitting the patient.

Regulatory Benefits

Regulatory agencies have recognized the link between how an organization leverages knowledge and a product's quality. In 2008, ICH Q10 *Pharmaceutical Quality System* (ICH, 2008) for the first time formally established KM as an enabler to the Pharmaceutical Quality System across the entire product lifecycle; existence of a formal knowledge management system is rapidly becoming a global regulatory expectation. The ultimate goal is to ensure safety, efficacy, and continuity of supply of medicines to the patient. Using an explicit framework such as that described is one way of enabling the industry to fulfill this requirement while also deriving other benefits. The recent inclusion of knowledge management within ISO 9001:2015 (released in October 2015) marks a positive development within the world of KM. For the first time one of the global business standards explicitly mentions knowledge as a resource, and specifies expectations for the management of that resource. This will likely lead to further development of formal systems, structures, and processes for managing knowledge across many industries.

Operational Benefits

The lack of free flow of knowledge across the variety of boundaries in the biopharmaceutical industry can lead to significant waste in how work gets done. The waste manifests itself in various forms including repeat deviations, search time for existing knowledge, wholesale recreation of preexisting solutions, infrequent sharing of best practices across the network, and more. By consistently managing the flow of knowledge using a framework such as the one described in this chapter, much of this waste can be reduced or eliminated to enable rapid, reliable, and sustainable flow of knowledge when and to where it is needed. In our experience at Merck, this flow of knowledge has resulted in many operational benefits including:

- Faster problem resolution to minimize risks to reliable supply for the patient.
- Lower inventory levels with positive impact on working capital.
- Improved efficiency and effectiveness of tech transfers through sharing of best practices and continuous improvement of the associated standardized work.
- Sourcing of materials, supplies, and equipment for commercial and clinical production across the internal and external manufacturing nodes.
- Support for business development and merger activities enabling more effective and efficient integration and supporting synergy targets.
- Improvements in how we develop regulatory filings, leveraging prior knowledge resulting in faster, more efficient, and more robust filings.
- Faster, efficient, and more informed regulatory audit responses by using knowledge across the network, leading to improved compliance posture.
- Sharing across the network of common issues with certain unit operations, equipment leading to proactive interventions and in time, to fewer market actions.
- Sharing of safety incidents and near misses leading to proactive interventions and in time, fewer people getting hurt.
- Enabling higher levels of engagement and faster onboarding of new personnel.

These and other examples enable stronger connections and integration across the global supply network consisting of internal and external manufacturing nodes. Ultimately, this robust connectivity and standards for knowledge sharing drives out waste and has pushed MMD to higher levels of sustained performance.

Workforce Benefits

The knowledge management activities associated with the framework described in this chapter have resulted in significant technical and leadership development opportunities and have led to routes for personnel development that did not previously exist (Guernard, Katz, Bruno, and Lipa, 2013). The most obvious opportunity is for the stewards of the different communities and platforms. These stewards greatly expand their personal networks, have shown significant growth in their subject matter expertise associated with their community, and have enjoyed a unique forum for demonstrating global leadership *in real time*. On numerous occasions, the stewards have driven rapid global problem-solving in responses to urgent issues. The community stewardship has also provided them with leadership of a community outside of the normal vehicles of organizational lines and project teams. As one steward noted, running his community made him feel like he was running his own business.

Workforce benefits extend well beyond the stewards. Non-steward participants in the various facets of the KM framework have taken advantage of the increased opportunities for networking and development of new capabilities. As noted previously, participants in the VTN create globally dispersed networks and are likely to meet colleagues whom they have never met before. Self-association with subject matter experts further increases skill levels. Being part of communities that provide up-skilling opportunities and enable the members to link their day-to-day work to the broader context increases the level of engagement for employees and contributes to discretionary effort. For new employees, we have seen several examples of KM facilitating rapid onboarding and engagement of employees by connecting them with expertise and resources to efficiently execute on the daily work. The mode of learning corresponds heavily to hands-on learning modes that feature prominently in learning models (Lombardo and Eichinger, 1996), such as the 70:20:10 model for learning beyond the classroom (Figure 1.3). In fact, capabilities such as the VTN help to catalyze the 70% (learning on the job) by establishing channels that accelerate learning through increased connectivity to subject matter experts (SMEs) and the organization at large.

Generational differences also create challenges for the modern workforce. The emerging workforce will be largely composed of workers who expect instantaneous access to information and use social media tools as main means of communication and networking. In a world where Millennials turn to Google, YouTube, or Mom and Dad when they need help with a problem—how do we replicate that in a manner inside the workplace which provides speed, trust, and satisfaction, leading to higher levels of engagement. Feedback from Millennials in our workspace suggests that the ability to translate experiences from social networking to the knowledge management approach, framework, and the tools that we have employed is very intuitive and allows them to navigate complex structures within the

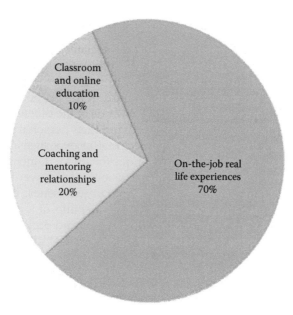

FIGURE 1.3
The 70:20:10 learning model.

company, enabling their rapid engagement and productivity. This ability to effectively access knowledge by linking to people and systems continues to rapidly become an expectation of our future workforce and can be viewed as a requirement for attracting and retaining talent.

Financial Benefits

Although it is difficult to predict a specific return on investment (ROI) for KM, it is common to capture savings through success stories and the operational benefits described earlier. A recent *Harvard Business Review* (*HBR*) article (Myers, 2015) estimated >$30 billion waste among Fortune 50 companies as a result of not sharing knowledge. During the initial deployment at Merck, from 2011 to 2014, the KM program delivered approximately a 3:1 ROI, which is consistent with benchmarking that suggests similar ratios for companies that have reached a maturity level of *standardize* (Level 3) in the APQC KM maturity model (Figure 1.4) (Hubert and Lemons, 2010). In addition, we have previously cited that we have quantified a $20MM benefit and believe that the indirect benefit—for example, through higher connectivity of the organization—is a several-fold factor of the benefit we have quantified, perhaps five times the amount of quantified benefits. Another example—use of KM has saved vials of lifesaving cancer therapies when their supply was limited, and another example is of development timelines that have been preserved. All of these have tangible benefit but are difficult to quantify in

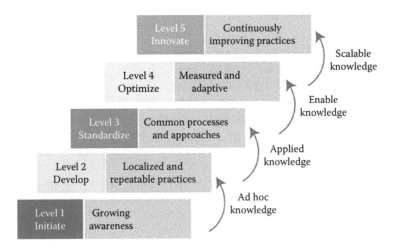

FIGURE 1.4
The KM maturity model.

financial terms. Obviously, further work could be done in order to quantify the financial benefits of our current framework; having demonstrated repeated value of the framework outlined above, we are now only seeking to quantify benefits for the development of new KM business capabilities.

Overall, managing knowledge has deep positive impact in the areas of top line (revenue driven by reliable supply of products, proactively preventing reliability issues, and rapidly responding to any issues), the bottom line (efficiency improvement across the product lifecycle), and the pipeline (enabling rapid development and introduction of new products into the market).

Leadership and Change Management as Key Enablers

As described above, the results from the initial KM experience in MMD are very promising. During the implementation of the different elements of the framework it was clear that strong leadership and change management would be critical factors for success (Lipa, Bruno, Thien, and Guenard, 2013) (Guernard, Katz, Bruno, and Lipa, 2013) (Koberstein, 2014). In this section we describe some of these factors based on our learning during the initial implementation of our KM framework.

1. Culture change: Although people, process, content, and technology (PPCT) are all important elements of the KM construct (Lipa, Bruno, Thien, and Guenard, 2013), there is a tendency to focus on technology and build the other elements around the technology solution.

However, our experience has been that the creation of lasting KM business capabilities has to start with people and behaviors and thinking differently about capturing, sharing, seeking, flowing, and using knowledge. For example, in case of the VTN, inclusion of the users from the different functions on the design team ensured that the value of the VTN got translated to the specific functional areas. Investment in up-skilling the VTN community stewards was key to ensuring that they could create an inclusive environment within the virtual communities and provide energy to the participants (Guernard, Katz, Bruno, and Lipa, 2013). This also helped overcome barriers such as the initial inertia toward participation in virtual discussions and moves the needle from *Why should I?* to *This is how we work*. Similarly, the role of leaders in enrolling sponsors and people managers to model the required behavioral changes and applying positive and negative consequences was a key success factor. Access to KM capabilities across the organization also empowers every employee as a knowledge worker that over time leads to a truly empowered and learning organization.

2. The sponsorship spine: Having clear sponsors at all levels of the organization, who can legitimize a change and provide meaningful consequences (positive and negative) is a key to successful installation and realization of any major change, and so it was with our KM program. During the implementation at Merck, the initiating sponsor for the program and the sustaining sponsors in each function played a key role by setting clear expectations, regular communications, advocacy, helping remove barriers to progress, and driving for results. This sponsorship was then expected down throughout the management hierarchy. It was made clear that managers should be asking questions like *What did you find in TK on that? Did the technology platform mention this possibility? What was the response on the VTN?* Creation of this spine of sponsorship ensured that all members of the organization knew what was expected and that the organization would not condone recreation of knowledge.

3. KM in the flow: In order to manage the flow and use of knowledge where and when needed across the network, it is important that this should not be viewed as work over and above the day-to-day work. Core work processes may need to be reconfigured to ensure that knowledge is captured, shared, and sought using the existing framework and platforms. Sponsorship by leaders is important to ensure that the core work process is not complete until the steps for managing the knowledge derived are completed. For example, the technology transfer for a new product is not complete until all the learnings from the work have been incorporated into the TK (Tech Knowledge) platform for that product. Building these expectations

into performance objectives is a tactic to help address this at the individual level, along with clear expectations outlined in standard practices and processes.

4. Alignment with business priorities: The work of managing knowledge should be closely aligned with the business priorities and is linked to the outcomes of the business unit. After the creation of the framework and platforms, the decision on the focus areas should be driven by the business. This will enable rapid transition from *KM as an initiative* to *managing knowledge as how work gets done*. The role of sponsors and leaders in the organization is a key to achieving this alignment and prioritization.

5. Stewardship: Stewardship roles are critical to provide energy and help people connect. Each of the KM framework elements described in this chapter features a stewardship role central to its success and sustainability. Leaders in the organization need to sponsor the selection and appointment of these stewards. Stewardship roles are great development opportunities for future leaders in the organization, as they become knowledge brokers who understand how to connect people to people and people to knowledge. In addition, a community of practice consisting of knowledge stewards allows for sharing and adoption of best practices.

6. Success stories: During the early days of KM deployment, it is often difficult to quantify the value generation that can be attributed to KM. Our experience has been that sharing success stories for each element of the framework is highly impactful. These examples of people connecting with people and knowledge to eliminate waste from the day-to-day work and adding value to the business are truly inspiring for others in the organization to understand KM and make it relevant for the broader organization.

Conclusion

In this chapter, we have described a practical approach to managing knowledge using the initial phase of KM deployment within the manufacturing division at Merck & Co., Inc. as an example. Although each organization will need to tailor their approach to KM based on their respective business needs, culture, and practices of that organization, the general approach to creating a robust framework, a long-term strategy, and sustainable execution is applicable across and beyond the industry. The business benefits in the areas of reliable supply to the patient, proactively fulfilling regulatory requirements, operational improvements, workforce development, and employee engagement are evident and tangible. The ROI derived from improved quality, rapid response time, internal efficiencies, cost reductions and cost avoidance, improved employee engagement, and the ability to leverage a diverse, global, and interconnected network is recognized by senior leadership.

In any situation, strong leadership and change management are a key to embedding KM in the flow of work to maximize the benefits to the business. As KM is scaled and deployed across different areas of the company, the corresponding business benefits for the top line, bottom line, and ultimately the patient. Hence, we believe that deployment of KM across the biopharmaceutical industry is good for the business.

The need for a strong KM framework is reinforced by the rapidly changing environment of the biopharmaceutical industry. Subject matter expertise and best practices are likely to shift as the industry develops new and different types of manufacturing. Supply chains are fragmenting and growing more complex, increasing the challenges for those who technically create and support such supply chains. Regulatory expectations are shifting, seeking stronger ties between risk assessment and knowledge. The emerging workforce has greater expectations with respect to knowledge infrastructure and the ability to be knowledge workers. Until recently, the need for a fully functioning set of KM business capabilities has been desirable, but an option. With the industry changes already in motion, a well-articulated and realized set of knowledge management capabilities will soon become a requirement.

In a rapidly changing world with increasing pressures, given the multifaceted benefits of effectively managing knowledge to the patient and the business, how could one not leverage knowledge management as a competitive advantage?

Acknowledgments

This work summarized the ideas, experiences, and efforts of many colleagues over several years. Special appreciation to Jean Wyrvatt and Anando Chowdhury for their pioneering efforts. Special thanks to our current and past colleagues in the Global Science Technology & Commercialization Leadership Team at Merck for their strong and continuous sponsorship and support of this effort, including: Gary Hoffman, Tom O'Brien, Parimal Desai, Jim Robinson, Maria Wirths, Jim Stephanou, and Wayne Froland.

References

Calnan, N. (2014). The 80/20 rule of knowledge. *Knowledge Management Supplement to Pharmaceutical Engineering*, 54–58.

Guernard, R., Katz, J., Bruno, S., and Lipa, M. (2013). Enabling a new way of working through inclusion and social media—A case study. *OD Practitioner*, 45(4), 9–16.

Hubert, C., and Lemons, D. (2010). *Using APQC's levels of KM maturity.* www.apqc. org. Houston, TX: APQC.

ICH. (2008, June 4). *ICH Q10 "Pharmaceutical Quality System".* Retrieved May 27, 2016, from International Conference on Harmonisation: www.ich.org. Geneva, Switzerland.

Koberstein, W. (2014). Merck—making it, no boundaries. *Life Sciences Leader*, 6(12), 24–29.

Lipa, M., Bruno, S., Thien, M., and Guenard, R. (2013). A case study of the evolution of KM at Merck. *Pharmaceutical Engineering*, 45(4), 94–104.

Lombardo, M. M., and Eichinger, R. W. (1996). *The career architect development planner* (1st edition). Minneapolis, MN: Lominger.

Myers, C. (2015). Is your company encouraging employees to share what they know? *Harvard Business Review*.

PWC. (2016). *Pharma 2020: From vision to decision.* Retrieved June 21, 2016, from pwc. com: http://www.pwc.com/gx/en/industries/pharmaceuticals-life-sciences/pharma-2020.html.

2

Theory of Knowledge Management

Stefanie N. Lindstaedt, Paul Czech, and Angela Fessl

CONTENTS

> The Know-Center, based in Austria, is a well-established and award-winning Big Data Analytics institution serving multiple services and process industries for their knowledge management research and application needs. To set the scene for the rest of the material to follow, they outline several important concepts and theories involved in knowledge management in a pragmatic and succinct way here.
>
> **Editorial Team**

The most important concepts, as well as new approaches involved in knowledge management are presented in a pragmatic and concise way. The Know-Center is Austria's leading research center for data-driven business and big data analytics. Their data-driven methods and technologies contribute to value creation and benefits for their customers in a sustainable way among various industries and use cases. Through their cognitive computing-based approach, which combines the strength of man and the machine, they are setting standards within both the local and the international research community. With their expertise in the field of knowledge management they are describing the basic approaches in this chapter.

Introduction

The business importance of knowledge management for the life sciences industry has been mapped out in Chapter 1. The goal of this chapter is to provide the reader with some definitions that are crucial for engaging in an informed discourse on knowledge management, its different types, and their advantages and disadvantages. We start out by examining the spectrum of knowledge, information, and data, and then dive into a short overview of relevant theories on knowledge management and organizational learning. With this background we then shortly examine the relationship between knowledge management and Quality by Design. This chapter closes with an outlook into a new emerging type of knowledge management: data-driven business. As the authors are convinced that this new paradigm will substantially alter the way we see knowledge management and the way business will be performed, a discussion of advantages, disadvantages, and the relationships between the concepts are provided. We close with four major challenges that need to be tackled in order to bring value to organizations and close with some practical examples on how they can be applied.

Definitions

Knowledge management has a long historical background and deals typically with capturing, storing, organizing, and managing organizational data in order to be able to use them effectively for companie's daily business (Beijerse 2000; Wong 2004; North 2011) and therefore also in the pharmaceutical industry. However, before being able to discuss how knowledge is managed, it is crucial to first agree on a common understanding of the term *knowledge*. Defining knowledge has kept humanity busy over the last thousands of years, starting with Aristotle and Plato or even earlier, and would warrant a whole book to itself. As within this book we want to explore different approaches to knowledge management in the pharmaceutical industry, we take a pragmatic approach.

When it comes to defining knowledge we distinguish between two major *schools of thought*: In the *human-centered view*, knowledge is defined as *something that we have between our ears* and does not exist outside the human brain. Everything else is referred to as information or data. In order to turn information into knowledge it needs to be *internalized* (learned) and then can be utilized for concrete tasks. This view on knowledge emphasizes the capacity for action, which is inherent in the human being. Knowledge management approaches that are based on this human-centered view, build heavily on supporting and enhancing the exchange between people in order to exchange knowledge.

In the *documentation-centered view*, knowledge is seen as something that can exist outside the human brain in the form of written documents, rules, process descriptions, and so on. The process in which a human documents his or her knowledge is referred to as *externalization*. Below we will further extend on this definition and its relation to information and data.

Nevertheless, in both *schools of thought* there is a common agreement that there exists knowledge, which cannot be completely (or at all) externalized. Examples for such *tacit knowledge* are skills such as bicycling but also situational knowledge (e.g., knowledge in the world). Knowledge management approaches that embrace this documentation-based view build heavily on externalizing knowledge of domain matter experts (e.g., through knowledge engineering practices), representing this knowledge in (ideally) a machine-readable form, performing computations on this knowledge (e.g., inferences), and making the resulting knowledge available to people in order to support their actions and decisions.

Clearly, knowledge management theories combine these two viewpoints in meaningful ways, however in most cases one of them is predominantly present and characterizes the approach. In the remainder of this section, we will discuss a number of different knowledge management theories, each with their own advantages and disadvantages.

Data, Information, and Knowledge

Data, information, and knowledge (Kendal and Creen 2007) are not static things in themselves but are stages in the process of using data and transforming them into knowledge—or the other way around. As with most knowledge-related concepts, there exist no universally accepted definitions of *data*, *information*, and *knowledge*. However, one widely used explanation utilizes an example such as the following (see Figure 2.1).

One sunny, wintery morning you look at the thermometer outside your bedroom window. It shows –10 degrees Celsius. Just reading the temperature makes you shiver and realize that it is very cold outside. Thus, when you leave after breakfast you naturally reach for your coat, gloves, and hat in order to stay warm on the way to work.

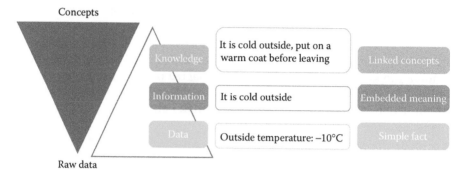

FIGURE 2.1
Data, information, and knowledge. (From Kendal, S. and Creen, M. *An Introduction to Knowledge Engineering*, Springer-Verlag, London, UK, 2007.)

This short description can be visualized as shown above.

In our example, −10 degrees Celsius measured with the thermometer can be seen as a *data point*. This data point is captured on a specific day and time. Collecting such data points every minute during the day can provide a good approximation of the temperature development throughout the day. In order to understand that it is cold outside you need to have an understanding of the Celsius temperature scale and know that temperatures below 0 degrees are perceived as cold. By putting together the data point and the temperature scale we can *infer* the *information* that it is cold outside. This information can be used by an agent as a basis for making decisions. If you want to take action based on this, for example, if you want to leave the house, you use the information that it is cold to decide that it will be better to put on a coat. In this example, the combination of the information *it is cold* and the context information *I want to go outside* provides you with the knowledge (inference) that you should put on a coat.

Being a little more formal, we can use these following definitions:

- *Data* are commonly assumed to be raw facts and recorded symbols. There is only little or no amount of semantics (meaning).
- *Information* is data in context, meaning that these data are presented in a form that is significant for the recipient.
- *Knowledge* is "The explicit functional associations between items of information and/or data" (Debenham 1988). Or, in other words, what someone has after understanding information.
- *Data*: It is −10 degrees Celsius (simple fact).
- *Information*: It is cold (embedded meaning).
- *Knowledge*: It is cold, and if it is cold you should put on a warm coat (linked concepts).

Interestingly, these three concepts are fluid, in that they do not hold in absolute terms but are depending on the context in which they are used. Let us reconsider our example: We interpreted the −10 degrees Celsius as a simple raw data point. However, in order to understand this data point a number of additional information has to be present: the information that the sensor presents is a temperature (it could also be a wind sensor or pressure sensor) that the sensor is outside the house, and thus tells us the temperature outside as opposed to inside the house, and so on. In this perspective, the value −10 could be seen as raw data, combined with the information that it is a temperature of −10 *degrees* becomes a piece of information and combined with the information that it is the Celsius scale (as opposed to Fahrenheit) that results in the knowledge that *it is cold outside*. On the one hand, a single data point is just raw data, but combining it with further context information can lead to knowledge.

We therefore must view the concepts of data, information, and knowledge as being parts of a continuum, which needs to be adjusted to each situation and application. By linking more and more data and/or information to each other the more, we move upward in the spectrum toward knowledge.

Knowledge Representations

After having fostered a shared understanding on data, information, and knowledge, the subsequent discussion leads us to the challenge on how to represent knowledge. One of the oldest and most successful, universally usable knowledge representation is *speech*. Since centuries, people have used words and sentences to explain relationships between concepts, tell stories, and even to dream up new worlds and ideas. The *written word* is a powerful extension of speech and allows knowledge to be exchanged between people who have never met. One of the oldest and most formal representations of knowledge is *mathematics*. With numbers and formulas, we can represent abstract attributes of real things and express their relationships between each other. For example, the Pythagorean Theorem expresses the relationships between the lengths of the sides of a right-angled triangle. In addition to these commonly known knowledge representations, different disciplines have developed their own specialized representations: (1) building plans for architects, (2) technical drawings for mechanical engineers, (3) electrical circuit drawings for electrical engineers, and (4) music notation for musicians, and so on.

In the context of artificial intelligence computer scientists have been busy in designing knowledge representations that enable knowledge to be represented in great (formal) detail, are universally applicable (domain independent), and most importantly, can be used for further computation such as inferencing. One early knowledge representation, which provides the basis for expert systems, is rules. Rules are expressions in the form of IF–THEN statements (if condition x holds then y follows). Other widely used knowledge representations are for example semantic networks, topic maps, and ontologies, which represent concepts, their relations, and thus serve as basis for semantic systems. Along with the different knowledge representations computer scientists also developed knowledge engineering processes and techniques that support the externalization of knowledge from individual domain experts or groups of domain experts.

Knowledge Management Theories

The essential principle of all knowledge management theories are *feedback loops*. When new knowledge is acquired it needs to be learned and applied by people. Through this learning-by-doing, new experiences are gained and out of these new knowledge can arise. To establish these feedback loops with regard to knowledge management in organizations, they need to provide structures, processes, and tools that at least allow this feedback loop to take place or even better, promote such feedback processes. However, knowledge management activities are always grassroots activities, which can be supported and encouraged but not mandated or even demanded. People's motivation for knowledge management has to be intrinsic and therefore the management of knowledge will always rely to a large extend on the knowledge sharing culture within the organization.

In order to highlight different perspectives on knowledge management, we will introduce six widespread knowledge management theories out of literature. Two of them examine Communities of Practice (Lave and Wenger 1991) and the role of knowledge brokers (Hargadon 1998) (Ericsson), represent the human-centered approach, whereas the theory of double-loop learning (Argyris and Schön 1978) and the Experience Factory (Basili et al. 2002) substitute the externalization-centered perspective. Furthermore, we will explore two theories, which try to bring together both, the human-centered and content-centered view, in order to achieve a higher level of knowledge excellence in businesses: Organizational Learning (Nonaka and Takeuchi 1995) and "Knowledge Maturity Model" (Hofer-Alfeis 2003). All theories provide a holistic view on knowledge management across disciplines as well as across people, processes, content, and so on.

Communities of Practice

The concept of communities of practice (CoP) is built on the learning theory Legitimate Peripheral Participation (LPP) (Lave and Wenger 1991). LPP describes and explains the process in which an apprentice within a certain community (e.g., Vai and Goa tailors, U.S. Navy quartermasters, meat cutters, and others) learns about the domain, the community, its tools, and its practices, and over time becomes a part of the community (e.g., apprentice and fellow) or even a master—and therefore earning authority. It is a learning process that is based on peripheral participation (helping, preparing, and watching) that is legitimately without responsibility for the outcome and over time turns into full participation, comprehensive responsibility, and social standing. The focus in the context of knowledge management is, that a community of practice develops tools, which crystallize domain knowledge in them and practices, which enable apprentices and masters to work together and mutually benefit from each other. The concepts described in this book have then been applied to modern organizations. The challenge in this context is to initiate the building and development of CoPs for different aspects interesting to the organization as a whole. For facilitating the emergence and development of Communities of Practice (CoP), organizations can leverage the BarCamp or Open Space Technology format (Dennerlein et al., 2013).

Key Characteristics: Human-centered. The theory describes the practices and tools, which are used in a CoP to share knowledge.

Business Application: In order to be used in organizations, the theory had to be extended from understanding and describing the knowledge sharing practices into developing approaches that help to instantiate CoPs in organizations and have them thrive there.

Advantages: Helps us to understand and within limits support the building of CoPs. Considers the holistic development of people within social settings and the importance of expertise.

Disadvantages: Requires a lot of experience and sensitivity to apply the approach successfully within an organization.

Technical Support: A large variety of computational tools have been developed to support CoPs. Typically, they are based on community spaces and communication support between the actors of the CoP.

Examples: Lave and Wenger describe how one person enters a new community (e.g., tailors) where the main activity is to cut clothes in a certain way. The new person starts to learn basic shapes and step-by-step new knowledge is created. This knowledge is then transferred into more complex cutting patterns. In the end, the new one will be slowly turned into the knowledge keeper and teacher, which provides her with extra authority.

In the late 1990s, Chrysler employed the CoP approach extensively within its development departments. The idea was to bring people with similar expertise but working on different car platforms together and to help them to develop their practices and tools and ideally to document them. The CoPs provided an orthogonal structure to the platforms (matrix). A group of knowledge management people had the sole task of helping motivated domain experts to establish CoPs and to run them.

Knowledge Brokers (Ericsson)

The simple idea is to hire people (*knowledge brokers*) whose specific task is to wander around the engineering and production sites, talk with engineers and managers about their problems and projects, and bring people with similar interests, problems, and approaches together. The underlying understanding was that people are generally so deeply involved in their own projects and busy in solving their own problems that they are not aware of other people working on similar challenges or even developing solutions that could be used to the problems at hand.

> *Key Characteristics*: Human-centered. A few dedicated knowledge brokers are responsible for sharing knowledge across teams and organizational boundaries.
>
> *Business Application*: Straightforward application.
>
> *Advantages*: If knowledge brokers have enough authority within the organization this approach is highly effective and leads to quick wins.
>
> *Disadvantages*: If knowledge brokers are perceived as nuisance this approach will not work.
>
> *Technical Support*: This approach does not call for specific technical support. Often the knowledge brokers develop knowledge maps of the organizations. Specific tools were developed to support this activity.
>
> *Examples*: In the 1980s, Ericsson employed this strictly human-centered approach to knowledge management and reported successes for more than a decade. After that, the knowledge brokers disappeared. It is unclear who took over the task.

Double Loop Learning

This knowledge management theory assumes the execution of a work process or task by one or several people (Argyris and Schön 1978). Learning in this model is based on experiences, by analyzing, rating, and considering work activities in their existing context. Before the execution of a work process certain preconditions are thought to exist and after the execution certain postconditions should hold.

Argyris and Schön (1978) defined two types: learning and a metalearning condition. In a *single-loop learning* situation, the person looks at a work process step, therefore gains feedback. If the postcondition does not fit the model she changes her behavior within her existing behavior repertoire. If she runs out of strategies to cope with the situation, she can evaluate the preconditions again and create new ways of thinking and acting. This more complex approach is called *double-loop learning*. This model can help people to successfully conquer challenging situations because one has to look at her own behavior that is based on learning experience. Clearly double-loop learning is especially important when the preconditions in the market change rapidly. Then it might be essential to realize in an early stage that it will not be sufficient to produce better rubber boots, but that, with the same competences, one can produce mobile phones. The third way of learning described in this theory is *deutero learning* or metalearning. The main activity for organizations in this type of learning is to develop capabilities that enable it to do better learning (learn how to learn). It is of major interest to change defensive routines that result in ignoring errors.

Key Characteristics: Human-centered. This theory provided the basis for the learning organization (see below). It stresses that within organizations different types of learning needs to occur and that specific learning capabilities have to be created.

Business Application: The theory itself needs to be extended in order to become applicable. This was done by Nonaka and Takeuchi in the learning organization (see below).

Advantages: Theory stresses the importance of questioning preconditions and postconditions and not accepting them as *given*.

Disadvantages: The theory does not explain how to establish the three different types of learning within an organization.

Technical Support: Tools that help to document the current practices and to reflect on them are helpful. Here the novel approaches around quantified self for learning and experiential learning are applicable (Fessl et al. 2011; Rivera-Pelayo 2015).

Examples: During car manufacturing, a car door has to be attached to the chassis. The precondition might be, that the door is already fully assembled and in a certain place. The postcondition should be that the door is securely attached and closed. During assembly, one or two operators attach the door to the chassis using bolts. Applying single-loop learning in this scenario means that the operators learn through repeated assembly that it is more efficient to attach the lower bolt before the upper bolt. By repeated acting, they become better at their specific task. Double-loop learning in this scenario occurs when an operator realizes that the assembly could be greatly improved, if the inside panel of the door was only included after the door was attached to the chassis. To reach this level of understanding he reexamines the preconditions of his execution step and thus can start a deeper improvement process. In doing so, he is not getting better at his specific task but helps to achieve the overall goal more efficient. The metalearning strategy in his car producing company can be, that in internal education programs the workers learn how to approach problems—learn how to sufficiently use single-or double-loop learning.

Experience Factory

The basic assumption of this theory is that an organization needs to be able to learn from the combined experiences of all its employees (Basili et al. 2002). The *natural resource* are the experiences of individual employees. Whenever an employee has an experience that triggers a positive or negative *eureka!* emotion, she should document this experience. The experience of all employees is collected and over time, experiences concerning the same process/task/subject are brought together and are *refined* into one *superexperience*. During the refinement process, contradictory experiences might be discovered and need to be discussed and consolidated.

Key Characteristics: Externalization-centered. Individual experiences are documented and together with similar experiences are *refined* into valuable insights.

Business Application: Basili worked with a number of businesses (e.g., DaimlerChrysler) as well as applied research institutes (e.g., Fraunhofer) and collaboratively developed guidelines and practices for establishing experience factories within organizations.

Advantages: The approach is based on the knowledge as well as experiences of all employees and includes all employees in the generation of insights.

Disadvantages: Documenting experiences is difficult and adds to the tasks employees have to do. In addition, refining experiences without turning them into noninteresting, seemingly obvious statements are challenging.

Technical Support: At the time as Victor Basili developed this theory, natural language processing capabilities were not yet as prevalent as today. Thus, all the refinement and consolidation steps had to be performed by a group of people (Experience Managers) in close collaboration with the experienced authors. Today this theory can be supported efficiently with modern text-mining approaches. Quantified self for learning and experiential learning are described in Fessl et al. 2011; Rivera-Pelayo 2015.

Examples: In the 1990s, this experience factory approach was embraced by DaimlerChrysler within the engineering departments. The goal was to help the distributed software developers to collect experiences about how successful different software engineering approaches could be applied.

Learning Organization

The SECI (socialization, externalization, combination, and internalization) model (Nonaka and Takeuchi 1995) has become one of the most important models with regard to knowledge creation and knowledge transfer theory in the organizational context. They describe a way on how to continuously transform tacit knowledge to explicit knowledge, in order to bring individual to organizational knowledge. The model encompassed four ways of how knowledge can be combined, converted, shared, and created within an organization. *Socialization*: a tacit-to-tacit knowledge transfer takes place by sharing tacit knowledge through face-to-face experiences. Similar to CoP, socialization typically takes place in a traditional apprenticeship relation where the apprentices learn from their masters. *Externalization*: making tacit knowledge explicit for example in written documents, manuals, and images. In this phase the knowledge is crystallized, and can be easily shared within the organization, however some knowledge is not possible to be codified. *Combination*: external knowledge is combined with other external knowledge to create new knowledge. *Internalization*: explicit knowledge is internalized by learning-by-doing; the external knowledge will become a part of the individual and as consequence new knowledge for the organization.

Key Characteristics: Organization-centered: continuously transforming tacit knowledge to explicit knowledge and vice versa and as a result create new knowledge for the individual and finally for the organization.

Business Application: This theory can be directly applied in business organizations using the mentioned model. Nonaka and Takeuchi explained the approach with Honda's Brainstorm Camp.

Advantages: This type of knowledge creation can lead to rich experiences, and knowledge becomes permanent for the individual and the origination. In addition, new ways of doing the job and business process improvements can be achieved.

Disadvantages: Converting tacit knowledge into explicit knowledge is often very difficult and sometimes impossible. Learning by sharing, observing, and imitating is time consuming.

Technical Support: Tools to detect, interpret, and analyze knowledge do vary between four forms of knowledge (e.g., video camera for recording situation for a postanalysis).

Examples: Nonaka and Takeuchi used this approach for developing the *perfect* bread maker. Their goal was to develop a bread maker baking the same good bread as in a five-star hotel. Although using the same recipe as in the hotel, the bread was not that good. It was possible to improve the bread maker in the desired way, after the bread maker developers have baked the bread together (learning-by-doing) with the baker of the hotel. The key was to identify how to knead the bread dough. The process of kneading could be seen as the implicit knowledge of the baker, which could be transferred to the developers through socialization.

Knowledge Maturity Model (Siemens)

The knowledge maturity model (KMM) was developed within the knowledge management group at Siemens, Munich, Germany (Hofer-Alfeis et al. 2003). It is a model and process, which guides the introduction of knowledge management initiatives into an organization based on the strategic goals of the organization and on its knowledge capabilities. The first step in the KMM is the analysis of the knowledge capabilities within the business unit at hand. Which domain expertise exists? Is the expertise only present in a few people or is it a widely available competence? Is the competence externalized (e.g., documented in how-to-descriptions) or is it mainly communicated from person-to-person? Each knowledge area is analyzed accordingly and a knowledge capability map is produced. This map is then contrasted with the strategic objectives of the business unit and a knowledge gap analysis is performed. Based on the identified knowledge gaps activities can be defined that can help to overcome them.

> *Key Characteristics*: Organization-centered. The KMM is more an analysis, planning, and controlling methodology. Based on the business strategy and the knowledge expertise within the organization knowledge gaps are identified and fitting knowledge management activities are selected. In turn, their outcomes are evaluated.
>
> *Business Application*: This more holistic approach is implemented best in businesses that do not have any knowledge management strategy yet, or will do a complete makeover on their old ones.
>
> *Advantages*: Very systematic approach with clear guidelines and practices on how to embed it within an organization.
>
> *Disadvantages*: Can take long before first concrete knowledge management activities are launched and produce a value.
>
> *Technical Support*: Planning tools for KMM were developed. Other tools depend on the concrete knowledge management activities chosen.
>
> *Examples*: Siemens used this approach in their own engineering departments.

Knowledge Management along the Product Life Cycle

Many of the introduced theories mentioned above emerged directly out of business environments. Specifically in the 1980s and 1990s high-tech and automotive companies invested a lot of money into the development of knowledge management strategies. From a historical perspective, Peter Drucker contributed significantly to this development by emphasizing knowledge as a *fourth* factor of production (besides land, labor, and capital).

One very interesting effect of the early attempts to employ knowledge management approaches within organizations was the establishment of a *knowledge management line of command* in parallel to the traditional management line. A common perception was that different people were needed to manage organizational knowledge assets (see also example in Chapter 1). This was practiced for one to two decades and then the tasks of knowledge management and skill development were merged back into the responsibilities of the line.

The vision for knowledge management in all producing industries is a fully cycled approach of knowledge through the entire product life cycle and supply chain. Feedback loops are essential for passing information and knowledge back and forth. Knowledge is produced in each single step of the supply chain, from research, procurement, development, production, marketing, and so on up to the customer contact. For example, the customer feedback can be used in the process of research and development (e.g., dosage form), the swaying and sometimes insecure pricing structure at procurement can have an influence on human resource planning (e.g., expertise) and on research (e.g., use of alternative substances). This vision is already implemented in some producing industries like cell phones or car production (Figure 2.2). Pharmaceutical knowledge management has a lot of regulatory limitation to slow down the process of implementing this vision.

The approach of the pharmaceutical industry toward this vision is often referred to as *Quality by Design*. These principles—first adopted by the automotive industry—focus the development process around the product, which is optimized toward the customer's needs. These can be derived in very different manners—in the case of car or mobile phone ownership the data and knowledge about customer needs is sent directly to the industry because of online connectivity. In pharmaceutical processes, this seems to be a challenge, as the customer (respectively the patient) is only integrated in the knowledge–loop–process during clinical trials. After the product has passed this stage, the knowledge flow from customer back to the industry

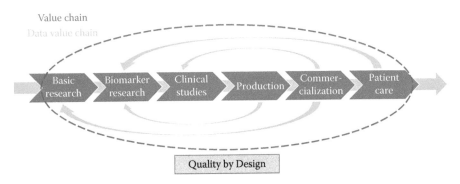

FIGURE 2.2
A product life cycle in life sciences with the underlying data value chain.

is almost completely missing. Modern supply chain management, Quality by Design approaches and all issues around personalized medicine have to come up with a way to integrate the customer knowledge into their management processes.

From Knowledge Management to Data-Driven Business

Peter Drucker postulated that the main challenge of the twenty-first century is to reach the same level of efficiency improvement in knowledge work than the industrial revolution has achieved with manual labor. Knowledge management initiatives and theories have made great advancements in this direction and already provided significant efficiency improvements in many sectors and industries.

Today we can observe a similar, ongoing discussion with respect to the importance of *data* for production and businesses processes (Meyer-Schönberger and Cuiker 2013). However, the rise of *data-driven business* (also referred to as *data-driven industry* or *industry 4.0* or *industrial Internet*) goes even one step beyond Drucker's vision: not only allow (big) data technologies to speed up business and industry processes, they enable to treat data as a resource that can be produced, refined, delivered, and sold as a product in its own right or as part of a data-driven service. The great advantage of the data-driven business approach over the traditional knowledge management approach is that data can be completely separated from humans, or let us put it differently: *content can be separated from interpretation and action*. Although in knowledge management, human motivation, and willingness to share knowledge and to make it explicit is key to success, the data-driven approach data can be captured completely automatically without any human involvement. Also, in order to process the huge amounts of captured data, automatic data analytics, and machine learning approaches can be utilized. The critical moment where *the rubber hits the road* is at the point of interpretation, sense making, decision making, and action. Here again (as with knowledge management) *humans are the critical agents to interpret the results of data analytics in the light of domain knowledge* (e.g., of logistics and agile manufacturing) and in context of the concrete situation and task at hand (e.g., supplier selection in a crisis situation). The advantage is that this interpretation only needs to happen in the moment in which a decision is needed and can be based on all data that has been captured until then by multiple agents and even from around the world. In contrast, knowledge management requires us to document knowledge and rationale for decisions long before a concrete decision needs to be made. This has two major disadvantages: first, humans need to invest significant resources (and brain power) in order to document knowledge that might never be used again; second, during time of documentation it is completely unclear which parts of the knowledge (e.g., situation parameters) will be of interest in the future. The first disadvantage leads to demotivation on the side of your employees who should provide the knowledge, second disadvantage leads to frustration on the side of your employees who should reuse the documented knowledge.

Again, data-driven business approaches offer solutions. With big data analytics approaches existing data can be reused in contexts and situations are completely different than what they were created for. That is, it is not important beforehand to know how a specific data set will be utilized for deriving insights. Instead, for each new task we can review our data storages and identify the data that might be useful to answer open questions. In addition, if we do not find any suitable data within the organization we can turn to outside sources (open or fee-protected) or even implement more sensors to generate the missing data. Thus data-driven business benefits from a cycle in which new technologies trigger new questions and hypothesis and these in turn trigger new data collection.

This data-driven approach creates a whole new market for data (and knowledge) because it can be largely separated from humans and thus sold freely. This idea is captured in the *data value chain* concept. It recognizes that data has a value of its own right and that this value increases with the proximity to the data consumer. Each refinement or interlinking of data with other data leads to an increase in value. This provides vast opportunities for companies—also small and medium companies—that specialize in niche data, applications, and services. Together with novel business models they have the potential to change knowledge work as much as the industrial revolution has changed manual labor, thus revolutionizing industry and business in an equally disruptive way.

In Europe this development is impeded by many different cultures and languages that lead to a high fragmentation of available data. For example, telephone registries in different countries are structured differently and are in different languages, thus hindering the common usage across the EU. Therefore, the commission has developed the *Digital Single Market* strategy and promotes it strongly throughout the member states.

Clearly data-driven business is not a silver bullet either. Traditional knowledge management approaches have their own advantages and prove useful in many situations. However, the combination of both can help us to reach our vision as described above more rapidly and provides us with flexibility. There are a number of challenges that have to be tackled before the full potential of data-driven approaches can be utilized:

1. *Combining domain knowledge with data analytics knowledge*: Currently (big) data analytics technologies still require a deep understanding of the algorithms, their suitability for application in specific situations, and their limitations. We need to combine this analytics knowhow with profound knowledge of the application domain in order

to identify interesting correlations and identify novel effects. *Visual analytics* is a computer science research field that tackles this challenge by utilizing visualization techniques to enable domain experts to navigate, browse, search, iteratively analyze, and discover relationships within large amounts of data.

2. *Establish feedback loops*: As discussed earlier, feedback loops are the defining element of knowledge management. Only by testing assumptions, evaluation, and taking feedback seriously can a system improve. When utilizing (big) data analytics approaches it is crucial that the domain experts with each interaction provide feedback to the analytics results and thus help to develop the system into a self-learning system. *User in the loop* is a concept that takes this iterative and cyclic approach seriously. With the help of domain-specific visual analytics tools (see above) domain experts can interpret the data and gain insights. In turn, these insights together with qualitative as well as quantitative feedback can be fed back to the analytics engine and help improve it (e.g., through improvement of algorithm selection in specific situations). We have seen the emergence of terms such as *doctor in the loop, engineer in the loop,* and *controller in the loop*.

3. *Integrating knowledge-based models with data-driven models*: Knowledge-based models such as engineering models of combustion and business models of production processes have been developed and refined over decades and thus embody significant amounts of domain knowledge. With the help of these models domain experts develop abstractions of the real world and strive to predict outcomes in novel situations. However, by abstracting from the real world they typically represent a *clean* reality (e.g., one without inferences between different machines in the same room). The massive availability of data on the other hand allows us now to build data-driven models directly from *dirty* reality. The data captured represents both effects explained in knowledge-based models as well as unintended side effects. However, it is difficult to assess which aspects of those data-driven models are relevant, are generalizable, and provide interesting insights. Thus, both types of models have their advantages and disadvantages. The challenge of the next years will be to combine both types of models in order to better understand *dirty* reality.

4. *From descriptive models to predictive models*: Many knowledge-based models are able to predict how the change of one parameter will affect another and how this will affect the results. This prediction though is only valid as long as all assumptions and preconditions about the situation in which the model operates hold (clean reality). If we want to transfer knowledge-based models to different situations the data-driven approach might be able to help. One typical

approach is to capture data in situations in which the knowledge-based model holds and then train a data-driven model that exhibits the same results and predictions as the knowledge-based model. Afterward the data-driven model can be used on data that represents situations for which the knowledge-based model does not hold (or in which it is unclear if it will hold). By training the data-driven model on this new situation, an integrated knowledge and data-driven model can be developed that allows us to predict results in the new situation.

In the automotive and high-tech industries the concept of data-driven industry (or industry 4.0) is already being implemented: both along the complete product life cycle as well as with specific focus on production processes. In the following we will provide a few small examples from this environment that might be useful to reflect and explore within the life sciences:

Predictive Maintenance: To accomplish batch size one and full automation, the production should be flexible and continuously up and running. Machine failures can disrupt the entire production and therefore have to be eliminated. The concept of predictive maintenance conquers this problem by analyzing sensor data streams and detecting irregular events before they occur (see Figure 2.3). Although engineering models have proven useful in this area, data-driven approaches have been shown to provide added value in cases where inferences between different machines and other exterior factors are suspected.

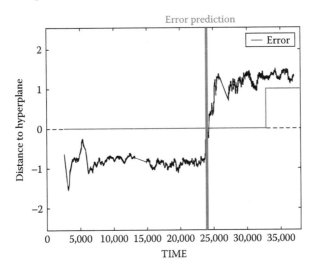

FIGURE 2.3
Predicting an error before it happens.

Hypothesis Generation and Validation: When analyzing chains of events and their effects on each other it is helpful to turn to visual analytics tools that visualize the data space and help domain experts to filter and zoom in on suspicious patterns. When analyzing failure mode and effects analysis (FMEA) within complicated engineering and production processes critical situations can be identified early and relevant actions can be triggered. These techniques can support research when no models regarding the correlation of the gathered data exist (Figure 2.4).

Automatically Linking Different Types of Data: Complex processes such as software development as well as pharmaceutical engineering share certain characteristics: there is not one reference structure but many equally relevant structures that are heavily interconnected (complex network structure), different roles and business units interact with different parts of this network, and generating information that needs to be accessible by many other roles and business units (constant change triggering unexpected inferences in the network). In order to make information and data available to all relevant stakeholders at the moment when they need it (without searching!) is crucial. Engineering models, knowledge from production machinery, analytic data from probes, and much more has to be taken into account when approaching pharmaceutical research with all aspects of Quality by Design. By automatically linking different types of data to each other and allowing users to explore and browse them along different dimensions and specialized for specific tasks supports effective information exchange and knowledge generation (Figure 2.5).

Summary

Within this chapter we have provided the background to engage in an informed discourse on knowledge management and data-driven business. Many of the concepts have stayed theoretical but will be taken up in the following chapters.

Take Home Message

The essential principle of all knowledge management theories is the existence of feedback loops. When new knowledge is acquired it needs to be learned and applied by people (internalized). Through this learning-by-doing, new experiences

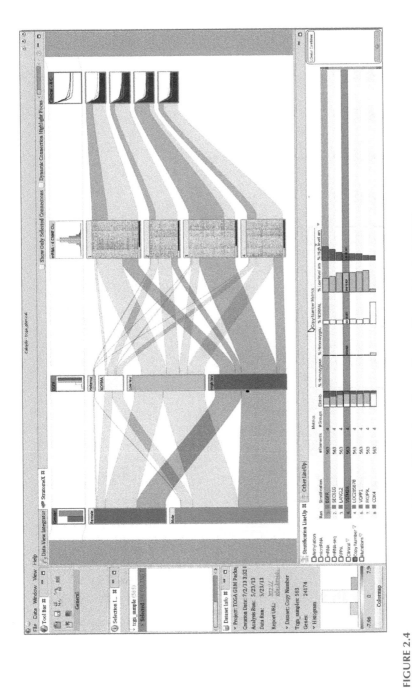

FIGURE 2.4
Using visual analytic tools to gain insights and define hypotheses. (Source: Dennerlein et. al. (2013)).

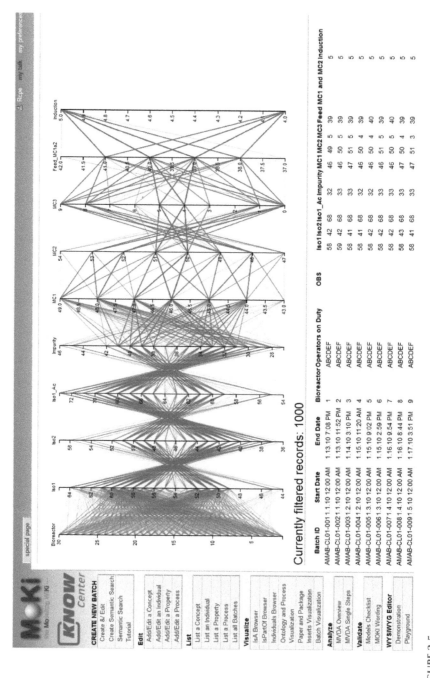

FIGURE 2.5

Interactive, visual tool to support the Quality by Design approach in pharmaceutical research. (Source: Dennerlein et. al. (2013)).

are gained and out of these new knowledge can arise and can be documented to a certain extent (externalization, tacit knowledge). Organizations need to provide structures, processes, and tools that encourage, structure, and reward such feedback processes. There exist many knowledge management (or organizational learning) theories and approaches that suggest how to do so. They distinguish themselves by how much emphasis they give either to the human-centered or the documentation-centered school of thought. However, knowledge management activities are always grassroots activities, which can be supported and encouraged but not mandated or even demanded. People's motivations for knowledge management has to be intrinsic and therefore the management of knowledge will always rely to a large extend on the knowledge sharing culture within the organization.

Data-driven business separates content from interpretation and action and turn data into a resource that can be captured, refined, and used. But since data, information, and knowledge are not absolute terms but represent a spectrum from simple data points to highly connected knowledge, this new paradigm will affect how we will approach knowledge management in the future.

References

Argyris, C. and Schön, D. (1978). *Organizational Learning: A Theory of Action Perspective.* Reading, MA: Addison Wesley.

Basili, V. R., Caldiera, G., and Rombach, H. D. (2002). *Experience Factory Encyclopedia of Software Engineering.* Hoboken, NJ: John Wiley & Sons.

Debenham, J. K. (1988). *Knowledge Systems Design.* Englewood Cliffs, NJ: Prentice-Hall.

Dennerlein, S. et.al. (2013). Assessing Barcamps: Incentives for Participation in Ad-hoc Conferences and the Role of Social Media. *In Proceedings of the 13th International Conference on Knowledge Management and Knowledge Technologies (i-Know '13),* Stefanie Lindstaedt and Michael Granitzer (Eds.). ACM, New York, Article 15, 8 pages. DOI=http://dx.doi.org/10.1145/2494188.2494208.

Dennerlein, S., Lindstaedt, S., Leitgeb, S., Lex, E., Trattner, C. (2013). Bringing ICH Q10 to life - *Knowledge Management for efficient life-cycle management in the context of QdB.* In: BioProcess International Conference. Düsseldorf: BioProcess International.

Fessl, A., Rivera-Pelayo, V., Müller, L., Pammer, V., and Lindstaedt, S. (2011). Motivation and user acceptance of using physiological data to support individual reflection. *2nd MATEL Workshop at European Conference for Technology Enhanced Learning* (ECTEL 2011).

Hargadon, A. B. (1998). Firms as knowledge brokers: Lessons in pursuing continuous innovation. *California Management Review,* 40(3): 209–227.

Hofer-Alfeis, J. (2003). Measuring knowledges and KM in an organization with an explicit top-down knowledge strategy. In Reimer, U., Abecker, A., Staab, S., and Stumme, G. (Eds.), *Wissensmanagement,* Springer: Berlin, pp. 433–442.

Hofer-Alfeis, J., van der Spek, R., and Kingma, J. (2003). The knowledge strategy process. In *Handbook on Knowledge Management: Knowledge Directions*, 2, Holsapple CW (ed.). Springer: Berlin; 443–466.

Kendal, S. and Creen, M. (2007). *An Introduction to Knowledge Engineering*. London, UK: Springer-Verlag.

Lavez, J. and Wenger, E. (1991). *Situated Learning: Legitimate Peripheral Participation*. Cambridge, UK: Cambridge University Press.

Meyer-Schönberger, V. and Cuiker, K. (2013). *Big Data: A Revolution That Will Transform How We Live, Work and Think*. London, UK: John Murray Verlag: Wiesbaden.

Nonaka, I. and Takeuchi, H. (1995). *The Knowledge-Creating Company: How Japanese Companies Create the Dynamics of Innovation*. Oxford, UK: Oxford University Press.

North, K. (2011). *Wissensorientierte Unternehmensfüuhrung*. Gabler Verlag, Wiesbaden.

Rivera-Pelayo, V. (2015). *Design and Application of Quantified Self Approaches for Reflective Learning in the Workplace*. Karlsruhe, Germany: KIT Scientific Publishing.

Uit Beijerse, R. (2000). Knowledge management in small and medium-sized companies: knowledge management for entrepreneurs. *Journal of Knowledge Management*, 4(2): 162–179.

Wong, Y. K. and Aspinwall, E. (2004). Characterizing knowledge management in the small business environment. *Journal of Knowledge Management*, 8(3): 44–61.

3

Knowledge Management—A Japanese Perspective

Yukio Hiyama

CONTENTS

The Japanese perspective on knowledge management (KM) comes to us by kind permission of International Society for Pharmaceutical Engineering (ISPE) as a reprint of an article authored by Dr. Yukio Hiyama originally published in May 2014 in an e-journal on knowledge management for the pharmaceutical industry.* Dr. Hiyama, a leading international regulator of long-standing, represented the Japanese National Institute of Health Science (NIHS) on the developments of significant International Conference on Harmonization (ICH) Q8, Q9, and Q10 guidance documents. This article presents a historical reflection on knowledge management issues relative to these ICH discussions and the 2005 Pharmaceutical Affairs Law change in Japan. Dr. Hiyama shares his unique insights on the importance of managing knowledge for our industry and our patients.

Editorial Team

I am delighted to see detail emerging with expanded presentations on Knowledge Management, particularly the case studies from industry. The International Conference on Harmonization (ICH) Quality Implementation Working Group (QIWG) team, which included myself, could not provide such practical advice on the topic at the time of our work.

---

* Supplement to pharmaceutical engineering, May 2014, http://www.pharmaceuticalengineering. org/pharmaceutical-engineering-magazine/pe-supplements

Early ICH Discussion and Japanese Regulation Change

First, let me present my personal reflection on *knowledge management* related issues at ICH and in regard to the regulatory framework development for 2005 Pharmaceutical Affairs Law (PAL) change of Japan.

In July 2003, the ICH GMP workshop adopted the vision; "Develop a harmonised Pharmaceutical Quality System applicable across the lifecycle of the product emphasizing an integrated approach to risk management and science." The U.S. FDA, who proposed the workshop, suggested knowledge sharing and transfer models, as a basis of efficient postapproval change management and defined optimal knowledge content and knowledge sharing as agenda items for discussion in their proposal [1]. The MHLW presented the new Pharmaceutical Affairs Law framework to become effective in 2005 [2] and the outcome of the 2002 MHLW study [3]. At that time, MHLW expected ICH to take on *technology transfer*, as the MHLW study in 2002 had identified poor communication between Research and development, and manufacturing as one of the significant problems. The study group sorted key information that should be transferred from R&D to manufacture and issued a technology transfer guideline.

The PAL change in 2005* was intended to allow the (Japanese domestic) pharmaceutical industry to contract out manufacturing activities. Very often contract givers are R&D-based organizations, whereas contract recipients are of course manufacturing organizations. This was one of the reasons why the Japanese authorities had significant concerns over the effective communication between R&D and manufacturing.

Having those concerns in mind, I participated in the ICH discussions in the following years. The first ICH Q10 meeting in November 2005† produced a proposed structure of the quality system guideline. The initial structure contained four chapters:

1. Introduction
2. Pharmaceutical quality management system
3. Management responsibilities
4. Lifecycle models

---

* Before 2005, manufacturing contracts were not allowed under the manufacturing authorization framework. The 2005 law change introduced the market authorization framework where manufacturing contacts are possible. The framework before 2005 was seen as discrimination against Japanese industry because industry outside of Japan was allowed to contract manufacture under the importing authorization framework that coexisted with the manufacturing authorization.

† PMDA conducted the first foreign GMP inspection in fall of 2005. As significant concerns were expressed earlier, discrepancy between manufacturing practices and the content of submission is often cited by PMDA foreign inspection.

The chapter on lifecycle models had a subchapter called *technical transfer/ knowledge management* with a note; "resolve terminology Knowledge Management: intent manage knowledge through lifecycle." The subchapter had an additional heading of *organizational learning* (i.e., learn from one product to next). This represents the early thinking about KM by the Q10 team.

In October 2006, the team produced draft version 8.0 that went outside the team for the first time. The draft expanded the *lifecycle models* chapter into two separate chapters for Product Lifecycle and for Quality System Lifecycle. Knowledge management and quality risk management were then described as principles and tools in the product lifecycle models chapter. At that time, there was NOT consensus on the difference between the quality system's elements (or functions) and tools that should be used in quality system. After extensive discussion, the team reached a conclusion that quality risk management and knowledge management are the most important tools that should be used in the quality system and declared that they are not Pharmaceutical Quality System (PQS) functions. In the step 2 document for public consultation issued in May 2007, the two tools are finally identified as *enablers*. The four PQS elements (monitoring system, CAPA, change management system, and management responsibilities) are required directly as tasks in the PQS, whereas QRM, KM, and others are tools to ensure the performance of the PQS. This was confirmed by extensive discussion at Q10 meetings between draft 8.0 and final step 4 document.* Later in order to reconfirm this, QIWG wrote the Q&A document† stating that KM is not a system and that there is no regulatory expectation to see a formal knowledge management approach.

More Recent ICH Discussion

In 2008 at the QIWG first meeting in Portland (OR), there were three breakout sessions for Quality by Design, Pharmaceutical Quality System, and knowledge management. The team decided to write Q&As on the three topic areas and to invite case studies from outside. The knowledge management subteam, including myself, struggled in obtaining practical case studies and Q&A proposals. As a result, the subteam was not as productive as the others in terms of writing Q&As. However, during the course of QIWG training material development, QIWG was able to write recommendations on knowledge management in various parts of the training documents.‡ (See box below)

---

* ICH Q10 guideline.
† ICH QIWG QAs.
‡ ICH QIWG Training material in 2010 are available from ICH website.

It should be noted that during the ICH discussions, only explicit knowledge was discussed. At one time in a QIWG meeting, there was a proposal to take up tacit knowledge for discussion. However, others did not support that proposal. This may be because there was a view that explicit knowledge is the only knowledge that can be actually formally used; tacit knowledge may be useful to connect knowledge to create new explicit knowledge but cannot be used directly (formally) for actions.

KNOWLEDGE MANAGEMENT PLAYS VERY VITAL ROLES IN THE PHARMACEUTICAL QUALITY SYSTEM

In 2009 and 2010, QIWG wrote extensive training materials that included six presentations (titled: *Introduction, How ICH Q8, Q9, Q10 Work Together, Case Study, Regulatory Assessment, Manufacturing/PQS, and Inspection*) and four breakout session slides (titled: Design Space, Control Strategy, PQS, and QRM). Below are extracts from these training documents on knowledge management.

- Prior knowledge to support the understanding, risk assessment, and scope of DoE in development. (*Work Together slide 14*)
- Maintain and update knowledge management in commercial manufacturing stage. (*Work Together slide 17*)
- List of prior knowledge for the case study. (*Case study slide 14*)
- Manufacturing have a key role to play; Using knowledge gained during development; Using current site knowledge (e.g., similar products); Building on knowledge through transfer, validation, and commercial manufacturing activities; Feedback of knowledge to development. (*Manufacture slide 4*)
- General on PAI Drug Product; Is there a process for acquiring and managing knowledge? (*Inspection slide 21*)
- Information from technology transfer activities, scale up, demonstration, and process qualification batched is *particularly* valuable. (*Inspection slide 38*)
- DS development–prior knowledge. (*Design Space session slides 8, 9*)
- Assess prior knowledge to understand materials, process, and product with their impact in the process for defining the control strategy. (*Control Strategy session slide 11*)
- Expand body of knowledge for continual improvement of product and PQS (*PQS session slides 16, 17*)

- Linkage between QRM and KM; Risk assessment in relation to knowledge management can be linked to identifying data to be collected (risk identification), analyzing raw data
- (Risk analysis), evaluating the results from measurement will lead to information (risk evaluation); New information should be assessed and risk control decision captured; Knowledge management facilitates risk communication among stakeholders. (*QRM session slide 14*)

Feedback from the training sessions, which were held in the three ICH regions, showed that there were not significant questions about knowledge management at that time. As a result, knowledge management is not among the six topics included in the *Points to Consider* document[*] issued by QIWG.

Some Thoughts on Knowledge Management to Conclude

Yakushi-Ji Pagoda Rebuild Story

It may be appropriate to bring up the 10 year long (2009–2019) disassembling and rebuilding project of East Pagoda of Yakushiji Temple[†] (Yakushi is Medicine Buddha) in Nara, Japan. The Pagoda was built in 730 and it retains the original structure with original materials that have survived earthquakes, typhoons, and war fires. The last rebuilding project was finished in 1900 and the one before was in 1644. Major building components include wood pole, wood beams, and Japanese nails (Wakugi) that are expected to last for thousand years. So selecting components is very challenging. Knowledge transfer for rebuilding is even more challenging. Training of shrine/temple carpenters is difficult because they have rare opportunities to use their expertise. Techniques or the craftsmanship they use would be extremely difficult to document. In a recent (only two decades ago) rebuilding project at Horyuji Temple, the head of carpenters conducted an assessment of the existing structure during the disassembling process in order to identify the previous building process and the tools. Compared to the challenges shrine/temple carpenters face, the challenges pharmaceutical manufacturing professionals have in terms of knowledge management seem to be straightforward. However, there are common challenges between the two different tasks. That is to obtain and develop explicit knowledge that can be used.

---

[*] ICH Quality Implementation Working Group Points to Consider (R2), December 6, 2011.
[†] Yakushiji Temple http://www.nara-yakushiji.com/

Lessons from Yamamoto Science History

If you look carefully at the history of science and technology (e.g., Yoshitaka Yamamoto [4]), they have been developed through the dynamics between strong belief (even religious) and observations. Among them is the modern scientific breakthrough of the seventeenth century, based on Johannes Kepler's laws of planetary motions, the theory of which heavily relied on the precise and comprehensive Mars orbit observations by Tyco Brahe.

Recent technology development have a tendency to use a <*Develop theory (hypothesis) first and conduct experiment (observe)*> approach rather than <*Observe first and interpret the result*> approach that was historically employed. Although the <*develop theory first*> approach may provide the quickest solution, one cannot discover something that has not yet been thought of. So, do not abandon the <*observe and interpret*> approach totally.

I also learned from Yamamoto's masterpiece that scientific knowledge gained by humans is very limited compared to the natural rules that govern universe. So the value of "20" is discussed in 80/20 rule of knowledge management (Calnan N., 2014) might indeed be overestimated.

Publications to Share Knowledge and Build Common Knowledge Base

Based on my personal experience as an NDA reviewer at NIST, techniques and approaches found in dossiers are commonly used between companies. So those techniques are unlikely unique know-how to one company. In order to use prior knowledge more effectively, by every party including the regulatory authorities, I would like to encourage industry to publish more on the learnings gained from actual development.

References

1. Calnan, N. (2014). The 80/20 Rule of Knowledge. Pharmaceutical Engineering: Knowledge Management E-Supplement Edition. ISPE.
2. FDA Proposal. *ICH Workshop,* July 16–18, Brussels, Belgium, Circulated on June 30, 2003.
3. Isozaki, M. MHLW's view on the quality regulations for the 21st century, *ICH GMP Workshop,* Brussels, Belgium, July 2003.
4. Hiyama, Y. Studies on quality assurance supported by health sciences grant (H14-Iyaku-04), *ICH GMP Workshop,* Brussels, Belgium, July 2003.
5. Yamamoto, Y. *Jiryoku to Jyuryoku no Hakken (Discover of Magnetism and Gravity),* Vol. 1–3, Misuzu. Tokyo, 2003. ISBN 4-622-08031-1 C 1340 http://www.msz.co.jp/book/author/14051.html.

4

Accelerating the Opportunity for the Pharmaceutical Industry through KM

Lauren Trees and Cindy Hubert

CONTENTS

> APQC has more than 20 years experience in researching and advising many industries on knowledge management best practices. In this chapter APQC describes best practices that could accelerate implementation of KM in the pharmaceutical industry.
>
> **Editorial Team**

APQC (formerly the American Productivity and Quality Center) has been leading research and working in the field of knowledge management (KM) for more than 20 years. This—along with our work in quality[*] and process improvement—has given us the foundation to develop a standard roadmap for KM program development, collect and disseminate thousands of best practices and lessons in KM, establish a framework for KM maturity assessment, and publish two books as strategic and practical tools to guide

---

[*] APQC is the co-creator of the Malcolm Baldrige National Quality Award.

47

organizations in designing and implementing their own knowledge management initiatives.

After working with countless KM programs over the years, our key message to the pharmaceutical industry is, "Do not reinvent KM best practices!" This is one of the most important guiding principles for KM. Industries ranging from oil and gas to professional services have spent the past two decades building effective enterprise KM capabilities and learning what it takes to sustain activity and value over the long term. Proven practices and methods can be readily adopted to help organizations accelerate their efforts and standardize how knowledge moves through and enables their business processes. Yet the pharma industry, staffed by some of the brightest engineers and scientists in the world, is off trying to solve the KM problem on its own, instead of benefiting from the lessons and experiences of those who have gone before them.

This is not to say that the pharmaceutical industry does not have good activities going on in the name of knowledge management—in fact, APQC sees evidence of many localized or business-unit–specific KM capabilities. But when each pocket of the business is pursuing KM according to its own rules, local capabilities can feel random and end up duplicating efforts. Not only does such duplication represent an inefficient use of resources, it also hampers the free flow of knowledge and expertise across teams, functions, and business units. Standardization, by contrast, promotes the strategic alignment and integration required to coalesce around an enterprise approach to KM.

Pharma has a great opportunity to accelerate its learning. That being said, there are several factors impacting the industry's ability and readiness to adopt KM best practices. Over the years, APQC has received continuous feedback from its members and clients on what is impeding adoption. Highlights of the common feedback follow:

- APQC observes a lack of common understanding of KM as a holistic practice among many in the pharma industry, including the regulatory bodies. Interactions between industry and regulators are time consuming and iterative. Although regulatory agencies attend pharma events to listen and learn, they are driving the need to document everything. Documented evidence of pharma activities is important to demonstrate compliance—*if it is not written down, it did not happen* is a common mantra used across the industry. This certainly may be true from a regulatory view, but when organizations get so focused on capturing explicit knowledge, they often ignore the importance and impact of sharing tacit knowledge, tapping into expertise, and encouraging collaboration.

- The pharmaceutical industry has a strong belief in its own uniqueness. *We are so different in our operations (compared to other industries) that we cannot even adapt, much less adopt, proven practices in KM* was a statement shared with APQC at a recent pharma association meeting. This led to a lively discussion of leadership's role in KM. Insights shared highlighted the lack of direction and engagement from the top leadership in pharma organizations. Many leaders have not fully grasped the opportunity, which means that grassroot efforts are proceeding without the benefit of an enterprise-wide vision. Without top-level support, KM capabilities are unlikely to align with the organizational strategy or be capitalized on for a broader purpose.

- The pharmaceutical industry is doing well. When the bottom line looks good, it is difficult to focus on the wasted time that comes from repeated reinvention and searching for needed content and expertise, especially because this waste is difficult to measure directly. Alternatively, the pharmaceutical industry may be a victim of the external forces hitting it, including globalization and the continuous churn from mergers and acquisitions. These forces result in the creation of huge organizations with massive stores of knowledge and expertise to draw on, but where knowledge flow is hindered by structural and cultural silos, incompatible technology systems, and a deficit of trust among colleagues who have little history with one another and limited opportunities for face-to-face collaboration.

With the recognition of these challenges comes optimism that the pharmaceutical industry can and will apply best practices in order to embrace KM on a broader scale. This book highlights the opportunity by providing insights and examples from the early adopters who are recognizing the gains from KM efforts. In the remainder of this chapter, APQC will provide a point of view about the current state of KM in pharma, why the timing is important for accelerating adoption of best practices and lessons, and what the industry can do to maximize the learnings from others.

The Current State of KM in Pharma

APQC's research provides ample evidence of the pharmaceutical industry's stalled progress on the road to KM maturity. Since 2007, APQC has used a 146-question diagnostic called the KM Capability Assessment Tool to evaluate more than 200 KM programs against a five-level KM maturity model. APQC codeveloped both the levels of KM maturity (Figure 4.1) and the assessment with a working group of advanced KM practitioners, which helped validate the tools and ensure that they reflect the reality of KM within a wide swath of industry and government.

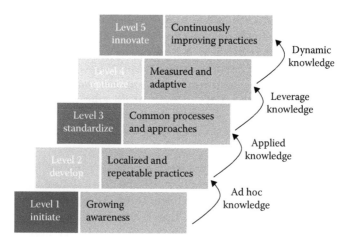

FIGURE 4.1
APQC's Levels of Knowledge Management Maturity®.

The assessment process allows KM teams to reflect on their progress and prioritize improvement opportunities, but the aggregated data also reveals trends in KM program development. Pharmaceutical's overall KM maturity ranks in the bottom third of participating industries, with an average maturity rating of 1.19 [1]. This means that a majority of pharmaceutical participants are aware of the need to support and enable the flow of knowledge, but have not yet put in place formal KM strategies or developed standard, replicable KM processes.

As KM programs frequently reach out to APQC to complete an assessment as part of initial strategy development, it is not uncommon to see a majority of responses clustered at the lower end of the maturity scale. What *is* noteworthy about the assessment results from the pharmaceutical industry is the scarcity of top performers. Only 4% of pharmaceutical KM programs have reached maturity level 3 or above—compared to 18% across all other industries—and no pharma program has yet achieved the top two maturity levels (Figure 4.2).

This relative lack of maturity is reflected in many facets of knowledge flow, but most especially those related to the transfer of tacit knowledge and expertise. For example, 79% of pharmaceutical respondents to an APQC survey ranked increasing STEM (science, technology, engineering, and math) expertise as a significant or urgent business priority—the most urgency expressed by any of the 20 industry groups represented [2]. Nearly four in five pharmaceutical industry representatives said that it was critical for their organizations to build employees' technical competencies to keep up with changing knowledge domains and product mixes,

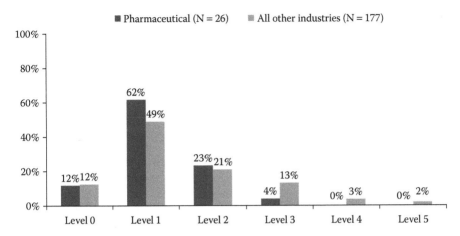

FIGURE 4.2
Distribution of assessed KM programs by KM maturity level.

and half said their employees were in critical need of technical leadership and program/project management skills.

However, when compared with other industries that cite STEM expertise shortages as a key issue, pharmaceutical firms are not investing in standardized, integrated efforts to accelerate knowledge transfer and learning for novice *or* midcareer employees. Industries such as petroleum and software, which face similar challenges related to knowledge retention and upskilling, are significantly ahead of pharma in developing approaches to combat expertise shortages and prevent the loss of critical product, project, and process knowledge (Figure 4.3).

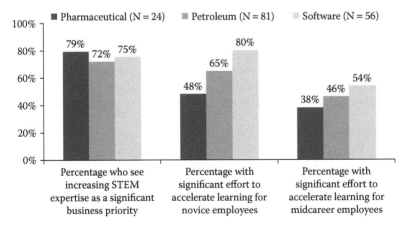

FIGURE 4.3
Growing STEM expertise: Urgency vs. action in select industries.

Pharmaceutical firms should increase learning opportunities for all parts of the workforce, but a focus on midcareer employees may yield the biggest and most immediate payoff. As demand for expertise threatens to outstrip supply, these midcareer professionals provide the best hope to address knowledge gaps and preserve know-how related to key business processes. The transfer of deep tacit knowledge from experts to midcareer professionals is particularly crucial because pharmaceutical's long product lifecycles mean that the current experts in a particular drug are likely to retire or move on before that drug goes out of production.

In short, if pharmaceutical firms want to safeguard knowledge across the product lifecycle and ensure that a sufficient pipeline of technical leaders are able to manage projects, innovate, and make tough decisions, they will need to focus more attention on the transfer of tacit knowledge, allocate additional resources to the development of midcareer employees, and give these employees more time to engage in formal and informal learning activities.

Why This Is a Pivotal Moment

Despite its relatively low level of knowledge management maturity, the pharmaceutical industry has seen an upsurge of interest in KM in recent years, fueled in part by regulatory emphasis on quality management. The implementation of Quality by Design (QbD) [3] and Quality Risk Management (QRM) [4] have brought additional rigor to the information generated across the product lifecycle. With this, processes and tools to better manage the complexity of information and knowledge are needed.

In addition, many companies have grown (either organically or through acquisitions) and are operating with more dispersed teams; this makes it less feasible to share knowledge informally around the water cooler. Accelerated timelines provide the opportunity for faster ROI, but they also increase the risk of insufficient knowledge transfer leading to costly errors and poor quality.

It is insightful that pharmaceutical regulatory guidance ICH Q10 [5] has included KM as an *enabler* to a Pharmaceutical Quality System (PQS). Not only does this focus regulatory agencies' attention on KM and the expectation to leverage prior knowledge, it also makes good business sense. With the release of ICH Q10, organizations have started directing more attention to the flow of knowledge through the product lifecycle and how teams working on the same product capture knowledge, communicate, and share with one another throughout the end-to-end lifespan. ICH Q10 calls for knowledge to be systematically captured, shared, and transferred throughout the product lifecycle to reduce the risk of knowledge loss. Although this definition is more data- and content-centric than APQC would like, it has created aware-ness and discussion of the importance of tacit knowledge flow. Most impor-tantly, it helps to highlight why KM is crucial for the industry to adopt and implement now.

APQC recommends beginning your KM journey by adopting a holis-tic view of risk that takes knowledge flow into account. It is important to understand how knowledge moves through your organization's product lifecycle, paying special attention to moments where responsibility for the product shifts from one team to another. Looking at these transitions can help uncover potential disconnects or bottlenecks that impede knowledge transfer and have the potential to compromise quality.

Avoid the Mistakes of Other Industries as You Move Forward

Below are three of the most common mistakes we see companies make in terms of their KM approaches. Recognizing and avoiding these pitfalls will help you shape an effective KM strategy and put you on the right path to improving the flow of knowledge and expertise throughout your company's product lifecycle.

Mistake 1: Treating All Knowledge as Equal

A lot of knowledge gets created as a product moves from research through commercialization to market, but not all of it is important. Consider, for example, only 1 in 5,000 drugs that enter preclinical testing are ultimately approved for market—and for the meager 0.02% that make it through this pipeline, the process takes an average of 12 years [6]. Mountains of informa-tion get generated at each stage of development, and for products that do not make it to market, only a small subset will be relevant to future potential projects. Knowledge related to successful products is more critical to capture and share, but still must be winnowed down to a manageable volume.

When a company embarks on a KM effort, a common early misstep is attempting to document and transfer *all* the knowledge and expertise asso-ciated with a product's evolution, without a concerted effort to isolate details with a direct bearing on quality, compliance, and other key issues. If employ-ees are encouraged to share everything learned during a particular develop-ment stage, they are likely to throw up their hands in despair. In addition, even if they manage to wade through the tedium (note: they will not), it would be impossible for the next team in line to absorb this level of detail.

So what constitutes critical knowledge, and how does a team decide what to share? Common examples of knowledge that should be prioritized for transfer—some of which are explicitly called out in ICH Q10—include technology transfer activities, process validation studies, manufacturing experience, raw material testing data, stability reports, product quality reviews, complaint reports, adverse event reports (patient safety), and product or manufacturing history. But there is no one-size-fits-all answer. What is critical depends on the product itself, the process under which it is being developed, the skills and expertise of the groups involved, and how you intend to apply the knowledge in the future.

For this reason, we recommend bringing together the key stakeholders from R&D, regulatory, and commercial manufacturing to create a knowledge map of what should be documented and shared. A knowledge map acts as a snapshot in time to help a company understand what critical knowledge and expertise it has and what it lacks. It highlights and organizes knowledge assets so that managers and employees can quickly see gaps in knowledge, pinpoint key resources, and learn where to find in-house experts who can offer advice.

A good knowledge map starts with a detailed view of the underlying processes that make up the product lifecycle (if product development does not follow well-documented processes, then no amount of knowledge transfer will be sufficient). Then, for each step in the process, the stakeholders should collaborate to answer the following questions:

- What knowledge is vital to this process step?
- What do we learn during this step that we need to preserve or impart to teams further down the value chain?
- Who possesses the knowledge, or where is it located?
- Is the knowledge already documented, or does it exist only inside someone's head?
- Who will need the knowledge in the future, and in what context?

After mapping its knowledge, the product lifecycle team should apply various risk filters to zero in on the specific knowledge that is most likely to be overlooked and for which the consequences of loss will be most severe. This further refines the priorities for knowledge flow and lets employees know what they should emphasize when communicating with adjacent groups working on the same products and processes.

Once the product lifecycle groups perform this exercise a few times, they will start to get a feel for the knowledge and expertise that need to accompany the product through its lifespan. In many cases, an organization with mature KM and collaboration practices can let this knowledge flow naturally without explicitly calling out each type of knowledge to be transferred. However, it is still a good idea to recalibrate at key intervals and make sure that new processes and technologies have not shifted the organizational definition of critical knowledge.

Mistake 2: Paying Insufficient Attention to Transitions

In addition to identifying the types of knowledge that should be prioritized for transfer, a good knowledge map helps expose points in the product life-cycle where knowledge is most likely to fall through the cracks. Usually, these are moments when one function passes the baton to another—for example, when a product moves from pharmaceutical development to commercial manufacturing or a product moves from one manufacturing site to another. When the individuals are working on the project change, there is a risk that information and insights will not shift seamlessly to the next group. Business disruptions such as mergers, acquisitions, internal reorganizations, and changes in leadership can also open up knowledge gaps, especially when these disruptions lead to significant or sudden turnover in the teams associated with a particular product.

Obviously, the team relinquishing responsibility must provide documentation to satisfy regulatory requirements and ensure project continuity—and structuring these records so that they are easy to navigate and search comes with its own set of challenges. But experts who are passing on a project after years of work have deep expertise and contextual knowledge that is difficult to capture in a technical report or database. Some companies give teams handing off responsibility a very limited time in which to share tacit knowledge associated with the project. Allowing these knowledge chokepoints to develop is a critical error with the potential to negatively influence quality metrics. A well-structured database where the project team documents its lessons learned can help convey past experiences and current-state understanding, but it does not eliminate the need for person-to-person transfer, whether it occurs face-to-face or virtually.

In short, do not expect a day or two of meetings to satisfy the knowledge transfer requirements when a project shifts to the next phase or group. Instead, treat these handoffs as extended transitions, where a core team stays engaged in order to share its expertise in context. Sometimes knowledge that does not seem particularly relevant during one part of a project (and is therefore not shared during handoff) becomes crucially important down the road. Any overlap in the people responsible for successive lifecycle stages can prevent nascent knowledge gaps from going unnoticed and unchecked.

Another option to retain critical knowledge and ensure it flows between functional groups is to establish technical networks focused on the end-to-end lifecycle for a particular product. If key stakeholders from each phase in a product's development are arranged into a knowledge-sharing community, members will be in a position to surface their tacit knowledge if and when it becomes relevant. Such communities may hold face-to-face meetings, or they may rely exclusively on virtual exchanges through community discussion boards, listservs, or enterprise social networks. The key is to ensure that significant developments are shared throughout the product lifecycle and to allow people to chime in with expertise related to their unique experiences on the project. When the company is able to bring its collective product knowledge to bear on challenges as they emerge, it is more likely to avoid costly errors and optimize the product and associated processes at each successive stage of development.

Mistake 3: Underestimating the Support Needed for Change

Although it is important to understand how knowledge flows through the product lifecycle and to target bottlenecks where they occur, the biggest reason why KM efforts fail is that they do not recognize the need to nurture and communicate the change. People are busy, and they may not understand why they are being asked to do something *extra* to share knowledge with colleagues outside their team. Similarly, those on the receiving end of such exchanges may not want to spend time learning about the previous stages of a project, or they may not have the absorptive capacity to take in the knowledge in a meaningful way.

All industries struggle with KM change management, but the pharmaceutical industry is significantly behind the cross-industry mean. In fact, KM programs in pharmaceutical firms report low levels of maturity in all people-related capabilities, including governance/leadership, resources, change management, and communication (Figure 4.4). Furthermore, the gaps on resources, communication, and overall people capabilities were rated as highly statistically significant, indicating that they are likely to persist beyond the specific survey group.

To truly improve the flow of knowledge, companies need to treat their KM implementations as they would any significant organizational change initiative, securing buy-in from all levels and addressing barriers that prevent new approaches and behaviors from taking root.

When building a business case and change strategy for KM, the first step is to determine the value proposition for the enterprise. Talk to business leaders,

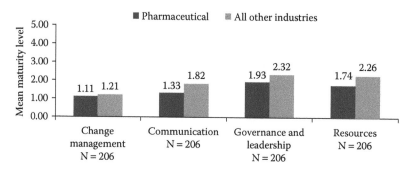

FIGURE 4.4
Knowledge management maturity on people-related capabilities.

find out what is keeping them up at night, and then articulate the role that KM can play in mitigating those concerns. Leaders are more likely to actively endorse the effort—and encourage their employees to take it seriously—if they see a direct impact on staff development, quality, and business results.

APQC suspects that the frequency of mergers and acquisitions within the pharmaceutical industry has impeded the development of cohesive, leadership-driven KM business cases. But companies that have recently undergone a merger or acquisition have a particularly urgent need for—and are in a unique position to benefit from—enterprise KM. Effective knowledge transfer and collaboration approaches can help these newly formed companies break down legacy knowledge silos, assimilate disparate groups of employees, and ensure that everyone is aware of the expertise and intellectual property available to them post-merger or acquisition.

Armed with executive-level support, the next step is to answer the *What is in it for me?* question for the individuals expected to participate. Often, this comes down to a good communication strategy. Employees have many of the same goals as the companies they work for: to ensure the success of their projects and become better at what they do. If KM is positioned as a way to build expertise and improve long-term outcomes, people are more likely to get engaged. Moreover, KM can provide networking opportunities and act as a vehicle for greater visibility and a shared sense of belonging, all of which are powerful drivers of participation. Knowledge sharing can help people get noticed and counteract the feelings of invisibility and insignificance that some experience in large, globally dispersed organizations.

Once the business case is established, the company should work to address logistical or cultural obstacles that impede effective KM. APQC CEO Dr. Carla O'Dell likes to point out that people do not hoard their knowledge; they hoard their time. Often, the biggest barrier to KM adoption is that people think they are too busy to share or that in-depth knowledge exchange will slow down their projects. In APQC's experience, the best solution to such challenges is to explicitly call out sharing and collaboration as part of people's jobs, emphasizing the importance of effective knowledge flow for broader business goals. When employees are allotted specific hours to transfer their knowledge and the expectation that they will do so is built into business processes, job descriptions, and performance goals, it becomes easier to move people beyond the *I am too busy* mentality.

Other barriers stem from the nature of global operations, where the teams that need to share knowledge:

- work in different locations and time zones;
- do not share a primary language;
- belong to different subcultures within the organization that are not bound together by the same processes, norms, and business principles.

New technologies—including video conferencing and automatic translation tools—can help bridge some gaps, but clashes between internal subcultures are more difficult to deal with. Usually, it requires sustained executive support, combined with a hard look at the incentives that drive behaviors, in order to get divergent groups working together effectively toward enterprise goals. The value of knowledge transfer must be communicated from the top and reinforced through the performance management and reward structure in order to drive meaningful change in employee attitudes and behaviors.

Conclusion

APQC is confident that, using the best practices described here and in the rest of the book, pharmaceutical companies can jumpstart their knowledge management efforts and accelerate to the next level of KM maturity. Use industry developed guidance to your advantage. ICH Q10 provides the industry with a visible platform to raise awareness of the importance of knowledge flow, get leadership engaged, and build the foundation for an enterprise approach.

For pharmaceutical organizations that are launching new KM initiatives (or reenergizing current ones), changing how knowledge flows between groups—along with getting people to adopt new attitudes toward sharing and collaboration—can feel daunting. However, other organizations have already blazed this trail, and you can improve your odds of success by applying their proven practices and learning from their mistakes. A clearly communicated value proposition, a knowledge map to guide your efforts step-by-step, and a comprehensive change strategy to reinforce the desired behaviors will put you well on your way to effective knowledge management across the product lifecycle.

APQC helps organizations work smarter, faster, and with greater confidence. It is the world's foremost authority in benchmarking, best practices, process and performance improvement, and knowledge management. APQC's unique structure as a member-based nonprofit makes it a differentiator in the marketplace. APQC partners with more than 500 member organizations worldwide in all industries. With more than 40 years of experience, APQC remains the world's leader in transforming organizations. Visit us at www. apqc.org and learn how you can make best practices your practices.

References

1. APQC. Knowledge Management Maturity Industry Report: Pharmaceutical. APQC Knowledge Base, November 30, 2015: https://www.apqc.org/knowledge-base/documents/knowledge-management-maturity-industry-report-pharmaceutical
2. Carla O'Dell and Lauren Trees. How Smart Leaders Leverage Their Experts: Strategies to Capitalize on Internal Knowledge and Develop Science, Engineering, and Technology Expertise. APQC Knowledge Base, March 14, 2014: https://www.apqc.org/knowledge-base/documents/how-smart-leaders-leverage-their-experts-strategies-capitalize-internal-kno
3. ICH Q8. ICH Harmonised Tripartite Guideline: Pharmaceutical Development Q8 (R2), 2009.
4. ICH Q10. ICH Harmonised Tripartite Guideline: Pharmaceutical Quality System Q10, 2008.
5. ICH Q9. ICH Harmonised Tripartite Guideline: Quality Risk Management Q9, 2005.
6. Drug Approvals – From Invention to Market ... A 12-Year Trip. *MedicineNet.com* (retrieved February 22, 2016).

Section II

Perspectives on Knowledge

This second section of the book brings together several key perspectives on knowledge on behalf of the patients, the regulators, academia, and an editor-in-chief of a leading industry newsletter. A special contribution is also included from outside the biopharmaceutical industry shared by the former Chief Knowledge Officer (CKO) at National Aeronautics and Space Administration (NASA).

This section opens with the patient perspective, shared by Dave deBronkart who puts forward a personal and passionate case for more proactive collaboration between patients, the medicinal products scientific community, and clinicians. There follows two interesting KM case studies from within the Swiss and the Irish medicines regulatory agencies, proving that managing what is known about the safety of medicines is not just the responsibility of the R&D or manufacturing organizations. NASA then shares how the International Space Station (ISS) project involved 21 international partners, and that 80% of every program or project undertaken at NASA is an international collaboration. What are the lessons that biopharma can apply to our complex partnership networks on managing knowledge amidst complexity? There follows two perspectives from academia, the first of which questions if knowledge management is the *orphan* enabler of ICH Q10, and the second introduces an exciting new knowledge assessment model, which has been developed following rigorous academic research, known as the managing individual knowledge (MinK) framework.

This section closes by looking forward to what ICH Q12, *Technical and Regulatory Considerations for Pharmaceutical Product Lifecycle Management*, might hold for KM; the role knowledge will play in the effective lifecycle management of products and the impact on both the regulators and the regulated to embrace *knowledge excellence*.

5

Who Moved My Facts?: Patient Autonomy and the Evolution of Infrastructure Mean Best Available Knowledge Is Not Where It Used to Be

Dave deBronkart

CONTENTS

> In his TEDTalk, Dave deBronkart tells of when he learned he had a rare and terminal cancer, he turned to a group of fellow patients online—and found the information that helped to save his life. Here he outlines how proactive collaboration between patients, the medicinal products, scientific community, and clinicians can lead to better outcomes for all.
>
> **Editorial Team**

Who Moved My Facts?

We used to know with certainty where to get the best medical advice: you asked the leading authority, or its proxy, the peer reviewed literature. That is still a good source (usually, as we will see below), but today's smart patient communities know that other sources may hold better information than the available literature, with additional knowledge that the published literature does not include, containing greater wisdom about what matters to patients.

This represents an earthquake of uncertainty, a profound shift in reality that every knowledge manager must master, or else run the risk of being an expert in an expired model. This risk is heightened by a sociological change progressing in tandem, which is the move toward greater patient autonomy, toward *e-patients*: patients who are empowered, engaged, equipped, and enabled.

From a KM perspective, empowerment is driven by access to information. The World Bank says* empowerment is "increasing the capacity of individuals and groups to make choices and to convert those choices into effective actions." Empowerment, therefore, springs from information in action, where knowledge is power.

From a business perspective, *information in action* plays an important role in optimizing many challenges. But from a patient perspective, the stakes are higher: when the context is *my mother has cancer* or *my baby is dying, action* takes on urgency on a different scale, far more than the publication of a quarterly report.

As a survivor of stage IV renal cell carcinoma (RCC), I speak from experience when I say such circumstances give patients a burning desire for the latest facts. It also gives patients a keen sense of what is important *to us*—which may or may not match science's current goals. As patients become more empowered, we learn to *speak and seek*, that is to speak up about it and go hunting on our own. At the Connected Health conference in 2008, Internet guru Clay Shirky said about my patient community, "The patients on ACOR† don't need our permission and they don't need our help."

In times of change, a historical perspective can be helpful. This is not the first time the goals people must seek have changed due to a radical change in the environment.

---

* http://web.worldbank.org/WBSITE/EXTERNAL/TOPICS/EXTPOVERTY/EXTEMPOWER MENT/0,contentMDK:20245753~pagePK:210058~piPK:210062~theSitePK:486411,00.html
† ACOR.org: Association of Cancer Online Resources.

Who Moved My Cheese?

Who Moved My Cheese?* is a classic 1998 book about how past methods no longer work in a changed world, leaving us thinking, *Hey! I was doing fine—but things are not the way they used to be! What happened? Why?* That book was about profound changes in labor markets, and urged readers—who were typically newly jobless—to adjust to the new world, and find *cheese* where it is now, not where it used to be.†

In terms of knowledge management, the birth of the web has caused a comparably seismic shift in how knowledge arises and travels. The methods we formerly relied on to know whether something is or is not trustworthy are no longer sufficient. Simply put, reliable facts are no longer reliably present where we used to look for them: as we will see below, some facts in the scientific literature are acknowledged to be suspect, and good facts show up in unexpected places outside the literature.

We could paraphrase the book title for KM: *Who moved my facts? I used to be good at this—why did it change?* It raises the question of how we are to keep operating as competent professionals on this altered planet.

One answer is to apply scientific thinking, to understand what has changed and what has not.

Thomas Kuhn's 1962 masterpiece *The Structure of Scientific Revolutions* (Kuhn 1962) proposed that when too many undeniable cases arise that cannot be explained by a profession's current model, a crisis develops that leads to revolution, for instance the Copernican revolution. This type of revolution may be what we are now seeing in terms of medical knowledge: clinicians' and researchers' skills remain essential, but something has changed in the environment—in the flow of knowledge—and it is altering what is possible.

New opportunities are opening up, which may seem like an existential threat to the traditional discovery, research and development model for medicines. Indeed, if one stays in the pre-Copernican paradigm it may well be the case, but those with an analytical mind can embrace the new world and adapt.

---

* amazon.com/Who-Moved-My-Cheese-Amazing/dp/0399144463

† To be sure, many of *Cheese*'s readers had reason to feel bitter, having sometimes been abandoned by employers to whom they had long been loyal; this is reflected to this day in the book's Amazon reviews. The change in knowledge management is less dramatic in its personal impact, but every bit as important to understand.

The Transformation of Patient Knowledge Access

The change in knowledge access is particularly well illustrated by a visual model created in 2010 by Lucien Engelen at Radboud University Medical Center in the Netherlands with his colleague Marco Derksen. It suggests that knowledge is like a nutrient that enables a more robust response, and that the Internet (or social media) acts as its capillaries. Knowledge flows through a series of networks, which makes it far easier to obtain optimal information on any clinical case, and thus gain insights on potential optimal outcomes (Figure 5.1).

The *capillaries* metaphor can be extended in a telling way: e-patients do not just passively receive what floats by; they are highly motivated to *grow* new networks to bring what they need. One could call it *information angio-genesis*, akin to the physiological process through which new blood vessels form from preexisting vessels. But as a new stream of facts starts flowing to us—patients and scientists alike—we are faced with an important new question: which ones are reliable?

In my own case my patient community quickly told me that some practicing clinicians were very out of date about current RCC treatments, because the literature they were taught to trust was out of date as compared to actual current clinical practice. To a scientist this may seem like "that's to be expected," but to patients this is a *big* problem. How can you find information you can count on, when your life is at stake? What makes knowledge reliable?

At the core of the medical inquiry is the question, *What is the best thing to do in the case of this patient?*

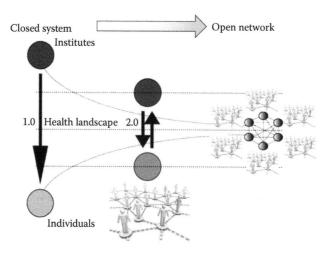

FIGURE 5.1
Transformation of knowledge access. (Adapted from Engelen and Derksen 2010.)

Our ability to answer this question has improved as science has matured in areas such as personalized medicine and the development of companion diagnostics, and as our understanding of which information we can count on has also improved.

Centuries ago physicians had little or no idea what they were doing: their practice was based on their belief that there were four humors (blood, yellow bile, black bile, and phlegm). Then in 1761, an Italian anatomist Giovanni Morgagni connected symptoms with pathologies discovered at autopsy—a major step forward. Meanwhile, bloodletting was still believed valid by learned men into the 1800s and it was not until the Flexner Report in 1910 that it was decided it would be good for medical students to see a lot of patients—supervised—before being licensed to practice on their own.[*] Even a half century later, brutal practices such as radical mastectomy were practiced en masse by the "science" of medicine with little or no evidence to support them.

It was not until the 1980s that evidence-based medicine began to take hold. Perhaps ironically, this advance toward more scientific thinking also led to an artificial and rigid belief that if it is not in the literature, it is not true and should be rejected. This scientific method, widely taught in schools and universities throughout the world, conflicts with evidence from the academic and publishing worlds that has acknowledged the occurrence of publication bias and inherent failures in the peer review process.

Meanwhile today, patients report that their suggestions are often greeted with *"There is no literature supporting that,"* as if it must therefore be false. However, we know that lack of evidence does not mean it is not true—if all truth has been discovered then governments and industry can save a bundle by closing the labs and sacking all the researchers.

Just imagine how the e-patient community feels about statements like this from the editors of three major journals, all issued in the past decade:

> "After 30 years of practicing peer review and 15 years of studying it experimentally, I'm unconvinced of its value.... Evidence on the upside of peer review is sparse, while evidence on the downside is abundant.... most of what appears in peer-reviewed journals is scientifically weak."

Smith (2009), 25 year editor of the BMJ

> "It is simply no longer possible to believe much of the clinical research that is published ... I take no pleasure in this conclusion, which I reached slowly and reluctantly over my two decades as an editor of *The New England Journal of Medicine.*"

Angell (2009), NEJM

---

[*] Flexner called the hapless patients "clinical material."

"The case against science is straightforward: much of the scientific literature, perhaps half, may simply be untrue. Afflicted by studies with small sample sizes, tiny effects, invalid exploratory analyses, and flagrant conflicts of interest, together with an obsession for pursuing fashionable trends of dubious importance, science has taken a turn towards darkness. As one participant put it, 'Poor methods get results.'"

Horton (2015), Editor-in-Chief, *The Lancet*

The questions remain, what knowledge *is* reliable? And who gets to say?

This is not just a philosophical pursuit. To seek the answer to the question: *What should we do for our dying baby?* requires new ways of viewing the issue of reliable knowledge. Table 5.1 outlines a summary of the traditional view versus the e-patient perspective.

The Problems of Knowledge Flow

The pharmaceutical industry considers knowledge management as an internal matter. However, can KM be outbound, too? Can industry—through their knowledge management programs—advocate for proactive, purposeful dissemination of new medical knowledge, so that less value is spilled and more reaches the point of need, that is, the patient and clinician working on a case? If not, what does that say about the priority of maintaining a patient focus right across the medicinal product lifecycle?

In terms of knowledge flow, at least four classes of problems exist that challenge the paradigm of how information reaches the patient: three L's (latency, liquidity, and liability), and information blocking. When lives are at stake these are *real* problems.

Latency

A learning health care system continuously and reliably captures, curates, and delivers the best available evidence to guide, support, tailor, and improve clinical decision making and care, safety, and quality.

Best Care at Lower Cost*

Balas and Boren (2000) established that on average it takes *17 years* for *half* of the physicians to adopt new knowledge. Throughout this time many patients continue to suffer, while available knowledge and treatments go largely unharvested in decade-size gulps. Information, knowledge, and wisdom

---

* Institute of Medicine, *op. cit.*

TABLE 5.1

Traditional View and e-Patient View of Trustworthy Information

Traditional View	e-Patient Perspective
Science is hard—what could less trained people know?	• Patients and families have lots more access to information than they used to, on demand or as needed • Enabled by this and 20 years of the Web, autonomy is on the rise: people are accustomed to going looking for something even if they do not know what it is—and finding it!
Peer-reviewed literature is reliable; nothing else has been vetted so well.	• The editors of several leading peer-reviewed journals have raised their own concerns about publications being: *scientifically weak …. no longer possible to believe much of the research that is published … much of the literature may be simply untrue* (see above for citations) • Publication bias and *the decline effect* (Ioannidis 2010) exist • Many negative trials go unreported,[a] even though they would add to the body of scientific knowledge • Even if a trial is conducted perfectly, who gets to say what should be studied and what outcomes should be measured?[b]
The latest published literature is your best source.	• Publication delay occurs • Dissemination delay (see *information latency* below)
If it is not in the literature, do not trust it.	• Who gets to determine whether something is worth studying or sharing? • *Gray knowledge* has value, that is, valid information that will never rise to the level of a clinical trial[c]
Resulting belief: Reliable information comes from, and only from, the peer reviewed literature.	*e-Patient reality*: The literature is one source, but it is not at all guaranteed, nor is it the boundary of what is *valid or not*. We must think, explore, and revalidate constantly.

[a] AllTrials.net.

[b] The BMJ patient initiative has solved this: "Authors of research papers are being asked to document if and how they involved patients in defining the research question and outcome measures, the design and implementation of the study, and the dissemination of its results." In (Richards and Godlee 2014).

[c] Example of gray knowledge: side effects of the treatment that saved my life, high-dose IL-2, sometimes kill patients, but the literature contains no information on how to endure them. My patient community gave me 17 firsthand stories, and today my oncologist says "I'm not sure you could have tolerated enough medicine if you hadn't been so well prepared." Dave.pt/davebmj1.

can die on the vine—or worse, from the patient's perspective, patients die—because the "knowledge ambulance" never got to the point of need.

The cure for scurvy was discovered in the 1500s, but was not disseminated throughout the British Empire for a further 264 years (Hey et al. 2009). Three centuries later, Semelweiss's hand-washing discovery, which cut maternal deaths by 90%, languished for a half century before being fully accepted. Imagine how many mothers died from this failure of knowledge management and how many families felt that pain.

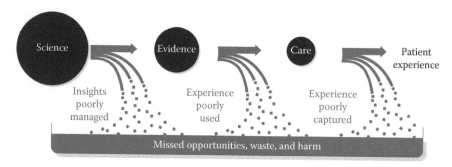

FIGURE 5.2
From science to patient experience. (Adapted from Best Care at Lower Cost, IOM 2012.)

Here is how the Institute of Medicine (IOM) depicts the missed opportunities that can arise between scientific discovery and the patient experience (Figure 5.2).

Notice that the knowledge and value that reaches the patient, at right, suffers from every shortfall in this chain: The missed opportunities arising from scientific insights that are poorly managed, coupled with evidence that is poorly used, compounded by experience that is poorly captured. In a manufacturing or retail context such waste is referred to as rejects, breakage, or shortage, and it is countered with *loss prevention* strategies and quality improvement programs. But whether we consider medicine, manufacturing, or retail, such spillage is a clear loss in potential value.

When patients have a voice in all aspects of medicinal science—including governance of information—those days will end. Meanwhile, in activated patient communities today, we take matters into our own hands—even if governments (and doctors) advise us to stop googling.*

Liquidity

Knowledge liquidity is a simple concept. Imagine two paper cups on a table: one filled with coffee beans and one filled with coffee. If the coffee bean cup tips over, the beans do not travel far; however, if the liquid cup tips over, the coffee goes everywhere.

In the Internet age, the liquidity of information flow is the reason why facts can show up where clinicians do not expect them. If this is not well understood, public and private policy in regard to knowledge sharing will continue to be suboptimal.

---

* dave.pt/belgiangoogle2, dave.pt/belgiangoogle3, dave.pt/belgiangoogle4, dave.pt/googlemug

Lability—The Rate of Change

Many of us were trained in an age where once you learned something, it was valid for most of your life. To understand how wrong this is in today's society, take a look at the book *The Half-life of Facts: Why Everything We Know Has an Expiration Date* (Arbesman 2012).

e-Patients know it is essential to check whether any advice they receive has expired.

Information Blocking

Francis Bacon famously said that knowledge is power. In health care, knowledge can be a matter of life or death. Dissemination delay is one acknowledged failure mode; however, from the patient's perspective, another failure presents a moral crisis: intentional blocking of information flow.

There are two principal modes of information blocking. One is sociological—policies that keep information penned in—and the other is commercial. From the e-patient's perspective it is bad enough that the system does not proactively push information out to the point of need but it is even worse when patients seek information and are actively blocked. Consider this anecdote:

> "Mr. Murphy ... wanted to know more about the potential risks and benefits of the proposed treatment and had repeatedly asked Dr. Blakely to help him obtain a copy of the definitive review article, which had recently appeared in a major medical journal.... Dr. Blakely had not done so. Finally, in desperation, Mr. Murphy had decided that there was only one way to obtain this vital medical information he needed: He would have to impersonate his own physician. As he picked up the envelope with Dr. Blakely's name on it, a hospital security guard stepped out of the doorway where he had been waiting."
>
> *e-Patients: How they can help us heal health care* (Tom Ferguson 2007)

This story actually took place in New Jersey in 1994, the same year the Netscape browser arrived on the market. The browser—arguably the biggest change seen to date in information infrastructure—was such a threat to the information establishment that the following year Bill Gates made his browser free of charge, declaring that he would make browsers a *zero revenue business.**

---

* Web Week, Vol. 2 (1996), No. 15 (October 7), p. 19. As quoted in Eliason and Karlson (2001). In the same chapter, Eric Schmidt (now Google CEO, then at Novell) is quoted as saying that at Sun Microsystems his customers had lamented, "I can't take this much innovation this quickly." Note: he was there 1983–1997—20 years ago.

Today open access to journal articles is a hot topic, with many leading scientists supporting it, especially where research is funded by government grants.

Yet in the arena of personal health data, such as that held in electronic medical records (EMRs) databases, the term *information blocking* has assumed a new life. In 2015, the U.S. National Coordinator for Health IT used this term in a report to the Senate stating that some hospitals and some system vendors were *knowingly interfering* with the flow of patient data to other systems or service providers, for commercial reasons. The conclusion drawn was that information blocking was occurring to keep patients from seeking care elsewhere or to keep hospitals from receiving data from a competing vendor. In this age of big data analytics, practices such as these may also represent a missed opportunity to systematically study postmarket trends for patient outcomes in populations much greater than those typically seen in a clinical trial.

At the 7th annual Health Datapalooza event in Washington in May 2016, Vice President Joe Biden gave a heartfelt speech about his son Beau, who died a year earlier of glioblastoma, a devastating brain cancer. He showed himself to be a highly engaged patient/family member (that makes him an e-patient), noting that "When someone dear to you is in trouble, you want to learn as much as you can."

He forcefully expressed his view of how unacceptable it is that important patient information does not currently move freely from hospital-to-hospital, and went further to emphasize how equally unacceptable it is for researchers to think they do not have to share their data with others. In doing so he called out the *New England Journal of Medicine* January 2016 editor's contribution on this subject, noting that some call such thinking "data parasites."

Noting the importance of sharing the data, Biden states, "we don't do this [hide the raw data] in NASA and we don't do it in DARPA. We only do it in medicine." Biden's passion in that speech illustrates an aphorism in my speeches: "Patient is not a third person word. Your time will come."

The patient view: How can information in action produce value, when it is leashed and bound?

Patient Needs and Perspectives: A Culture Change

In 2012 the Institute of Medicine wrote:

> A learning health care system is anchored on patient needs and perspectives.

> **Best Care at Lower Cost**[*]

---

[*] Institute of Medicine, 2012. Table S-2.

Note the term *anchored* on the patient's needs. This infers a starting point, an essential element, not that you commence your work then step back for another view. Furthermore, your anchor point must hold true even when the seas get stormy.

A movement is afoot to promote patient collaboration in medicinal science development. In a 2015 essay in the *British Medical Journal* (*BMJ*) I wrote:

> A growing movement, exemplified by the Society for Participatory Medicine and the annual Stanford Medicine X conference, asserts that patients and clinicians must collaborate. Central to it's belief is that whereas the physician brings training and clinical expertise, patients bring their life experience, their deep investment in the outcome of their case, their skills and resilience, and a unique perspective on needs and priorities. The movement recognizes patient autonomy as a valid priority and patients' hearts and minds as essential contributions to the best possible care. We believe that medicine cannot achieve its potential if it ignores the voice of thinking patients.[*]

Some medical schools are beginning to integrate patients as lecturers, or even as participants in the design of curriculum. As chronic pain e-patient Britt Johnson[†] recently said to a class of medical students, "Patients, by default, learn a version of medical education that then needs to be valued and learned by practitioners." In addition, when Radboud UMC (which created that knowledge flow diagram) spent two years completely revising its medical school curriculum, patients were involved in creating every clinical course.

But like all culture change, this will take time, partly because not everyone in the patient community has asked for change—yet. Paradigms die hard, and so do cultural beliefs about roles and responsibilities. So it is with the changing role of patients, including their pursuit of information, assessment of its value, and about putting it into action. Ideally, this change will take place in a participatory partnership with the pharmaceutical industry, the scientist, and the clinicians, but if not, patient empowerment will happen from the ground up. It is especially ironic when an informed patient is aware of weaknesses in a given study and their clinician is not. A vested interest in outcomes is a powerful motivator.

In the final chapter in his manifesto, *The Autonomous Patient and the Reconfiguration of Medical Knowledge*[‡], Ferguson writes:

> In the late nineteenth century, Oliver Wendell Holmes assured a group of medical students that, "Your patient has no more right to all the truth you know, than he has to all the medicine in your saddle bags ... he should get only as much as is good for him."[§]

---

[*] dave.pt/davebmj2
[†] theHurtBlogger.com; Twitter @HurtBlogger
[‡] Ferguson, *op. cit.*, chapter 7
[§] "Why do patients need info?" Filler/endnotes, BMJ 2000;321: October 28, 2000. The Holmes quote is dated 1871.

The twin presumptions here, of course, are that the patient could not possibly know what is good for them and the physician certainly does. Today we know neither is necessarily true. Yet as recently as 2001 the American Medical Association proposed this advice to patients:

> "Only your physician has the necessary experience and expertise to diagnose and treat medical conditions. Trust your doctor, not a chat room" (AMA 2001).

American Medical Association

As with many cultural revolutions there can be a well-intentioned side to the resistance of change. Ferguson illustrates this point below with regard to the training of clinicians:

> ... distinguished medical educators repeatedly reminded me that I must strive to convince each patient I saw that my colleagues and I knew everything we needed to know about his or her condition and its treatments—even when I felt this was not the case. Should I fail to do so, I was warned, my patients might "lose confidence" in me.

Care meant being responsible for the patient's state of mind and this was the era when many would not tell patients they had cancer, because there was little they could do, and they felt—with the best of intentions—that the best care meant not burdening the poor souls with sadness and fear any sooner than necessary.

You might protest that this was long ago and that we are better now. However, remember, change travels slowly. In 2010 at a gathering of patients who had my disease (RCC) and received the same medicine, about a third said they had been given no warnings about the possible side effects of their medication. Their doctors said that as there was nothing they could do about side effects, why burden the sufferer with fear of things that might not happen? That is well intentioned, but should not I be the one to decide if I want to know this information or if I would rather not?

In the Society for Participatory Medicine blog, heart patient Hugo Campos, who has an implanted cardioverter/defibrillator wired into his heart, wrote about the importance of patient autonomy:

> "Autonomy is true empowerment. It promotes patient responsibility and holds the promise to lead us to more engagement and better health. We must move beyond participatory medicine and focus on educating, enabling, and equipping patients with the tools necessary to master autonomy and the art of self-care" (Dawson 2014).

On this note I close this chapter by sharing the conclusions from my 2015 BMJ essay:

Patient powered healthcare is no insult to clinicians any more than home thermometers insult pediatricians or home glucose tests insult endocrinologists, nor any more than a drowning swimmer insults a lifeguard by climbing onto a raft.

In the era when patients could not make use of advanced information, it was the job of science to digest it before delivering it. In today's world patients want the raw facts, promptly, so we can help solve *our* problems according to our priorities. We know firsthand that the result will be more lives improved and more lives saved, and to us that is even more important than it is to you.

Please, let patients help.

References

Angell, M. 2009, January 15. *Drug Companies and Doctors: A Story of Corruption.* The New York Review of Books 56. Available: http://www.nybooks.com/articles/archives/2009/jan/15/drug-companies-doctorsa-story-of-corruption/. Accessed 20 September 2010.

AMA suggests resolutions for a healthy new year. Press release, 20 December 2001. http://www.e-patients.net/AMA_Dec2001.pdf.

Arbesman, S. 2012, 27 September. *The Half-life of Facts: Why Everything We Know Has an Expiration Date.* New York: Penguin Group.

Balas, E.A. and Boren, S.A. 2000. Managing clinical knowledge for health care improvement. *Yearbook of Medical Informatics 2000.* 65–70. https://scholar.google.com/scholar?hl=en&q=%22Managing+clinical+knowledge+for+health+care+improvement%22&btnG=&as_sdt=1%2C30&as_sdtp=

Dawson, N. A case for autonomy and the end of participatory medicine, 5 December 2014. http://e-patients.net/archives/2014/12/a-case-for-autonomy-and-the-end-of-participatory-medicine.html.

Eliason, G. and Karlson, N. (eds). 2001. *The Limits of Government: On Policy Competence and Economic Growth.* New Brunswick, NJ: Transaction Publishers. p. 105.

Hey, A.J.G., Tansley, S., and Tolle, K. 2009. *The Fourth Paradigm: Data-Intensive Scientific Discovery.* Redmond, VA: Microsoft Research.

Horton, R. 2015. Offline: What is medicine's 5 sigma?. *The Lancet 385.* 9976:1380.

Ioannidis, J. 2010, December 13. The truth wears off. *The New Yorker.*

Kuhn, T. 1962. *The Structure of Scientific Revolutions.* University of Chicago Press: Chicago.

Richards, T. and Godlee, F. 2014. The BMJ's own patient journey. *British Medical Journal.* 348:g3726.

Smith, R.W. 2009, October. In search of an optimal peer review system. *Journal of Participatory Medicine.* 1(1):e13.

6

Valuing Knowledge at Swissmedic—
A Regulatory Agency's Perspective

Michael Renaudin and Hansjürg Leuenberger

CONTENTS

To further emphasize the importance being placed on the management of knowledge from the regulators perspectives, here the Swiss medicines agency (Swissmedic) shares their journey toward valuing the knowledge held within their knowledge intensive organization. They discuss how knowledge management can often be difficult within an *expert*-based organization, where the structure of the organization may be characterized by the autonomy of the individual experts coupled with the potential for lower levels of interest in the concerns of the organization as a whole. This drove them to develop a strategic program where the focus is on people and culture first and foremost, then on the organization and finally on the technology, IT, and tools.

Editorial Team

Where We Are Today

Swissmedic is the Swiss agency for the authorization and supervision of therapeutic products including both medicinal products and medical devices. We fulfill our legal mandate to ensure that the authorized therapeutic products are of high quality, effective, and safe, and in doing so work with partner authorities on a national and international basis. We endeavor to make a considerable contribution toward protecting the health of humans and animals, and we also participate in safeguarding Switzerland as a location for industry and research.

Swissmedic is currently based across three separate locations in Bern. This includes an office and laboratory building on Freiburgstrasse that has been operational for two years, a modern office building on Erlachstrasse that has recently come into service in 2015, and our main headquarters on Hallerstrasse, which is not very new but still a fine building.

In 2015 the management board decided that the HQ building on Hallerstrasse should be modernized and brought up to the technical and architectural standard of the other two buildings. The head of infrastructure, who was responsible for this upgrade project, set up a working group comprising representatives from each of the organizational units concerned and the Swissmedic knowledge manager. This working group was given the task of developing two or three design concepts for using the available space. The scope of work for this project acknowledged that Swissmedic is a knowledge-based organization, and the premises therefore needed to be adapted to the needs of its knowledge workers. For this reason it was necessary to ensure that the concerns and suggestions of the knowledge manager were taken into account in the design concept. The key evaluation criteria for the concept were as follows:

- Provide benefit for the organizational units, that is, meeting their logistical and functional concerns and requirements.
- Management of costs, including one-off refurbishment costs and changes to ongoing operating costs.
- Support for knowledge management goals and enable effective knowledge exchange.

As you can imagine, leadership support of this approach endorsed the role of knowledge management within the organization and was a sign of the great estimation for the people responsible for knowledge management at Swissmedic. No doubt, like us you know of knowledge managers who spend years trying to get approval for a concept, who put enormous effort into attracting attention and interest for their concerns, all to no avail. Here at Swissmedic we are lucky enough to work in an area that has a substantial impact on the everyday working lives of employees and that allows us to provide a sustainable benefit for the organization.

How did we get this far? Why does knowledge management enjoy such high standing within Swissmedic today?

We think that this has a great deal to do with the organization itself, and more particularly with the culture within this organization and the people who work there. Senior management recognizes the advantages that can be gained through specific KM activities and acknowledges that knowledge management at Swissmedic is organized along very pragmatic and unbureaucratic lines.

What follows seeks to explain how this positive situation arose. We describe how knowledge management at Swissmedic is both organized and positioned, and provide an overview of the people involved, the tools we use, and the approaches we take. We end with a summary of the most important lessons learned.

Swissmedic and Knowledge Management

About Swissmedic

As noted, Swissmedic is the Swiss authority responsible for the authorization and supervision of therapeutic products. Swissmedic's activities are based on the *Law on Therapeutic Products*. As a Federal public law institution with its headquarters in Bern, Swissmedic, the Swiss agency for therapeutic products, is autonomous with respect to its organization and management, and has its own budget.

Swissmedic is mainly financed by means of fees and to a smaller extent by payments from the confederation in return for providing services of public utility.

Swissmedic is attached to the Federal Department of Home Affairs. The Agency Council, Swissmedic's strategic decision-making entity, represents its interests to both the Federal Department and the Federal Council. The Agency Council also approves Swissmedic's budget, annual accounts, and annual report.

A wide range of activities are carried out in accordance with both our legal mandate and the needs of the various stakeholders. These include patients, the therapeutic products industry, health care professionals, authorities, and organizations in Switzerland and abroad and the media.

Knowledge and Swissmedic

Swissmedic is *knowledge-intensive*; this means that much of the work it does is intellectual in nature. Its employees are well-educated knowledge workers. The knowledge of our employees is the major factor that determines the success of our organization; it is an elementary resource of the kind listed among the classic factors of production: labor, capital, and land. The special feature of this intangible resource *knowledge* is that it is not depleted if it is used; in fact, it grows with use. It can, however, diminish, for example, if employees leave the organization, taking their knowledge with them. Or if knowledge lies idle and goes to waste through lack of use.

Given the importance of knowledge for our organization and for the general (societal and organizational) development toward a knowledge society, it is elementary for Swissmedic to perceive knowledge management not only as a central challenge but as an opportunity too. It is a challenge that necessitates a consistent approach adapted to the specific needs of the organization. In addition, as we will show later, cultural aspects play the dominant role here.

Swissmedic as an Expert Organization

As an organization, Swissmedic has a major interest in making its knowledge available to its employees, irrespective of the role of the individual. This raises the question of what options the organization has for accessing the available knowledge. Within an expert organization, knowledge management is a more difficult task than in other types of organizations because the structure of the organization is characterized by the pronounced individual autonomy of the experts who work there and the frequently low level of interest in the concerns of the organization as a whole.

Features of an expert organization[*]

- High level of specialist qualification, knowledge as a factor of production
- Provision of complex, nontrivial products, and services
- Identification with the profession, not with the organization
- Orientation to the professional reputation system
- Little interest in coordination tasks
- High level of individual autonomy

Knowledge Workers

Knowledge worker is a term coined in 1959 by Peter Drucker (2011) in his book *The Landmarks of Tomorrow*. In its original meaning, it referred to workers who are not paid for doing physical work or because of their manual skills, but for using the knowledge they have acquired. Knowledge workers are highly qualified people who contribute to a company's value generation through their knowledge. A look at the current literature on this topic is a useful way of gaining a wider and better understanding of the term: self-determination of individual organization, flexibility in terms of time and place, and autonomy are terms that crop up time and again. The management summary of a study by Hays[†] on the subject of knowledge work describes the cohort of knowledge workers as follows:

- Knowledge workers do not define themselves in terms of fixed rules and processes; they need freedom to organize their work and are not tied to fixed times and places.
- From the knowledge worker's viewpoint, self-determination and flexibility as regards time are fundamental prerequisites for productive working.
- Knowledge workers are aware of their market value and are very willing to change jobs. Their loyalty is to the work they do, rather than their employer.
- Companies often still fail to provide an environment that is conducive to dialog and enables their knowledge workers to network.
- Databases continue to be viewed as an important tool in knowledge work, while social media often still meet with skepticism.

---

[*] After Grossmann et al. (1997, p. 24)
[†] http://www.wissensarbeiter-studie.de/das-projekt/, retrieved on November 5, 2015.

- Knowledge workers draw primarily on conventional conferences and specialist fairs in order to build up networks and exchange information.

- Knowledge workers demand a high level of support from their companies to help them create a balance between family and work life. Many employers have so far done little to meet this need.

- Permanently employed knowledge workers have a positive attitude toward collaborating with external specialists and view mixed teams as a way to boost productivity.

We therefore need to consider how to support these employees and tie them to the company. The crux of the matter is the need to personalize and multiply knowledge and how to show appreciation for those who hold this knowledge.

Background

Swissmedic's strategic plan consists of its *guiding principles* and the *strategic objectives*. The *strategic objectives* show the way in which we want to achieve our vision. Measures for implementing the strategic objectives are defined using the balanced scorecard (BSC) approach. The objectives are formulated from various perspectives: our social responsibility, stakeholders, finances, processes, and employees. Specific measures for implementing the strategic objectives are defined.

It was a measure of this kind that marked the official start of institutionalized knowledge management at Swissmedic. Five years ago, in pursuit of the strategic objective *Complete task on time and to a high standard*, the management board initiated a measure related to the development of knowledge management. The intention was to develop a concept for knowledge management, with the laboratory selected as the pilot unit. At the time, the management board assigned the task to the human resources department.

Approach and First Steps

Realizing that knowledge management already had to exist in an expert organization, our first step was to identify and catalog the existing tools within Swissmedic. We talked to managers and employees, visualized the information we had compiled, and organized it systematically using the building block approach developed by Probst et al. (2006, p. 25) This model, shown in Figure 6.1, structures the management process into logical phases, suggests approaches to intervention, and provides a tried-and-trusted system

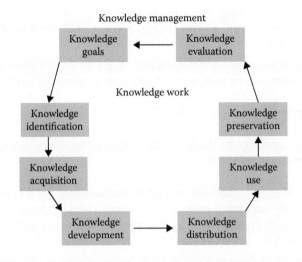

FIGURE 6.1
Knowledge management and knowledge flow – A structure approach. (From Probst, G. et al., *Wissen managen - Wie Unternehmen ihre wertvollste Ressource nutzen* [Managing knowledge - How companies use their most valuable resource], Wiesbaden, Germany, Gabler, 2006.)

for identifying the cause of *knowledge problems* in an organization. We found it to be a suitable tool for structuring knowledge management measures.

We were soon pleasantly surprised to see how many tools already existed. We heard, for example, from organizational units that had already given systematic thought to the topic of using knowledge, and from project teams that had experimented successfully with lessons learned or after-action reviews. We *found* databases and identified knowledge transfer processes. A lot of measures had already been incorporated into human resource and management processes, or integrated into internal project management guidelines, or they were already an integral part of organizational processes and communication platforms—albeit not under the name of knowledge management.

Through this early knowledge mapping process, we recognized that the high standing of knowledge management illustrated the level of attention already paid to the topic by management and employees (e.g., inquiries to the human resources department and participation in internal presentations).

However, at that point

- Knowledge management had not become a strategic topic at Swissmedic; rather it was a collection of individual, unrelated knowledge management initiatives.
- There was no coordination, communication, or internal metaknowledge.

- Many of the existing measures needed to be optimized. Knowledge management is not a project that is completed on date x; it is an ongoing task.
- We heard very few success stories likely to convince or motivate other employees or managers to promote knowledge work.

Defining Objectives

We took these findings as the basis for defining the major objectives of our knowledge management offensive.

- If we want to establish an all-encompassing, well-functioning, and widely accepted knowledge management system, it needs to be anchored strategically and supported by a commitment from management.
- A uniform understanding of knowledge management needs to be created. The system must meet the needs of employees and the organizational units.
- Low hurdles to acceptance: existing elements should be retained where possible, new elements should only be introduced if appropriate and useful.
- Good communication is decisive. Knowledge management should be a subject of interest to everyone.

We drew up a brief concept document (six pages) describing the approach derived from these objectives and the associated cultural and organizational initiatives and presented it to the management board.

...and the Benefits

Management naturally asked what benefit Swissmedic would derive from knowledge management—after all, even a streamlined concept and pragmatic approach requires an investment of time and money.

Furthermore, it is difficult to quantify the success of knowledge management, and we spent a long time testing various approaches to evaluating knowledge. We tested performance indicator systems and we evaluated the use of knowledge scoreboards, that is, we looked at knowledge from a strategic point of view. We ultimately found that the effort involved in setting up a knowledge scoreboard system would have far outweighed its benefit.

We therefore initially looked at the question of what we would get out of knowledge management in detachment from any quantifiable performance measures. We posed the following question, *Why use knowledge management?* and came up with the following answers on the basis of the work done by the knowledge management forum[*] and by focusing on the objectives we had previously formulated:

- To increase the learning ability of the organization
- To use knowledge resources that were lying idle within the organization
- To improve communication
- To avoid loss of knowledge
- To avoid unnecessary resource use and transactions (*reinventing the wheel*)

The management board approved the concept.

Principles for Action

Formulated objectives and expected benefits cannot on their own form a basis for action; at this stage they provided no indication of what specifically needs to be done. Taking our understanding of knowledge management and the above objectives as a starting point, we agreed on three principles as the actual basis of our future actions.

- The focus is on people and culture first and foremost, then on the organization, and finally on the technology (IT, tools, etc.).
- Knowledge management only works with a combined top-down/ bottom-up approach.
- Knowledge management needs a team of dedicated specialists if it is to function.

We would like to take this opportunity to mention a few specific examples to demonstrate what we mean.

Culture>Organization>Technology

Knowledge management and human resource management have a central overlap in the form of the employees who hold the organization's knowledge. The more knowledge-intense an organization is, and the more highly qualified knowledge-intensive work is done within that organization, the

---

[*] http://wm-forum.org/wissensmanagement/warum-wissensmanagement/#, retrieved on November 5, 2015.

more important it becomes to align human resource management with knowledge-oriented value-added processes.

This is particularly sensible in an expert organization because there are close theoretical and practical links between human resource and knowledge management. The acquisition of knowledge in an organization begins with the recruitment of suitable employees, who are preferably knowledge strategy-oriented. The preservation and evaluation of the knowledge that exists within the company is one of the core functions of human resource management in a knowledge society.

Ultimately it is people, in their role as users, who determine whether knowledge management is successful, or in other words useful, within an organization. It is therefore necessary to involve employees in the implementation process at an early stage and to deploy wide-ranging communication and/or consultation processes to overcome reservations, arouse interest in shaping the system, and to motivate these employees to use the system at a later stage. This is an elementary factor in the success of knowledge management, because a knowledge management system's existence is ultimately dependent on the people who use it, animate it, and refine it.

It has been our experience that knowledge workers are happy to share their knowledge and experience. Over the years we have not encountered a single employee at Swissmedic who was unwilling to pass on his knowledge or who showed no interest in his colleagues' knowledge. Employees do, however, expect feedback and recognition.

We are convinced that the culture of an organization plays a vital role. A living knowledge culture based on respect, transparency, individual responsibility, and flexibility is the only way to create an integrative and participative working environment. The approach taken by the human resources and organization department has been driven by this idea in recent years, whether in the context of rolling out the new mission statement or recruiting and developing managers and employees. This same idea guides us in our knowledge management efforts. We firmly believe that employees have a need to share and pass on their knowledge and skills in the interests of the organization. For this reason we initially focused on employees' needs, attempting through our ideas to promote appreciation, an active culture of constructive criticism, defined freedoms, and creative learning processes.

Interestingly, the success of these efforts can be quantified. Every two years we carry out a staff survey that provides information on employees' satisfaction and commitment. The results are more positive every year.

Top-Down/Bottom-Up

Fundamentally, we believe that managers play a particularly important role in determining corporate culture. Knowledge management that targets employees needs a management culture that motivates, nurtures, and creates a stable framework to empower employees (at the same time as allowing individual freedoms).

We also believe, however, that it can certainly be worthwhile to first allow individual ideas to grow within individual organizational units and then to initially test them there. Novel and creative ideas, in particular, are often driven by the enthusiasm of those involved, and we want to promote this enthusiasm, not regulate or forbid it. Guerrilla tactics can work well provided that management is tolerant of uncertainty and ambiguity. We are convinced that a management culture of this kind ties employees to the organization and can ensure their loyalty, commitment, and willingness. Communication is the central activity here, because communication is the core business of managing.

We outline below how we have promoted knowledge management from *above*—with the emphasis on management—and from *below*—with the emphasis on initiatives launched by individual employees or organizational units.

Managers: Top-Down

It was extremely fortunate for us that the Swissmedic management board initiated a strategy-oriented management development program almost at the same time that knowledge management was launched. The intention was for all managers to benefit from modular training, and the emphasis was on the ongoing development of qualified, socially-skilled, and motivated managers. Sponsored by the executive director, the goal was *not for everyone to manage the same way, but for everyone to manage equally well.*

We (in the human resources and organization department) seized this opportunity. In designing the training, we incorporated the principles of transformational management according to Burns (1978) (idealized influence, inspirational motivation, intellectual stimulation, and individual consideration), at the same time trying to keep the emphasis on the organizational context and the needs of the knowledge workers.

We held six training days in total. The managers began with an individual 360-degree feedback round followed by a plenary session, and then did the training in small groups. And it worked. The feedback from managers was very positive, and we also noted an improvement in the results for management quality in the staff survey. One aspect that we found particularly interesting was, that there is a connection between good management and knowledge management, albeit one that is almost impossible to underpin statistically. Figure 6.2 positions the managers at Swissmedic in terms of management quality and knowledge management. The y-axis shows the employees' assessment of management quality, the x-axis shows our assessment of the maturity of knowledge management in the organizational unit in which the manager works (criteria: number of enquiries to CoP, participation in information and training events about knowledge management, and a number of knowledge management activities [workshops, tools used, etc.]).

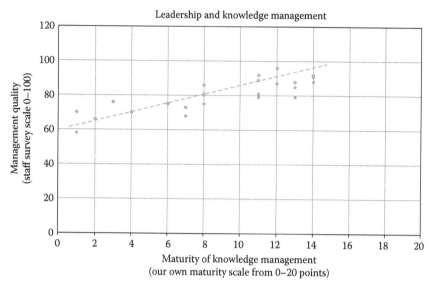

FIGURE 6.2
Integration of KM and leadership capability.

Laboratory: Bottom-Up

The laboratory's task is to analyze medicines that may be authorized, illegally manufactured and/or imported. The laboratory operates in an integrated management system. This means that, in addition to quality, topics such as management, risk assessment, health protection, and safety and environmental protection are documented and managed. The management system is structured according to the principles of a process-oriented organization, and the responsibility for the processes lies directly with the staff involved. This system, combined with the use of synergies, facilitates streamlined, and efficient management.

The laboratory operates in an environment characterized by a high standard of transparency, quality, standardization, and comparability. It is therefore important for the laboratory to have a system that functions well. Over the years, a powerful awareness of quality, quality assurance, and knowledge management has developed, which for a long time was unique within Swissmedic.

First Steps

The realization that knowledge is an autonomous factor has only recently become established, including at Swissmedic. It is only in the past few years that existing tools, performance indicators, and social platforms have been brought together under the heading of knowledge management.

The necessity of a coordinated approach to handling knowledge was first recognized in the laboratory. The initial step was for a small project team to design an approach and produce an overview with the assistance of external support.

In this process we discovered that, here too, a wide range of activities can be brought together under the heading of knowledge management. It was therefore not necessary in the initial stage to introduce new tools or methods, merely to consolidate existing elements or to publicize their existence.

In order to *put a face* to these activities, the managers created the additional function of knowledge manager in the laboratory, with an appropriate time quota, thus deliberately reinforcing the status of knowledge management.

In this way, a large number of different processes and tools developed over time, and with them greater awareness of knowledge management as something that needs to be as efficient and all-encompassing as possible.

The aim is to keep the topic fresh in the minds of everyone in the laboratory and to incorporate it into the context of everyday working life as fluidly as possible, thus enabling knowledge management to become a natural and self-perpetuating entity. The benefit of this approach is better *know-how* and *know-who* and, above all, better networking between employees—which is the basis of successful working. People in the laboratory are convinced that significant value added can be created if employees are motivated to work according to these principles.

Examples

- *Wiki*: In a laboratory, a lot of knowledge is codified and recorded in operating procedures, instruction manuals, and other types of document. This knowledge is abstract and written in as general a way as possible with the intention of maintaining clarity. It is hardly surprising that this knowledge is helpful only to a limited extent in resolving problems involving complex analytical equipment or analyses. A wiki was put together in order to compile and retain the valuable experiences of those that work in the laboratory more effectively. This content can be written and edited by anyone; it is governed by minimal formal or quality requirements, and can include both general documentation and a chronology of individual events.

- *Interdisciplinary working groups (involving the stakeholders)*: The process for writing a laboratory analytical method or an instruction manual for a piece of equipment requires the document to be written by an author, checked by a reviewer, and then released by an approver for use in the quality management system.

- The implementation of complex analytical techniques or subjects has frequently generated time- and resource-intensive queries.

- In order to reduce these negative *side effects*, all relevant groups are now involved more intensively at the start of a project, so that customer requirements can be clearly defined, and technical and quality standards can be agreed.

- *Science forum (specialist, cross-departmental forum)*: Isolated knowledge or failing to air the problems of individuals or departments frequently presents a barrier to the rapid and efficient resolution of everyday issues. There is substantial potential in achieving joint solutions by eliminating hurdles of this kind and promoting a routine exchange of scientific findings. This can include discussion of problems when working at the development stage and facilitating knowledge transfer from meetings or committees across departmental borders. Better employee networking is a further benefit of these scientific forums. One of the challenges though is to motivate employees to contribute and present topics.

A Special *Community of Practice* (Decentralization)

At an early stage we started to consider ways of nurturing knowledge management at Swissmedic and of enabling knowledge management to address and solve specific concerns and problems of the organization and its employees. Following the principle that knowledge management is a discipline *of the people, for the people* (bottom-up!), we decided to share responsibility and set up a small group that in future would be responsible for knowledge management at Swissmedic: a Community of Practice. We have to concede, though, that our CoP is more than a practice-based community of people who are linked informally and perform similar tasks. We prefer to see our CoP as a kind of centre of excellence. We see ourselves as a group of employees who also provide advice to the organization and who try to produce an effect by means of targeted initiatives.

The knowledge management CoP at Swissmedic operates unbureaucratically; it takes skills forward and brings the latest developments into the wider organization. Results are documented, decisions that need to be taken are passed on to the responsible unit in the line. The CoP reports twice yearly to the management board.

At the same time, the members of the CoP represent their organizational units and are thus point-of-contacts for questions or queries relating to knowledge management. Since the CoP was set up, we have consistently promoted the methodological and specialist skills of its members and in this way have been able to share the work associated with specific knowledge management services and small projects among a growing number of people. Although the CoP members are networked with each other, they provide advice and support on a largely independent basis.

What the CoP Offers

We have put together a brochure outlining the advice and support we offer in 18 different areas of knowledge management with the aim of providing managers and employees with an overview of our full range of knowledge management activities. These range from a basic course for interested teams, through workshops on knowledge development to the production of a knowledge scoreboard. We offer these activities without external assistance. The advisors and speakers are members of the CoP who have gained their skills in the course of continuous learning-by-doing.

The CoP members are called on particularly often to provide advice on the wiki and on databases and to act as internal moderators at knowledge transfer meetings. This latter activity receives a lot of recognition within Swissmedic, and in the past few years we have been able to develop knowledge transfer, arising from changes in position (internal rotation, move outside the organization or retirement), into a knowledge management best seller within the organization. Here we have developed a standardized procedure that is readily tangible for managers. The members of the CoP act as knowledge managers, supporting and assisting the line managers and structuring the process. They act as catalysts (are allowed to ask *stupid* questions, pick up on unclear points, clarify terms, etc.) and at the same time make sure that those involved take the time needed for an effective handover.

We also organize a different type of knowledge transfer internally after congresses and symposia such as the annual meeting of the DIA (Drug Information Association). Several Swissmedic employees usually attend these meetings, so we simply organize an internal *follow-up fair* without further ado. In this way the annual meeting of the DIA becomes a mini-DIA, a knowledge transfer congress for the people who stayed at home. Employees can sign up for the fair and attend short presentations (20–30 minutes) during which those who took part in the annual meeting pass on to what they have learned. The response is positive. The employees benefit from an opportunity to exchange information across a wide range of topics.

Success Factors

We are confident that we are fulfilling our objective of being useful to the organization by providing knowledge management capabilities. The subject meets with a very high level of acceptance and great interest at Swissmedic. In other words, while we no longer need to drum up business, we still need to invest a lot of time and resources in maintaining and developing our activities.

In conclusion, we would like to mention the factors that we feel are most important in achieving success with a functioning knowledge management system (in our organization).

The Value of Knowledge

One major condition for the implementation and relevance of knowledge management is that the resource of *knowledge* must represent a value for the organization and that this value must be acknowledged, that is, it has a firm place in the organization's value system. Knowledge is of course always bound to the person, the bearer of the knowledge, and for this reason there must be a cultural awareness of the fact that knowledge is valued within the organization, irrespective of the position that the person holds in the hierarchy. Whether the bearer of the knowledge is an employee on the lower rungs of the hierarchy or the executive director must make no difference. The development of a perception, a culture, of this kind must be exemplified, supported, and demanded by management. We believe that this can succeed best if this aspect features repeatedly in communication within the organization and is input into the management development.

Commitment: Knowledge Management Is Important!

The history of the implementation of knowledge management and the examples mentioned here show that we had no difficulty in establishing the relevance and benefit of knowledge management at a high level. On the contrary, the management board, managers, and employees have all proved to have a high affinity with the subject and provide a high level of support. We have established on many occasions that the term *knowledge management* is able to open doors, that it is a sales argument that helps to import new methods and tools into the organization.

Talking Is Worth Its Weight in Gold: *Marketing* and Communication

From the very start we flanked our knowledge management efforts with a media presence and work continually to make our ideas and successes visible within the organization.

The Swissmedic intranet features both the brochure mentioned above and something known as a *toolbox*, which is home to all relevant internal documents on knowledge management and a wide range of information on tools and techniques such as lessons learned, after-action reviews, and microarticles. All knowledge management activities are listed in the *agenda*, and the *media* section contains videos and a large number of links.

We have already seen that knowledge workers draw primarily on conventional conferences and specialist fairs in order to build up networks and exchange information. We can confirm this, as for the past three years we have kept a knowledge management blog and are more than familiar with the concerns and problems associated with online communication. Although we are sure that knowledge management enjoys a high level of acceptance at

Swissmedic, and we are confident that communication in our organization is an open and uncomplicated affair, we have not yet succeeded in motivating employees to post comments. We periodically post articles in the certain knowledge that they will be read, however, it is currently still not possible to initiate an interactive process through the medium of the blog.

We also organize our own fairs as a way of promoting dialog between specialists. In addition to the mini-DIA already mentioned, we have a WiMaMe (from the German for knowledge management fair), an internal platform for exchanging ideas, and experience relating to knowledge management activities at Swissmedic. This is where employees from all levels of the hierarchy present their cases, talk about the opportunities and risks associated with their projects, and answer questions. This is an indirect way for us to achieve an increase in the number of *visible* knowledge management experts and thus to increase our operational excellence.

Introduce Knowledge Top-Down and Bottom-Up

A bald directive to the effect that *We are now doing knowledge management!* is unlikely to lead to success. Ultimately, a knowledge management system always comes alive through the people who use it and develop it. Our experience—corroborated by reports from other organizations—shows that knowledge management cannot be imposed *from above* and that big (and, more particularly, long) concepts and loudly proclaimed initiatives are generally doomed to fail. This may be due to unfulfilled expectations, a lack of confidence, or too little involvement on the part of the employees.

Nevertheless knowledge management is above all successful if the topic receives support from the organization's management. In addition, there must be willingness to lay open habitual practices to scrutiny and to give employees the time and freedom to use and maintain the knowledge management system.

The *dual approach* that we deliberately chose from the outset—that of introducing knowledge management on two levels—has demonstrated its value. On the one hand we won over the Swissmedic management board with a streamlined concept and concrete ideas; on the other we generated benefit for employees and teams simply and directly by introducing tangible measures.

Gradual/Modular Introduction

Knowledge management activities initially tie up human and financial resources. Since both are often only available in finite amounts, it is usually more practical to introduce knowledge management in the form of projects focusing on limited topics or very specifically in individual departments. It proved to have been a good idea for us to focus on individual interested organizational units when developing the wiki, for example, and to let the tool grow where people were interested in growing it.

What Next?

Have we already achieved our objective? Will there still be a need for knowledge management in the future? And, indeed, how will we be working in the future? What will the set-up look like? And where?

To address these questions we put together a working group to consider how working at Swissmedic will look in the future and concluded that there will still be routine work, formal meetings, and concentrated work, and that the extent of these activities is not likely to change significantly. At the same time, though, we think that the following types of work will undergo some relevant changes in the sense of activity-based working.

- *Collaboration/informal exchange/ad hoc meetings*: The intensity and the importance of an informal exchange of information will increase. The era in which an expert kept his information to himself is history. The knowledge that exists within Swissmedic must become even more accessible in the future. Knowledge needs to be shared.

- *Maintaining relationships*: There is a growing need to maintain relationships, and the importance of doing so will increase too. Employees need to know each other well if they are to work effectively. It is extremely important to know what I can expect and ask of colleagues. Understanding each other can also help people to work more efficiently and, above all, more effectively—because misunderstandings and unclear instructions waste resources. Knowledge workers also need to feel comfortable, so it is not wrong, but in keeping with the age, to create a working environment that encourages good relationships.

This is the conviction that will accompany us as we move forward and continue our efforts to establish a culture that encourages knowledge management. Furthermore, we are determined to keep up our efforts to enrich the organization with good ideas and tangible initiatives.

References

Burns, J. M., *Leadership*. New York: Harper Perennial Modern Classics, 1978.

Drucker, P. F., *Landmarks of Tomorrow: A Report on the New*. New Brunswick, NJ: Transaction Publishers, 2011.

Grossmann, R., Pellert, A., Gottwald, V., Krankenhaus, Schule, Universität: Charakteristika und Optimierungspotentiale [Hospital, school, university: Features and potential for optimisation] (pp. 24–35). In R. Grossmann, A. Pellert, and V. Gottwald (eds.), *Besser Billiger Mehr. Zur Reform der Expertenorganization, Krankenhaus, Schule, Universität* [Better Cheaper More. On the reform of the expert organization, hospital, school, university], (iff texts, Volume 2), Springer Verlag: Vienna 1997.

Probst, G., Raub, S., Romhardt, K., *Wissen managen - Wie Unternehmen ihre wertvollste Ressource nutzen* [Managing knowledge - How companies use their most valuable resource]. Wiesbaden, Germany: Gabler, 2006.

7

Generating New Knowledge from Data: Protecting the Patient by Exploring What Is "Known" about Quality Defects and Product Recalls

Nuala Calnan and Anne Greene

CONTENTS

This chapter reports on the findings from a recent knowledge management (KM) research study undertaken at the Irish medicines regulator
to create new knowledge from the data available about sources of product quality defects and recalls in order to examine potential solutions
and prevention strategies for these market-based failures.

Editorial Team

The ever increasing complexity of health care products requires a data-
driven evidence-based approach to their regulation…that's understood
and recognized worldwide.

Pat O'Mahony
Former CEO, Health Products Regulatory Authority (HPRA)
RSI Knowledge Management Symposium Dublin, March 2015

Introduction

The expectation for high quality and safe medicine is an intrinsic requirement for patients. The responsibility for the assurance of high quality in the
manufacture and supply of these medicinal drug products lies largely with
the manufacturers and distributors of these products while the assessment,
oversight, and guidance for product safety and efficacy, is provided by the
international regulatory community.

The recently published EU Medicines Agencies Network Strategy to 2020
(EMA, 2015) identifies the handling of emerging events related to medicinal
products such as, quality defects or safety concerns as a major challenge for
authorized medicines. Enhanced integration between national competent
authorities on defect and product recall handling is crucial in addressing
these *emerging events* that relate to quality defects and product recalls. An
integrated approach could facilitate driving behaviors across the industry
toward prevention rather than cure.

This chapter introduces a recent Irish knowledge management research
study conducted within the Irish medicines regulator, the Health Products
Regulatory Authority (HPRA), which examined five years of medicinal products quality defect and recall (QDR) data, reported in the Irish market from
January 2010 to December 2014. The purpose of the study was to create new
knowledge from the available data about the sources of quality defects in order
to examine potential solutions and prevention strategies for these market-
based failures.

The study was the first of its kind in Europe and the initiative formed the basis of a pilot study to help inform a broader European QDR study currently being undertaken by the European Medicines Agency (EMA).

Research Position and Aims

This study was undertaken within the context of the current *paradigm shift in quality* for the pharmaceutical industry. The demands for high quality, innovative drug therapies increase daily yet, despite significant scientific advances in the development and approval of novel therapies, there continues to be cases of authorized medicines failing to deliver excellence to the patient. In the majority of the cases, these defects are reported to the relevant competent authority by the manufacturers themselves, and the resultant investigation may lead to a recall of the affected products from the market(s) in question.

However, the current transformation of the quality paradigm within the pharmaceutical industry is presenting real challenges for organizations, compounded by the impacts of globalization, in maintaining compliance with existing and emerging regulatory expectations. One route to overcoming these complex challenges lies in the ability to improve organizational culture, and influence key behavior change to improve product quality. Leveraging the knowledge already available within the organization is a key source of strength in this change process.

This research presented an invitation for reinterpretation of the available defect data, offering opportunities to reimagine outcomes that could positively impact the future for pharmaceutical supply. Furthermore, in order to encourage adoption by both the regulatory and industry communities, the primary focus of the theories presented in the study and recommendations drawn from the research was in providing practical solutions.

The actual premise of the research proposed that if the available knowledge in the pharmaceutical manufacturing sector could be shared and utilized more effectively then the quality of the decision making, and the actions taken relating to known failures and weaknesses in Pharmaceutical Quality Systems, could improve. This in turn should reduce the risks of defective medicines reaching the market and the patient.

The key aims for the research were as follows:

- Identification of trends and key learnings from the extensive quality defect and recall data that are currently available at the HPRA.
- Generation of new knowledge from the data by exploring what is *known* about the quality defects and product recalls that occurred over the five-year period.

- Development of a proof-of-concept report to support initiation of a pan-European project at the European Medicines Agency.
- Recommendation of strategies that could lead to higher quality medicines for patients with less likelihood of them receiving a defective medicine. See Figure 7.1 for the goals and objectives of the research project.

Quality defect and product recall research at the HPRA 2015	
Goals and objectives	Benefits and expected learnings
Establish do certain types of manufacturing processes/product generate more serious defects than others	Inform GMP and GDP inspectors where to focus their inspection time and resources on » Specific process validation activities » Quality risk assessment work » Change management activities » Corrective action effectiveness Such data could also feed into future *risk-based inspection planning*. Enable recommendations for *Pharmaceutical assessors*, on what parts of module 3 might benefit from higher levels of assessment, in particular » Justifications in critical attribute/parameter assignment » Validation plans presented for certain types of manufacturing processes. Such data could also feed into *risk-based assessment* approaches. This should provide useful information for developing *Annual sampling and Analysis plans* at OMCLs and NCAs. Products that are at a higher risk of having a quality defect could be targeted for more surveillance testing.
Identify what kind of issues are actually investigated as quality defects at NCA level	Understanding what percentage of QDR's are related to » Underlying quality failure arising at the manufacturing site » MA-Noncompliance issue involving the failure to implement a variation » Stability out-of-specification (OOS) » Distribution or supply chain failure
Identify if there are certain product lifecycle activities that contribute more frequently to failures that result in defects and recalls	Inform the *Pharmaceutical industry* about areas requiring increased attention, such as » Product and process development (e.g., for better process understanding and robustness) » Validation (e.g., Process, cleaning) » Qualification (e.g., Equipment, HVAC and other utilities, suppliers) » Quality control (e.g., more robust QC methods) » Corrective and preventative actions (e.g., effectiveness and root cause analysis) » Product quality review activities (e.g., use as a knowledge management tool)
Examine current QDR communications methods to identify improved knowledge sharing and transfer mechanisms	Improved knowledge sharing could assist *Regulators and industry* by leading to better prevention strategies, thereby reducing incidents and could also improve the performance of QDR investigations, for example » Knowledge of *recurring quality defect* issues could determine how much regulatory oversight should be devoted when assessing and agreeing CAPAs with companies » Knowledge of the *historical incidence rate* of a certain type of quality defect could inform regulators and industry about the likely extent of a defect in a batch

FIGURE 7.1
Key goals and objectives of the QDR research at HPRA, 2015.

In addition, a series of *key research questions* was drawn up to address the goals and objectives and to inform the data analysis phase of the research.

- Are certain defect types in medicines recurring?
- Where are the key trends?
- Are certain types of manufacturing processes, product types, and pharmaceutical forms more likely to generate/have defects?
- What kinds of root causes are being identified?
- How many QD investigations fail to determine the root causes?
- How often is *human error* cited as the root cause?
- What kinds of CAPA actions are generally taken?
- How effective are the CAPAs that have been implemented?

Reliable Drug Quality

> Patients require quick access to safe, affordable, effective and good quality medicines. (European Commission, 2015)

> **Fernand Sauer, European Commission—Directorate for Public Health**

The subject of reliable drug quality has received increased attention in recent years as a result of some highly publicized public health tragedies, such as the Heparin crisis and the New England Compounding Company (NECC) fatalities. There has also been an increased focus on the manufacturing quality problems underpinning recent drug shortage situations (EMA, 2012; FDA, 2013; ISPE, 2014; Kweder and Dill, 2013).

The ICH Q10 guidance outlines that an effective Pharmaceutical Quality System provides a pathway to "…enhance the quality and availability of medicines around the world in the interest of public health" (ICH, 2008). Although this statement has universally become known as the *desired state* for pharmaceutical products, almost ten years on from the publications of the guideline, the achievement of this desired state remains an on-going challenge.

U.S. Food and Drug Administration (FDA) Centre for Drug Evaluation and Research (CDER) director Dr. Janet Woodcock has cautioned on the potential for serious or critical consequences arising from quality problems in pharmaceutical manufacturing, "the consequences of quality problems such as sub-potency, lack of sterility, or product mix-ups can be so devastating" (Woodcock, 2012). Woodcock asks if other manufacturing sectors have successfully adopted quality management techniques such as Six-Sigma, in an area where the stakes are so high, why reliable, high quality manufacturing not also attainable in the pharmaceutical sector?

Understanding the Implications of Defective or Substandard Medicines

The World Health Organization (WHO) defines the term of *Substandard Medicines* as "pharmaceutical products that do not meet their quality standards and specifications" (WHO, 2009) and a recent article in the *British Journal of Clinical Pharmacology* (*BJCP*) asked if we are facing a potential public health crisis as a result of substandard medicines? (Johnston and Holt, 2014).

The article by Johnston et al. provides a very useful overview of the potential risks to public health associated with the supply of defective or substandard medicines. They acknowledge that much of the recent political and regulatory attention has been focused on the risks arising from the increasing availability of deliberately *falsified (and counterfeit) drugs* through illicit or unregulated online channels. However, their research also confirms that, as a result of some poor manufacturing or quality-control practices in the production of genuine authorized drugs, substandard medicines are also reaching patients through legitimate channels.

They point to the facts that incidences of failure arise in both branded and generic drug products. Categorizing the three primary contributing factors to substandard authorized medicines are as follows:

1. *Drug Content*: Cases where too much or too little of the active pharmaceutical ingredient (API) is present compared to the authorized formulation specifications.

2. *Impurities*: Cases where any substance found present in the product is neither the chemical entity defined in the specification nor an approved excipient.

3. *Pharmacological Variability and Stability*: Cases where drug efficacy or safety are compromised due to lack of therapeutic effect arising from issues such as lack of bioequivalence, differences in drug dissolution profiles, and altered drug metabolism actions due to formulation or excipient changes or accelerated stability failures due to inappropriate storage or packaging conditions.

Johnston and Holt, both eminent UK-based professors with research interests in the areas of therapeutic drug monitoring, clinical toxicology, and counterfeit drug detection, conclude that a concerted effort is now required on the part of governments and regulators, drug manufacturers, charities, and health care providers to ensure that only drugs of acceptable quality reach the patient.

Why Ireland Provided a Suitable Pilot Case Study
for This Quality Defect and Recall Research

Ireland's Foreign Direct Investment Promotion Agency, IDA Ireland reports that the life sciences sector in Ireland has grown to achieve global significance from humble beginnings in the 1960s (IDA Ireland, 2015). With the emergence of collaborative clusters across pharmaceutical, biotechnology, medical devices, and diagnostics manufacturers, Ireland has established a reputation for excellence in the manufacture of medicinal products. Consequently, Ireland has more than 120 biopharmaceutical companies established here, including nine of the top ten global pharmaceutical companies (PCI, 2014). These establishments are principally regulated by the HPRA and the EMA, with approximately one quarter of these facilities also approved by the U.S. Food and Drug Administration (FDA) for supply of medicines into the U.S. market. This sector accounted for just over €51.5 billion in annual exports from Ireland in 2014, represented 56% of the total exports (based on recent Irish Central Statistics Office figures), which sees Ireland with a ranking as eighth largest producer in the world for pharmaceutical products.

This leadership in the manufacture of high quality, safe medicines also relies on the high standing of the Irish regulator authority within the European and international regulatory communities. HPRA experts actively participate at a European and global level for all products under their remit and are key contributors to a range of important regulatory committees and working parties, both in Europe and internationally. The HPRA also works closely with their counterparts from other national competent authorities as an active member of the international rapid alert network.

These factors positioned Ireland as a particularly good environment to conduct this pilot KM research study.

Examining Quality Defect and Product Recalls of Medicines in the Irish Market—The Background

This research, undertaken in 2015, conducted a systematic review within the HPRA of all reported incidences of product quality defects and product recalls of medicinal product over the five-year period from January 2010 to December 2014. The proposal for this research was developed by the HPRA, in conjunction with the Pharmaceutical Regulatory Science Team (PRST) at the Dublin Institute of Technology (DIT), Ireland and Regulatory Science Ireland (RSI) granted the research funding.

The pilot study sought to promote the value to the patient, the regulators, and the industry of on-going collaboration and knowledge sharing in relation to quality defect and recall reporting.

The range of data examined was broad and included investigations into product quality problems encountered in branded and generic prescription drugs as well as over-the-counter products, affecting both traditional drugs formulations and the more innovative biologics. In terms of organizational complexity, those involved included large multinationals as well as smaller companies and third-party or outsourced partners. The analysis confirmed that incidences of quality defects were not limited to any one particular industry sector, and this highlights both the challenges and potential risks that are now present right across the globalized pharmaceutical industry.

The period under study also coincides with a global rise in the numbers of falsified or counterfeit medicines and unlicensed products found in the marketplace, fuelled by the rapid growth in Internet or online sales. Consequently, new challenges to medicinal product oversight now exists that increase the risk that a defective medicine may reach the patient.

An Overview of the Quality Defect and Recall Trends

Quality defect and recall reporting at many of the competent authorities has traditionally focused on a rate-based, annual reporting model. By undertaking a detailed systematic five-year analysis, this study was able to examine incidences and trends related to recurring quality defects throughout the industry in order to explore opportunities for incident prevention.

From the review of international activity, several competent authorities note an increasing trend in quality defects and this is matched by a rising trend of rapid alerts reported within Europe.

In Ireland, though the year-on-year trend from 2010 to 2014 of overall *Number of Quality Defects* reported to the HPRA from the market place rose only a moderate 8%, the number of *Critical Quality Defects* rose from 23% of all cases reported in 2010 to 45% reported in 2014.

In terms of *Causes of Defects*, parenteral medicinal products (administered by injection) are the most likely to be defective due to failures arising from product contamination and lack of sterility assurance. Other persistent recurring issues across all product types include stability failures, packaging, and labeling issues, noncompliance with the product specification and noncompliance with the market authorization (MA) for the product.

Three key elements of the proposed prevention strategy lie in harmonizing the *QDR Data Capture,* enabling *knowledge generation* to facilitate examination of recurrence, underlying factors, and root causes. This will also allow for enhanced *knowledge sharing* of the lessons learned across and between patients, the industry, and the regulatory communities.

Quality Defect Investigations 2010 to 2014 — Key Facts

Over the period from January 2010 to December 2014 there were 3,999 individual reports of quality defects investigated by HPRA's market compliance

	2010	2011	2012	2013	2014	5 Yr. Totals
Total QDRs investigated	751	917	741	774	816	3999
QDRs related to products manufactured in Ireland	278	299	216	193	165	1151
QDRs-Affecting Ireland	609	715	547	533	470	2874
Recalls required in Ireland	168	253	141	109	102	773
Critical QDR (total)	173	241	189	235	365	1203
Critical QDR (affecting Ireland)	77	100	58	55	45	335
Rapid alerts sent from Ireland	47	14	13	7	2	83
Rapid alerts received	126	215	194	289	198	1022
Rapid alerts received-Affecting Ireland	12	22	18	74	20	146
% of RA affecting Ireland	10%	10%	9%	26%	10%	14%

FIGURE 7.2
Quality defect reports (January 2010–December 2014)—Impacts on Ireland.

team. Approximately 71% of these investigations were undertaken because the defect report had the potential to impact on products available for supply on the Irish market and the remaining 29% involved investigations related to products either manufactured in Ireland, QP released, or exported from Ireland.

The analysis shows a rising trend in the number of QDR cases reported to the HPRA over the period 2010 to 2014 as illustrated in Figure 7.2.

The peak in the 2011 figures, showing a total of 917 QDR reports investigated, can be attributed to an additional 70 defect reports that arose due to a single cold chain failure event at one Irish primary wholesaler facility.

As seen from the figures, not all investigations of suspected quality defects will result in actions that affect Ireland. In many cases the report is reviewed and found not to have an impact on the product batches or lots currently available for supply in the Irish market.

The analysis shows that of the total 3,999 QDRs received over the five-year period only 773 (19%) of these cases required a recall of product from the Irish marketplace.

Furthermore, though the overall number of defects classified as *Critical* was 1203 cases, representing approximately 30% of the total QDRs reported in the period, only 335 of these critical defect cases affected Ireland (representing 8% of the total number of cases investigated in the period).

Finally, in terms of rapid alerts, the HPRA initiated a total of 83 rapid alerts to other national competent authorities within the network over the five years as compared to a total of 1022 rapid alert notifications received in that time.

Figure 7.3 provides an infographic with some additional facts arising from the analysis.

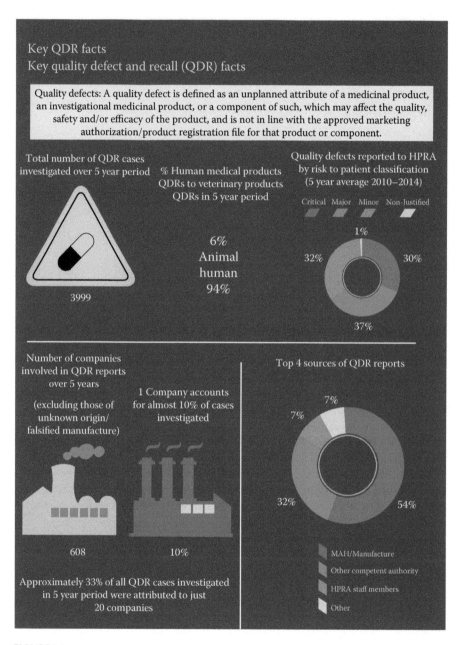

FIGURE 7.3
Key QDR facts.   (*Continued*)

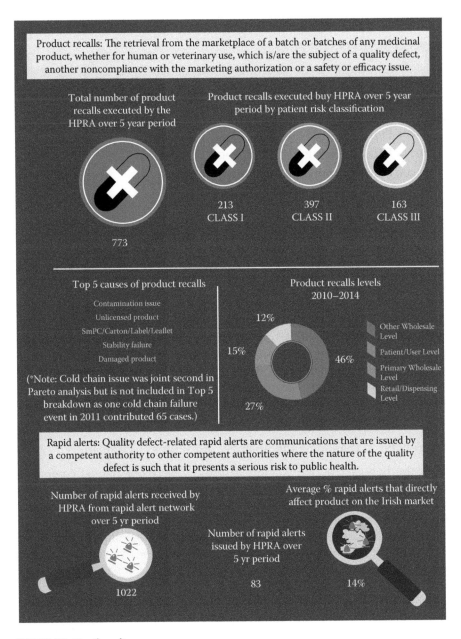

FIGURE 7.3 (Continued)
Key QDR facts.

Research Analysis Focus

Arising from this five-year data analysis four *Focus Points* areas were examined in greater detail to facilitate insight development for the research recommendations:

Focus Point 1: The rise in critical quality defects

Focus Point 2: Examining the anatomy of failures at *Company A*

Focus Point 3: Product recall analysis

Focus Point 4: Rapid alert coordination

Focus Point 1: The Rise in Critical Quality Defects

The five-year analysis showed that the number of cases classified as *critical quality defect* rose from 173 cases in 2010 to 365 critical defects reported by 2014. This represents a greater than two-fold increase in the number of reports that were deemed to have the potential to be life threatening or cause a serious risk to health. This worrying trend indicates that the nature of the quality failures now arising have very grave consequences indeed for patients. Noteworthy, from an Irish perspective only 45 of the total 365 critical defect cases reported to the HPRA in 2014 were found to affect products in the Irish marketplace. Figure 7.4 shows the breakdown of the top causes identified for these critical quality defects in 2014.

Focus Point 2: Examining the Anatomy of Failures at *Company A*

Over the five-year period 2010 to 2014 there were 608 named companies involved in the 3,999 quality defects investigated at the HPRA. In seeking to understand the *state of quality* at these companies a Pareto analysis of the top ten contributors by company was examined, as shown in Figure 7.5.

This Pareto analysis reveals that one company, *Company A* contributed almost 10% of the overall quality defect reports over the five-year period with a total of 303 cases directly linked to that company. Company A provides a wide range of medicinal products globally.

In an effort to gain further understanding of the types of challenges presented in the Pharmaceutical Quality System at Company A, a deeper analysis of the issues was undertaken during the *data analysis* phase of the pilot project. From the 303 cases of quality defects reported in the five-year period 75% were considered to be serious (i.e., classified either as a critical defect or a major defect). This means from a patient's perspective that the failures are either potentially life threatening or could cause a serious risk to health (critical), or

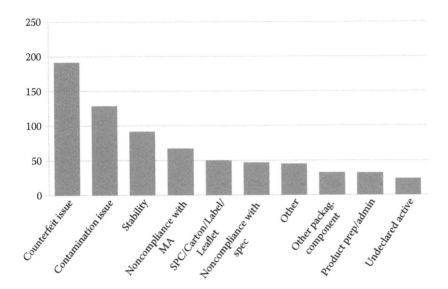

FIGURE 7.4
Critical quality defects by failure category (Pareto for Year 2014).

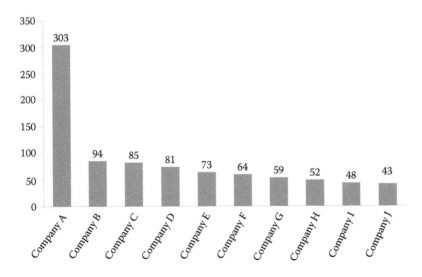

FIGURE 7.5
Top ten quality defects by company 2010–2014.

they could lead to illness or mistreatment (major). Items such as packaging issues, contamination, and sterility assurance are among the top causes of failure within the organization. However, there is also evidence of issues arising from adverse patient reactions, through noncompliance with their own specifications and approved marketing authorizations, stability, and cold chain problems to product mix-ups and supply of unlicensed products.

These failures led directly to a total of 35 rapid alerts being issued through the *rapid alert network* in relation to products supplied by this company. Twenty-four of which were issued at the highest *Class I* alert level and the remaining eleven were issued as *Class II* alerts, over the years 2010–2014.

The analysis points to a lack of effective knowledge transfer at Company A following a quality defect or recall event. It was observed that in the wake of the failures at one site, corrective actions implemented at that site were subsequently found to be absent in other parts of the organization at later dates, contributing to further recall events at those alternative sites.

Getting organizations to use what they know—right across the organization—as a proactive strategy to prevent reoccurrence of quality failures is critical to patient safety but also offers an opportunity for competitive advantage. The cost in terms of both financial and resource constraints is considerable for each of these failure events. The ability to learn from mistakes is fundamental to putting this right.

Focus Point 3: Examining Product Recalls—A Deeper Analysis

One aspect of the research design and methodology employed in this study involved a *deep dive* analysis of a number of case file records in order to undertake a detailed review of correspondence, decisions, and actions agreed between the companies involved and the HPRA. This facilitated a greater level of understanding of the capabilities and competencies demonstrated by the staff responsible for handling the failure event.

The research examined three cases associated with Class I and Class II recalls initiated by the HPRA in response to

- A contamination issue
- A packaging failure
- A stability failure

The detailed case file reviews of these events conducted in phase two of this pilot study provided further insights into

- The level of complexity associated with these products, processes, and supply chains.
- The robustness of the Pharmaceutical Quality Systems (PQSs) within the companies involved in detecting and responding to failures.

- The level of competency of the companies involved in terms of handling failure investigations, root cause analysis, patient risk evaluations, and business continuity planning.
- The timeliness and responsiveness of organizations to a market-place failure event.
- The effectiveness of the corrective actions undertaken to prevent reoccurrence.
- The evidence of learning within these organizations.

Focus Point 4: Rapid Alert Coordination

Rapid alerts are an "urgent notification from one competent authority to other authorities that a batch recall has been instituted in the country originating the rapid alert" (EMA, 2014). The rapid alert procedure covers the transmission of key information when urgent action is required to protect public or animal health relating to the recall of medicinal products due to quality defects or falsified products.

The procedure may be used also for transmission of other information such as cautions-in-use, product withdrawals for safety reasons, or for follow-up messages to previous alerts. In the period under study there were 1022 rapid alerts received by the HPRA and 83 rapid alerts initiated by the HPRA. In total 146 rapid alerts were assessed to have affected products available on the Irish market over the five-year period.

The *rapid alert* network represent a clear case of cooperation between the regulatory authorities in member states, including other interested parties such as the mutual recognition agreement (MRA) countries and members of the PIC/s inspections cooperation group. Systematic tools and processes have been established at each of the relevant NCAs, based on the compilation of community procedures, to facilitate this level of coordination. Notification through the rapid alert network enables an effective global response to threats discovered in the global supply chain with accountability for local market actions clearly assigned to each of the national competent authorities.

However, though the coordination of the rapid alert notification provides an effective pathway for the *transmission of urgent information*, the records of such events are not currently centralized within the network. Responsibility for the maintenance of these records resides with each local competent authority and therefore there is currently limited coordination of outcomes for products or patients arising from these events.

The potential for the development of a centralized database, even for those critical quality defect events that trigger a rapid alert, presents an opportunity to report on these outcomes as a means to gain insights regarding the *state-of-control* of medicinal product quality in order to share the knowledge and learnings gained. Coordination of the information gained from the investigations and subsequent regulatory actions related to rapid alerts would be a good starting point for further cooperation in this area. Indeed, the recently initiated *EU Telematics Project* has announced plans to develop a coordinated platform through which this might be achieved in the future.

Recommendations Arising

The recommendations proposed from this research have been grouped around three key themes of *discover, learn,* and *engage.*

1. *Discover*: Recommendations related to standardization in the *data capture* of records of quality defect events.
2. *Learn*: Recommendations related to improving *knowledge generation* capabilities arising from the quality defect events reported.
3. *Engage*: Recommendations related to improving *knowledge sharing* between patients and their caregivers, the industry, and the regulatory communities.

An outline of these recommendations follows.

Discover—Recommendations for Data Capture

To improve *knowledge discovery* it is recommended that

1. *A common taxonomy of terms* associated with quality defect and recall reporting is established.
2. *A standard EU classification system* should be developed for key data elements such as *dosage forms, defect failure causes, planned corrective actions, and defect reported by.*
3. *A systematic QDR tool* should be developed to capture and record quality defect and recall data for use by all EU member state national competent authorities.
4. *Data capture rules* should be enforced such that individual data elements, including key supply chain information, are captured in individual data fields to facilitate future comparison and analysis.

Learn—Recommendations for Knowledge Generation

To enable *knowledge generation* from the available data it is recommended that

1. The quality defect and recall data captured by the EU member states should be amalgamated and analyzed to produce *defect type and frequency reports.*

2. Develop a systematic process for completing an *End-of-Investigation review* to capture the insights, experience, and intuition of the QDR investigator at the close out of each market-based defect event.

3. The defect and recall data captured by the EU member states could be utilized to develop *a quality defect scorecard* to provide quality defect reports by organization, site, and affected products.

4. This quality defect knowledge could then be linked to the risk-based inspection planning processes.

Engage—Recommendations for Knowledge Sharing

To enable *knowledge sharing* among the main stakeholders it is recommended that

1. The leadership within organizations should be engaged in the drive for improvement by *formally communicating their risk rating/quality defect scorecard* and sharing the learnings gained about their performance and compliance history at the opening meeting of site inspections.

2. Examine the opportunity to implement *regular, scheduled meetings of the QDR/rapid alert network* to facilitate sharing of the valuable tacit knowledge available within this network.

3. *For cause inspection* represents an enhanced means of investigating certain quality defects directly at the site of origin of the failure. Establishment of clearer ways of tracking these inspections and feeding the outputs to the wider network should be explored.

4. Continue to promote the benefits of reporting quality defects through engagement with the reporting stakeholders.

5. Where warranted and based on risk, certain categories of defects, such as, *stability failures* and *MA Noncompliance* could represent one opportunity for regulators and the industry to agree a *flexible regulatory regime* based on the good standing (or risk-ranking) of an individual organization with respect to the reporting and management of appropriate resolutions to these incidences.

Conclusion

This pilot research study of the quality defect and recall data available at the HPRA has provided many useful insights of the potential risks to patients from substandard medicines. It offered an opportunity to examine a relatively small body of data and generate new knowledge in order to draw up recommendations that may provide benefits for the broader European network.

Ultimately, the ability to capture and share data about postapproval quality defects between national competent authorities conveys the opportunity to create knowledge that could enable regulators and industry to work together to deliver manufacturing quality improvements and ultimately enhance patient safety.

References

EMA. (2012). *Reflection Paper on Medicinal Product Supply Shortages Caused by Manufacturing/Good Manufacturing Practice Compliance Problems* (EMA/590745/2012), November 22, 2012. European Medicines Agency, London, UK.

EMA. (2014). *Compilation of Community Procedures on Inspections and Exchange of Information* (EMA/572454/2014), Rev 17, October 3, 2014. European Medicines Agency for European Commission, London, UK.

EMA. (2015). *EU Medicines Agencies Network Strategy to 2020 - Working Together to Improve Health* (EMA/MB/151414/2015), March 27, 2015. Heads of Medicines Agencies & European Medicines Agency, London, UK.

European Commission. (2015). Half a century of European pharmaceutical legislation. *Health-EU*, 159, 1–3.

FDA. (2013). Strategic Plan for Preventing and Mitigating Drug Shortages. Retrieved September 22, 2014, from http://www.fda.gov/downloads/Drugs/DrugSafety/DrugShortages/UCM372566.pdf.

ICH (2008). ICH harmonised tripartite guideline: Pharmaceutical Quality System Q10, Step 4 version, June 4, 2008. ICH, Geneva, Switzerland.

IDA Ireland. (2015). Bio Pharmaceuticals & FDI Opportunities. Retrieved September 20, 2015, from http://www.idaireland.com/business-in-ireland/industry-sectors/bio-pharmaceuticals/.

ISPE. (2014). *ISPE Drug Shortages Prevention Plan: A Holistic View from Root Cause to Prevention*, October 2014, ISPE, ISBN 978-1-936379-76-7.

Johnston, A. and Holt, D. W. (2014). Substandard drugs: A potential crisis for public health. *British Journal of Clinical Pharmacology*, 78(2), 218–243. doi:10.1111/bcp.12298.

Kweder, S. L. and Dill, S. (2013). Drug shortages: The cycle of quantity and quality. *Clinical Pharmacology and Therapeutics*, 93(3), 245–251. doi:10.1038/clpt.2012.235

PCI. (2014). PharmaChemical Ireland-The Pharma Factor in Innovation Ireland Review, Issue 9, December 2014.

WHO. (2009). WHO Questions and Answers on Counterfeit Medicines: Does Quality of Medicines Matter?, October 2009, retrieved from http://www.who.int/medicines/services/counterfeit/faqs/QACounterfeit-October2009.pdf.

Woodcock, J. (2012). Reliable drug quality: An unresolved problem. *PDA Journal of Pharmaceutical Science and Technology/PDA, 66*(3), 270–272. doi:10.5731/pdajpst.2012.00868.

8

A Perspective from NASA: Knowledge Services and Accelerated Learning in NASA—The REAL Knowledge Approach

Ed Hoffman and Jon Boyle

CONTENTS

All organizations live in a world of increasing complexity, perhaps none more so than at National Aeronautics and Space Administration (NASA). Here we learn that the International Space Station (ISS) project involved 21 international partners, and how 80% of every program or project undertaken at NASA is an international collaboration. What does it take to make a project like this successful? And what are the lessons that Pharma can apply to our complex partnership networks?

In this thought-provoking contribution, our NASA colleagues identify that one of the key issues in dealing with complexity is identifying the critical knowledge: that is, what is it that the organization needs to be the best at and what does it need to focus on? To address this challenge the NASA CKO Office has developed the *Rapid Engagement through Accelerated Learning* (REAL) knowledge model, which they share below.

Editorial Team

Disclaimer: This material is based on the work supported with resources and the use of facilities at the National Aeronautics and Space Administration (NASA), Office of the Chief Knowledge Officer, and is available as open source information and may be used by external individuals and organizations with proper author citation.

Introduction

This is an age of projects, in an age of entrepreneurship. For NASA, work often concerns project leadership issues such as the competencies of effective leaders, team development, knowledge services, and accelerated learning. This focus on leadership occurs because of the difficulty in achieving success without excellent leadership, especially when teams today are often global and work virtually. Knowledge is the lifeblood in terms of achieving success in this type of environment. This is also at odds with the prevailing wisdom in a highly technical environment, where NASA leaders often focus on science and engineering and not on the soft skills involved in leadership, knowledge, and learning.

In considering the International Space Station (ISS), it is easy to see why this bias happens. The ISS involved 21 international partners; this represents *normal conditions* in an organization where 80% of every program or project is an international collaboration. What does it take to make a project like this successful? What are the health implications when people are circling Earth for up to a year, and how is that monitored? What about power delivered through batteries and what types of battery technology is needed? What about food, exercise, psychology, work, sleep, and the myriad of issues that surround these and many more topics of functional concern? In this complex project environment there is little room or tolerance for diversions.

Complexity

In complex technical organizations, the processes are embedded in systematic ways of doing things, with a heavy emphasis on control and planning. It is about the customers, about the goals, about the measurements, and understanding the risks. For organizations without an entrepreneurial orientation, experience can become a disadvantage, where highly experienced practitioners may become resistant to change. The manifestation of this can be heard across public, private, government, industry, academia, and professional organizations, where practitioners often say it is increasingly difficult to bring ideas to fruition and projects to completion.

Recent research says that 44% of strategic initiatives fail in surveyed project organizations, and 58% of these failed projects were not highly aligned to organizational strategy (Project Management Institute, 2014). The second annual Aviation Week Young Professionals Study identified the top frustrations of the under-thirty-five (35) workforce as *bureaucracy and politics*, with over half of those surveyed believing that the pace of decision making, progress, and management of change are not what they could or should be (Anselmo and Hedden, 2011).

Organizations live in a world of complexity. At NASA, the driving motivation concerning project knowledge is ultimately—mission success. Complexity works against this focus on mission success, and it can take many forms:

- Confusing, vague, poorly defined priorities, strategies, lines of authority, governance, policies, and roles and responsibilities and support, characterized by iterative reorganizations, constant budget changes, constant resource level adjustments, a proliferation of administrative burdens, and endless requirements.
- A proliferation of customers, stakeholders, and strategic partner interfaces at multiple levels of interest, involvement, and responsibility.
- Technical complexity and system integration issues within and across multiple disciplines and multiple systems.
- Increased data and information amount and availability for process input, throughput, and output.
- Multiple overlapping, conflicting, outdated processes, and procedures that involve multiple points of contact distributed across multiple organizational levels and across multiple oversight and advisory entities, characterized by competing priorities, strategies, lines of authority, governance, policies, roles and responsibilities, and support requirements.

In a counterintuitive sense, when the budget is small, there is an increased ability to focus that in turn leads to a reduction in complexity. Large organizations often try to do too much and become overloaded and people can get confused.

In this age of knowledge, of projects, and of entrepreneurship, there are advantages in being agile and focused. There are approximately 100 countries involved in space, and the most successful countries seem to be the ones that figure out both their expertise and their niche skills so that no one competes with them, such as the current work in robotics by Italian craftsmen working with small budgets, very focused with great expertise. Another example of this approach is the Canadian Space Agency, a great partner with a very small budget, yet the Canadian robotic arm is their area of world expertise. NASA signs up Canada for missions involving this capability because they have an expertise that no one else in the world does.

Addressing Complexity through the REAL Knowledge Model

One of the key issues in complexity is identifying critical knowledge. What is it that the organization needs to be the best at and what does it need to focus on? This is fundamentally and empirically a knowledge question and it addresses the purpose.

The NASA CKO Office developed the *Rapid Engagement through Accelerated Learning* (REAL) Knowledge model (Figure 8.1)

- To promote the capabilities of how to more comprehensively and accurately define a problem.
- To encourage a pragmatic orientation that informs better decision making.
- To help address the issues of bias, ego, special interests, and personal agendas (Hoffman and Boyle, 2015).

At the core of the REAL knowledge model is the operational KM cycle activities of capture, share, and discover, but with an effectiveness measure paired with the knowledge activity. For example, capturing knowledge is the action and retaining is the measure; sharing knowledge is the action and applying is the measure; and discovering is the action and creating outcomes is the measure.

Surrounding the REAL knowledge core activities are the individual/team knowledge factors and the organizational/societal expectations that mitigate the journey of the challenge/opportunity from inception through the knowledge cycle to successful project outcomes. Note that the process arrows are bidirectional in terms of influence and input.

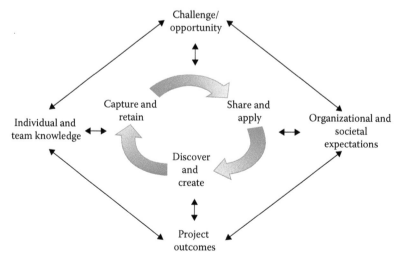

FIGURE 8.1

NASA REAL knowledge model. (Hoffman and Boyle, 2015.)

REAL Knowledge Model

In describing the REAL knowledge model, the following top-level generic application serves to illustrate a potential progression of activity:

1. A challenge/opportunity is selected and prioritized (characterized by *leadership, knowledge, project world, portfolio,* and *problem-centric* imperatives).

2. A learning project plan that compliments the project charter and project plan is initiated (characterized by *knowledge, accelerated learning, frugal innovation* and *governance, business management, and operations* imperative).

3. The functional communities of practice are recruited with points of contact identified (characterized by *leadership, project world, knowledge,* and *talent management* imperatives).

4. The core operational KM cycle is supported by specific KS learning strategies, methods, models, and technology tools to better define the opportunity; aggregate the data, information and knowledge; populate the alternatives for project decisions; provide appropriate online and traditional environments to spur and support innovation through discovery and creation; and support implementation through progressive and iterative knowledge support as the project proceeds through the lifecycle (characterized by *knowledge, technology, frugal innovation,* and *accelerated learning* imperatives).

5. Individual and team knowledge is leveraged, encouraged, supported, and enhanced through KS activities (characterized by *knowledge, talent management, accelerated learning, transparency, frugal innovation,* and *certification* imperatives)

6. External environment expectations in terms of the organization and broader society are identified and operationalized into objective definitions of performance over time and space (characterized by *leadership, knowledge, transparency, frugal innovation, accelerated learning, technology* and *governance, business management, and operations* imperatives).

The REAL knowledge model component definitions are provided along with associated keywords and concepts to aid potential future research in taxonomies and ontologies related to the narrower model and to the broader knowledge and learning disciplines:

- The challenge/opportunity is a problem-centered issue in terms of a product or service that presents a potential for action toward defined outcomes. Possible keywords and concepts include: vision and possibilities, requirements, and organizational capacity in technological, social, political, economic, and learning.

- Individual and team knowledge are formal and informal individual and collective education, professional development, and lessons from direct and indirect experience applied to a challenge/opportunity. Possible keywords and concepts include: assignments, abilities, formal education, professional development, and mentoring.

- Attitudes and values are predispositions based on learning, experience, and the challenge/opportunity to evaluate the environment in particular ways. Possible keywords and concepts include: personality and inclination, resilience, open-mindedness, curiosity and Skepticism, and tempered optimism.

- Heuristics and Biases are cognitive shortcuts and simplifications by individuals, teams, and organizations used to reduce complexity. Possible keywords and concepts include: normalization of deviance, problem solving, and decision making, fundamental attribution error, and culture of silence.

- Abilities and talent are learned or natural patterns of action for both individuals and teams that possess potential to achieve goals. Possible keywords and concepts include: critical thinking and creative thinking, problem solving, and decision making, creating alliances, and leadership and persuasion.

- Project knowledge is applied to existing and new data and information from a project context to a challenge/opportunity to gain efficiency and effectiveness. Possible keywords and concepts include: success stories and failure stories, learning through analogies, and organizational learning.

- Expectations are assumptions on probability of event occurrence for both individuals and groups based on learning and experience. Possible keywords and concepts include: adaptation to change, reputation, executive communications, and past performance.

- Organizational culture comprises a common set of values and assumptions that guide behavior in organizations that inform problem-solving and decision-making activity. Possible keywords and concepts include: organizational norms and mores, environmental context, and performance management.

- Knowledge capture and retention is the identification and storage of relevant content and skills. Possible keywords and concepts include: alliances, communities, and networks, cases and publications, risk records, mishap reports, organizational communications, and stories.

- Knowledge sharing and application is the representation, promulgation, and utilization of searchable and findable relevant content and skills. Possible keywords and concepts include: digital technology tools, informal learning, and best and emerging practices.

- Knowledge discovery and creation covers original content and skills derived and developed from previous relevant content and skills that result in project outcomes. Possible keywords and concepts include: searchability and findability, taxonomies, and innovation.

- Project outcomes achievement of original or improved products or services as defined by the project charter and validated by organizational expectations. Possible keywords and concepts include: value, improvement, innovation, and learning, knowledge, and growth.

Organizational and Societal Expectations

One particular component in the REAL knowledge model, *Organizational and Societal Expectations*, needs to be discussed due to its importance when addressing the topic of complexity. Human cognition is colored by inherent hard-wired preferences in thinking and in shortcuts that accompany decision-making processes, a product of choices and evolution. Biases and heuristics serve to reduce the amount of complexity, but also may introduce error. In addition, these biases and heuristics may differ across cultures. NASA represents a complex technical organization consisting of several divergent domestic and international cultures with different perceptions. Understanding these perceptions is important for the success of NASA's projects, especially since 80% of NASA programs and projects are international in nature.

Biases and heuristics are not just cognitive distortions that affect decisions, but also social biases that affect individual and organizational behavior as well as learning and memory tendencies that affect perceptions and explanations of the world. In our interview with Nobel Prize-winning scientist Daniel Kahneman on his recent *New York Times* bestseller *Thinking Fast and Slow* (2011), he clarified how humans address increasing levels of complexity in the project environment through heuristics that can introduce errors into decisions, a veritable catalog of fundamental biases, and heuristics that characterize human cognition.

System 1 thinking is fast, instinctive, and emotional, whereas System 2 thinking is slower, more deliberative, and more logical. Kahneman delineates cognitive biases associated with each type of thinking, starting with his own research on loss aversion, the unsettling tendency of people, and organizations to continue funding a project that has already consumed a tremendous amount of resources but is likely to fail simply to avoid regret. From framing choices to substitution, the book highlights several decades of academic research to suggest that people place too much confidence in human judgment, resulting in different outcomes, even given the same information input.

However, biases and heuristics should be viewed not only in a negative context, but one where these distortions and shortcuts can also provide positive outcomes. Many projects would not be started if executives waited until all the data and information were available to make a rational decision. Biases and heuristics serve in creating an environment where possibilities and vision can drive an idea toward reality. Busenitz and Barneyb (1997) found that there is a fundamental difference in the way that entrepreneurs and managers in large organizations make decisions, and that biases and heuristics drive entrepreneurial decisions and are used to reduce complexity in the project environment, simplifying decision making, and preventing data and information from overwhelming programs and projects, as well as serving to achieve buy-in and motivating practitioners. This often morphs into a tremendous disadvantage as projects mature from start-up activities to implementation and sustainability requirements. A brief set of examples from a rather extensive catalog of biases and heuristics are the following:

- Availability: making judgments on the probability of events by how easy it is to think of examples and their consequences.
- Substitution: substituting a simple question for a more difficult one.
- Optimism and Loss Aversion: generating the illusion of control over events and fearing losses more than we value gains.
- Framing: choosing the more attractive alternative if the context in which it is presented is more appealing.
- Sunk-Cost: throwing money at failing projects that have already consumed large amounts of resources to avoid regret.
- Mental Filter: focusing on one feature of something that influences all subsequent decisions.
- Fundamental Attribution Error: the tendency to overemphasize personality-based causes of behavior and underemphasize situational-based causes of behavior.
- Egocentric Bias: Recalling prior events in a favorable light to one's self rather than an accurate objective analysis.

The Four As: Ability, Attitude, Assignments, and Alliances

Another important facet of the REAL knowledge model is in what NASA refers to as the four As: *Ability, Attitude, Assignments, and Alliances* (Hoffman and Boyle, 2013). These components of the model are extracted from the interpersonal and team knowledge, attitudes and values, abilities and talent, knowledge capture and retention, and knowledge sharing and application

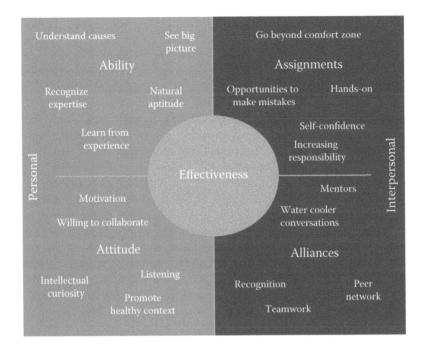

FIGURE 8.2
The 4A word cloud. (From Hoffman, E. and Boyle, J. *The Public Manager*, 42–43, 2013.)

components. They are represented in Figure 8.2 across a personal and inter-personal dimension of effectiveness.

Strategic Imperatives at NASA

The key challenge in addressing risk is integration at a systems level. The most successful organizations, countries, teams, and projects are the ones that know the system and know how to integrate in terms of both technical and people issues. The success or the failure of a mission often comes down to how effectively NASA works with their international partners because the success ultimately happens in terms of the effectiveness and the smoothness of the team.

Roughly a third of the success or failure of a project or a program comes down to political support. Practitioners are raised in projects to believe that control results in success. The reality is two thirds of the success happens outside of the project, at the headquarters, the teaming level, and the partner level. In this environment, being the second or third Project Manager is much better than being the first. The second Project Manager for the Hubble Space

Telescope is hailed as a great leader; the first was removed after six months. The background that is never revealed is that the first Project Manager received inadequate funding, did not get the team requested, and the project was not a priority to NASA initially because other things were happening. A new person comes on board and receives everything requested. This is addressed as a component of the REAL model.

At its core, NASA is a project organization fixated on mission success. There are 12 mutually reinforcing strategic imperatives that have emerged from interviews, studies, and experiences (Hoffman and Boyle, 2013). These guide the design, implementation, and evaluation of knowledge services (KS) for NASA. These are discussed in no particular order of priority.

The first imperative is *leadership*. It is ironic that one of the more fragmented disciplines provides valuable answers for the application of KS in organizations. Without effective leadership, KS and its results are at best serendipitous, at worst fail. The essence of leadership occurs with an insight that things should change, but also a realization that the reasons for change may be clear to leaders themselves but not necessarily to others. There is also an external stakeholder community as well as a core internal project team to lead, and both should be understood and managed. In addition, good leaders align projects with organizational strategy, mission and goals, and admittedly easier said than done in the modern environment of information overload and change. Successful implementation happens with a carefully articulated vision, leadership focus on that vision, and attention to detail on implementation.

It is a *project world*. Varied organizations worldwide require a methodology allowing for rigor in managing temporary, unique initiatives toward the achievement of defined requirements and project goals and outcomes that are aligned to organizational strategy in an era of constrained resources. In this context, project management (PM) is uniquely positioned as an adaptable discipline that fits these requirements and can maximize the use of learning to promote efficiency and effectiveness. Again, the alignment of project goals to organizational strategy through good leadership is critical.

Knowledge is the essential element for the creation of successful physical and virtual products and services. It can be viewed as an organized set of content, skills, and capabilities gained through experience as well as through formal and informal learning that organizations and practitioners apply to make sense of new and existing data and information. It can also exist as previously analyzed and formatted lessons and stories that are already adaptable to new situations. The ascendance of leaders that can validate the realities to which projects are able to apply knowledge and base decisions on is key.

Talent Management addresses the specification, identification, nurturing, transfer, maintenance, and expansion of the competitive advantage of practitioner expertise and competence. It encompasses the broad definition of diversity that goes beyond the classic categories of color, race, religion, and national origin to domestic and international variables important to geographically dispersed multicultural teams, such as multigenerational, cross-discipline, and cross-experiential variables. This allows diverse groups to bring a diversity of experiences, attitudes, knowledge, focus, and interests to the table, strengthening both inductive and deductive problem-solving approaches and nurturing innovation. Good leaders link talent management with executive sponsorship, organizational strategy, and the core work of the organization. They also achieve operational efficiencies by learning, working, and collaborating together at a distance independent of time and geography and leveraging smart networks that provide content, access and connection to project data, information, and knowledge. For NASA, talent management is represented as the variables of abilities, assignments, attitudes, and alliances (Figure 8.2).

Portfolio Management integrates projects with strategy and creates an organizing framework and focus that drives organizational purpose and activities. They provide a centralized function that promulgates a systems view of knowledge, where stove-piped disciplines and activities can transcend boundaries and discover and apply cross-disciplinary knowledge to increase competitive advantage and better achieve results. Organizational expectations can also be tested against reality at this level and adjusted and communicated accordingly to eliminate or mitigate errors and achieve better decisions.

Certification establishes objective, validated standards, and functions to benchmark achievement in defined categories of practitioner performance and capability. It also provides organizations and practitioners a way to establish trust with superiors, peers, team members, customers, and stakeholders, and provides a framework for adapting to change as well as a method to address emerging performance requirements. For practitioners, it provides a roadmap for individual development and serves to link organizational performance and individual capability. Since people are essential in projects (PMI, 2014), certification allows for objective definitions of abilities, assignments, attitudes, and alliances. An example of a discipline standard is the Project Management Body of Knowledge (PMBOK, 2013).

Transparency is an important consideration as the network of organizational portfolio sponsors, project team members, customers, stakeholders, strategic partners, suppliers, and other interested parties tie into organizational strategy and project operations through information and communication technology tools. In this environment, nothing is hidden for long and errors travel at the speed of light. Communications with each interface should be carefully defined across intensity and frequency dimensions, for example, where external stakeholder communities may expected to be informed about progress at a higher level, but not as frequently or in-depth as internal leadership. Transparency that is formally built into the strategic business process encourages innovation, translating economies of scale and a breadth of experiential lessons into innovation and flexibility.

Frugal innovation is a mindset that views constraints in an era of restricted and diminished resources as opportunities, leveraging sustainability and a focus on organizational core competencies to reduce complexity and increase the probability of better outcomes. Sustainability in particular has gained momentum as the cost to the planet and availability of resources increasingly impact business decisions. Organizational core competencies for a product or service involves what it must do in-depth rather than what it can do in breadth, ensuring that organizational capacity in areas such as technological, social, political, economic, and learning dimensions are part of the frugal-innovation process. In a mutually reinforcing perspective, imperatives such as *transparency* allow the broader team to share knowledge and experience to improve and innovate in terms of products and services, supporting the *frugal innovation* effort.

Accelerated learning is the tactic of employing state-of-the-art digital technologies, traditional knowledge-sharing activities, modern learning strategies, social media processes and tools, and cross-discipline knowledge into the broadest possible view of learning for an organization. The operational-knowledge process is closely linked to key internal and external knowledge sources and serves to clarify organizational expectations to optimize knowledge searchability, findability, and adaptability.

A *problem-centric approach* emphasizes a nonpartisan, nonbiased, nonjudgmental, and pragmatic orientation toward problems and solutions, keeping the focus on achievement, improvement, and innovation. Organizational expectations are kept pragmatic and constructive when a problem-centric approach is encouraged and expected. At the end of the day, it is about problems, communications, power, and building a community of support focused on credible challenges. This orientation serves as the fuel for change while addressing competing agendas and administrative barriers, and directly addresses the issue of bias and heuristics that may introduce error in decisions.

Governance, business management, and operations provide for pragmatic alignment, oversight, approvals, and implementation of project operations and establishes rigor and processes. In an era of frugal innovation, management of the budget and clarity of funding requirements that supports the overall effort must be visible and valued by the leadership and the workforce. Nothing brings trouble faster than mismanagement of funds and a lack of focus on funding flow, so the oversight, tracking, and implementation of project activities need definition. Defined governance addresses the issue of siloed implementation and raises executive awareness as well as formalizing successful localized grassroots efforts.

Digital technology makes it possible to examine new frontiers of potential knowledge and access multiple sources of data and information, but simultaneously causes organizations to be increasingly buried in data and information and have less time for focus and reflection. Technology is necessary but not sufficient for KS, but wonderful things can result from the application of technology, such as an open, social network-centric, nonproprietary, adaptable, and flexible framework that accelerate learning processes to deliver the right knowledge at the right time for particular needs while respecting context. The proper application of technology helps achieve learning results and better decisions at a lower cost.

Summary

So how is it possible that NASA has been doing talent management and leadership development, but practitioners still do not know what they need to know to succeed? The REAL model addresses this directly by providing a framework to communicate, to address pragmatic issues, to emphasize knowledge and learning in policies, and to develop accelerated learning in a complex unforgiving environment. Punishing practitioners because they did the wrong thing is the wrong approach; it basically shuts down the organization.

For NASA, knowledge services (KS) are developed with a steady progression of maturity influenced by the requirements of the missions over time. The agency today is not the same one that went to the Moon. Individual capability driven by internal experts fit the organization at the beginning, but that soon morphed into a team-based approach driven by diverse mission requirements as the purpose of the agency changed over the years. The complexity of the project environment addressed by this paper forces KS to adjust to the new realities of knowledge findability, searchability, and adaptability, highlighting the need for accelerated learning within a systems perspective, and reveals the synergy between the disciplines of knowledge management and organizational learning. The agency recognized this need by designating in 2011 the first NASA CKO to serve at the executive level for the agency.

The strategic imperatives that guide the development of NASA KS are a product of their times, addressing the realities and requirements for planning and action concerning leadership, complexity, limited resources, communication, knowledge, individual and organizational capability, and process. These imperatives can take different forms depending on specific organizational characteristics and needs at the strategic, operational, and tactical levels. For NASA, the federated approach allowed for the knowledge community to generate common definitions and purpose and develop reinforcing products and services that addressed both local and agency knowledge considerations, to include a new knowledge policy, an agency knowledge map, chairmanships of the federal knowledge community and the development of the NASA REAL knowledge model. This model allowed the agency to formulate KS activities that addressed the strategic knowledge imperatives, achieve buy-in across diverse communities, and accelerate learning to reduce complexity and ensure risks based on knowledge were identified and mitigated or eliminated.

The REAL knowledge model was presented as a descriptive model of how knowledge flow and knowledge services work at NASA. Future research can advance the understanding of the components of this model to achieve normative assumptions, definitions, and standards that promote effective and efficient knowledge practices that reduce complexity and accelerate learning to achieve successful outcomes.

Accordingly, the following future research initiatives should advance understanding and yield practical benefits for project organizations:

1. What are the characteristics of challenges and opportunities that achieve organizational and individual commitment, align individual and organizational agendas, and promote effective project management?

2. How should organizations systematically address talent development in terms of abilities, attitudes, assignments, and alliances?

3. What are the metrics and measures that best capture effectiveness and efficiency in the knowledge processes and outcomes of capturing and retaining, sharing and applying, and discovering and creating?

4. Can biases and heuristics that drive organizational and societal expectations be identified and addressed to inform how organizations can make better decisions and design better measures for the challenge/opportunity, the core knowledge processes, and project outcomes?

5. What are the operational definitions and certification parameters of knowledge behaviors for project practitioners and how does that address talent development and capability requirements?

6. How can the characteristics that make data and information searchable and findable, and result in adaptable knowledge in a systems approach to organizational knowledge and learning be operationalized to effective requirements and behaviors?

7. What is the nature of the relationship between knowledge services, accelerated learning, and reducing complexity?

In conclusion, there is still much work and research to be done in addressing how organizations and practitioners can best leverage project knowledge and knowledge services to get things done in the modern complex project environment. The potential mitigating and complicating variables that can reduce the power of knowledge and learning are too numerous to list, but a descriptive model serves as a framework to ensure that the breadth of relevant components that need to be addressed are present, as well as serving as a map for future research and development.

References

Anselmo, J. and Hedden, C. (2011, Aug 22). Up & down: A workforce in transition: Commercial companies hire as defense and space pare down. *Aviation Week & Space Technology*, 44–53.

Busenitz, L. W. and Barney, J. B. (1997). Differences between entrepreneurs and managers in large organizations: Biases and heuristics in strategic decision-making. *Journal of Business Venturing*, 12, 9–30.

Hoffman, E. and Boyle, J. (2013). Tapping agency culture to advance knowledge services at NASA. *The Public Manager*, 42–43.

Hoffman, E. and Boyle, J. (2015). REAL Knowledge at NASA: A knowledge services model for the modern project environment. Retrieved from: http://www.project management.com/white-papers/291288/R-E-A-L--Knowledge-at-NASA--A-Knowledge-Services-Model-for-the-Modern-Project-Environment.

Kahneman, D. (2011). *Thinking, Fast and Slow*. New York: Farrar, Straus and Giroux.

Project Management Institute. (2013). A Guide to the Project Management Body of Knowledge: PMBOK Guide (5th Ed.). Newtown Square, PA: Project Management Institute.

Project Management Institute. (2014). PMI's Pulse of the Profession(R): The High Cost of Low Performance 2014. February 2014.

9

An Academic Perspective: Knowledge Management: The Orphan Enabler— Enabling ICH Q10 Implementation

Nuala Calnan, Anne Greene, and Paige E. Kane

CONTENTS

> Despite the fact that the ICH Q10 *Pharmaceutical Quality System* guideline has been with us since 2008, the transformation of the traditional Quality Management Systems (QMS) in use within the pharmaceutical industry into an *effective* Pharmaceutical Quality Systems (PQSs) remains a work in progress. This chapter focuses on understanding the implementation essentials necessary to deliver an "effective quality management system for the pharmaceutical industry" [*Introduction, ICH Q10 p. 1*][1] and the crucial role that knowledge management plays.
>
> **Editorial Team**

Most of us by now are familiar with the stated aspiration of the pharmaceutical industry and regulatory authorities in the introduction to ICH Q10 Guidance (2008), which outlines that an effective Pharmaceutical Quality System provides a pathway to "… enhance the quality and availability of medicines around the world in the interest of public health." Yet, the achievement of this desired state, where top quality medicines are universally available, and affordable, to meet the current health needs of patients remains an ongoing challenge.[2]

In this chapter we examine the key role that *knowledge management* (*KM*), one of the two enablers designated within ICH Q10, plays in delivering the desired transformation. We propose that the means by which the pharmaceutical industry will achieve effective Pharmaceutical Quality Systems lies not in the management of knowledge, but in the utilization of that knowledge.

One important way of proactively managing patient risk and making better risk-based decisions, is to enhance the way we currently manage *and use* what we know. Knowledge must be evaluated by the quality of the decisions and actions that arise from it. Improving the quality of decision making within the industry is one important route to delivering an effective PQS.

> The best decision makers will be those that combine the science of quantitative analysis with the art of sound reasoning.[3]

The authors propose that the lack of maturity within the industry regarding the role that *knowledge* plays in delivering the necessary quality risk management, continuous improvement, and innovation is actually *disabling* the achievement of the ICH Q10 desired state. We look to leading international literature to examine strategies for how pharmaceutical organizations can improve the creation, management, and most importantly, utilization of the available explicit and tacit knowledge to deliver improved patient safety and product quality.

Introduction

> Knowledge derives from minds at work
>
> **Davenport and Prusak**[4]

Our exploration into the role that knowledge management can play in enhancing the effectiveness of the Pharmaceutical Quality System must start with the matter at the core of this challenge, *knowledge*. The ICH Q10 (2008) guideline has provided us with a precise definition of *knowledge management* as follows:

> Knowledge Management is a systematic approach to acquiring, analysing, storing and disseminating information related to products, manufacturing processes and components. [ICH Q10, p. 3]

However, it is useful to take a step back to develop an understanding of the body of knowledge that requires management.

ICH positioned *Quality Risk Management* (*QRM*) and *knowledge management* as the twin enablers for an effective PQS. However, although the topic of

QRM was assigned a full ICH guideline of its own (i.e., ICH Q9[5]), KM received somewhat less attention, with only a single paragraph included in Q10, along with some limited additional content in the related *Questions and Answers*[6] and *Points to Consider*[7] documents. Yet one important way of making better risk-based decisions and to engage in proactive management of patient risks is to enhance the way we currently manage *and use* what we know.

Developments in the area of knowledge management, within the fields of both management science and behavioral science, have made much progress in the intervening years since 2008. Indeed, it is difficult to pick up any newspaper or journal and not see some references to the importance of the knowledge economy, the role of knowledge workers, or the impact data scientists are delivering with recent advances in data analytics. Nevertheless, practitioners within the pharmaceutical manufacturing sector will freely admit that we have come to the knowledge table later than many other regulated (and nonregulated) industries.

Later in this chapter we explore the possible factors for this, but one immediate reason that resonates with many of the industry experts we have spoken to relates to the fact that many early initiatives focused more on data and information, and their *management* activities, rather than on the *knowledge* aspects. There was also less attention given to understanding the sources and variety of knowledge that may be important to the business. This emphasis may have arisen from the long tradition of managing records and documents (examples of data and information) within the GMP-regulated industry, and it led some of the early KM efforts to be mainly technology or IT platform-focused, designed to manage data and information, instead of being knowledge focused. Many of these efforts also consumed large resources but probably delivered little additional benefit to ensuring the supply of high quality, available medicines to the patient … because they were not utilized by the people in the organizations.

Therefore, let us start by examining what knowledge do we need to focus on, what is the impact it can have on patient outcomes, and why might the lack of maturity regarding knowledge management within the pharmaceutical industry have led us to controversially assign it the title of the *Orphan Enabler*.

What Do We Talk about When We Talk about Knowledge?

Let us commence with some basic definitions taken from the seminal work on knowledge by Davenport and Prusak from 1998, whose aptly named Chapter 1 title we have borrowed for this section. They adopted a simple three tiered approach as follows:

- *Data* is a set of discrete, objective facts about events.
- *Information* has meaning, that is, it is data that makes the difference.

- *Knowledge* is a fluid mix of framed experience, values, contextual information, and expert insight that provides a framework for evaluating and incorporating new experiences and information.

Most importantly, Davenport and Prusak identified the critical human element to knowledge, noting that it "originates and is applied in the minds of knowers." Missing this link to the human capital of an organization may account for why some of the earlier *systems*-based KM solutions failed to deliver their intended benefits. They go on to say that, if information is to transform into knowledge, humans must do virtually all the work (p. 6). They propose that this transformation happens through

1. *Comparison*: How does information about this situation compare to other situations we have known?
2. *Consequences*: What implications does the information have for decisions and actions?
3. *Connections*: How does this bit of knowledge relate to others?
4. *Conversation*: What do other people think about this information?

Considering these questions were posed in a 1998 book on knowledge management, they align almost exactly with the questions we might now pose when assessing risk during a pharmaceutical quality risk management activity. Note also, that the emphasis is firmly placed on the ability to make informed decisions *and* to take the necessary actions. This is referred to as *knowledge in action* and is linked to our main hypothesis.

> *Hypothesis:* The means by which the pharmaceutical industry will achieve effective Pharmaceutical Quality Systems lies not in the management of knowledge, but in the utilization of knowledge.

The desired goal in the area of KM is to have the ability to effectively use knowledge to make smarter, better, and risk-based decisions about the drug products we produce. Making better risk-based decisions, with the patient as the focal point, will also help the industry move away from the traditional *modus operandi* of relying on a purely compliance-led approach for decision making. In the traditional approach, while GMP compliance may be achieved, defective and potentially harmful medicines still get manufactured and released, and at an alarming rate. See Table 9.1[8] for a 10 year summary of the *Quality Defect and Recall* statistics for Ireland, other regional summaries reflect similar increases.

What is required therefore is a balanced approach of *doing the right things* rather than simply focusing on *doing things right*. Just to clarify, compliance is given, and it should (and must) be the foundation stone of any quality management approach. But a purely compliance-led approach, in the absence of

TABLE 9.1

Health Products Regulatory Authority (HPRA) Quality Defect and Recall Statistics 2006–2015

Year	2006	2007	2008	2009	2010	2011	2012	2013	2014	2015
Critical	84	173	127	105	173	231	189	235	365	213
Major	238	216	300	345	332	364	303	300	199	221
Others	49	84	128	164	246	322	249	239	252	325
Total	371	473	555	614	751	917	741	774	816	772
Recalls	58	97	141	98	168	253	141	109	102	113

Source: O'Donnell, K. Understanding the Context: Quality Defects and Recall Regulation in Europe—Role of the Irish Pilot Project, Regulatory Science Ireland Meeting 24th Feb 2016.

scientifically sound risk-based decision making, can lead to a static set of standards and controls being applied. This can lead to a situation where current or emerging risks are not identified, assessed, or properly managed. To compound things, the complexities presented by globalized manufacturing operations and supply chains, present an ever evolving array of sources of variation[*] that cannot easily be managed by a purely compliance-driven approach.

What is required are knowledge-led processes in organizations that enable the organization to take appropriate actions within this dynamic environment.

These organizations should have an accessible knowledge base that

- Continuously builds understanding regarding the sources of variation in their processes.
- Provides the necessary intelligence to promptly detect when a variation occurs.
- Facilitates a knowledge-led review of the risks posed by the variation.
- Enables proactive continual improvement activities.

And most importantly, it provides the

- Capability to take the necessary actions to protect the patient.

The justification underpinning any actions taken should come from a rigorous assessment of the risks utilizing "Knowledge … that provides a framework for evaluating and incorporating new experiences and information," as per the Davenport and Prusak definition.

---

[*] *Sources of variation may arise from People, Materials, Processes or Systems.*

The Evolution from Science to Knowledge within the *New Paradigm*

> Knowledge is experience, everything else is just information.

Albert Einstein (attributed)

In the context of understanding the body of knowledge we are referring to, it is worth considering the evolution of regulatory thinking on knowledge and its role in relation to risk management.

Early in the past decade, and up until the ICH Q8, 9, and 10 initiative, regulatory efforts at improving how medicines were manufactured and controlled focused mainly on enhanced science and risk-based approaches. Knowledge-led approaches were not given too much attention.[9] For example, when the FDA announced the Pharmaceutical cGMPs for the twenty-first Century Initiative[10] in 2002, the emphasis then was on *risk* and *science*. This was innovative and new at that time, and it led to many important developments and initiatives, such as those relating to PAT[*]. On rereading this document now, one cannot help but notice the absence of the term *knowledge* in it, whereas *science* is mentioned over fifteen times. Although the correlation between science and knowledge cannot be disputed, as evidenced by the definition for science from the Oxford English Dictionary:

> Science: *A systematically organized body of knowledge on a particular subject.*

If we limit the knowledge we value to that which is aligned with science (that one could argue is truly *explicit [documented] knowledge*), one is ignoring a whole range of *tacit knowledge* gained right across the lifecycle of the product. One might ask

> As the value of knowledge became more apparent throughout the past decade, have we come to recognize knowledge as the key and science to be a subset of that knowledge?

A review of key regulatory guidance documents published since 2002 for mention of either *knowledge* or *knowledge management* is informative.

The emphasis in the FDA's PAT guidance (2004) is on *understanding*, (product, process, and equipment understanding) and *continuous improvement*. The word *understanding* appears more than 25 times, whereas in regard to knowledge, the focus is on explicit scientific or mathematical knowledge. There is also reference to IT-based knowledge management solutions to analyze and store knowledge, but there is no real evidence of the importance of capturing tacit knowledge in the guidance.

---

[*] PAT: Process Analytical Technology.

But with ICH Q9 *Quality Risk Management* in 2005, we began to see the signs that knowledge was emerging as the overarching concept in itself, of which both science and understanding were a subset, albeit still with a leaning toward explicit knowledge. ICH Q9 leads off by stating that "The evaluation of risk to the quality should be based on *scientific knowledge.*" It does go on to mention *new knowledge, current knowledge* and *available knowledge* and the section on preliminary hazard analysis (PHA) identifies the role of *prior experience or knowledge* as a method of determining risks. However, most noteworthy from ICH Q9 are the details relating to uncertainty, where the concept of *knowledge gaps* is introduced. It suggests that uncertainty can arise;

> ... due to combination of incomplete knowledge about a process and its expected or unexpected variability. Typical sources of uncertainty include gaps in knowledge, gaps in pharmaceutical science and process understanding...

By linking the science and understanding gaps to knowledge gaps, ICH Q9 signalled, in a quiet way, that knowledge should be regarded as the overarching concept, with science and understanding a subset of that. ICH Q8 developed this concept further. The first publication of ICH Q8 *Pharmaceutical Development,* that same year saw the emergence of use of the term knowledge rather than understanding. The reference to *knowledge gained* in the quote below, underpinned the importance of knowledge gained through experience:

> It should be recognized that the level of knowledge gained, and not the volume of data, provides the basis for science-based submissions and their regulatory evaluation.

In the the following year, in September 2006, the FDA issued their guidance for industry on a *Quality Systems Approach* to pharmaceutical cGMP regulations, which provides a comprehensive overview of how pharmaceutical companies can implement robust quality systems based on a "science based approach ... and an understanding of the intended use of the product." This guidance does highlight the need for monitoring and analysis to enhance the body of knowledge about the product;

> Monitoring of the process is important due to the limitations of testing. *Knowledge continues to accumulate from development through the entire commercial life of a product.*

The above initiatives signalled the gradual emergence of knowledge as a key enabler, but it was with the publication of ICH Q10 *Pharmaceutical Quality System* in 2008 that we saw clear evidence of *knowledge* appearing at the forefront of regulatory thinking.

Although the complex question of *what is knowledge* may remain, for the first time we find a formal definition of what the international regulatory community considers *knowledge management* to be. Perhaps, the three-year time gap between the publications of ICH Q8/Q9 and the ICH Q10 guideline, allowed the thought process to crystallize and the role of *knowledge* and *knowledge management* to emerge. Within ICH Q10 there are also fine examples of what is considered knowledge, which infer the inclusion of experience and tacit knowledge, such as

> Development activities, using scientific approaches provide knowledge for product and process understanding

In the section on *process performance and product quality monitoring systems*, we see;

> Provide knowledge to enhance process understanding, enrich the design space (where established), and enable innovative approaches to process validation.

More significantly, rather than *science* being considered *knowledge*; Q10 suggests that scientific approaches *provide* knowledge, with knowledge rather than science being the key. Further sources of knowledge are identified in Q10 as:

- Prior knowledge (public domain, or internally documented)
- Pharmaceutical development studies
- Technology transfer activities
- Process validation studies over the product lifecycle
- Manufacturing experience
- Innovation
- Continual improvement
- Change management activities

From this list we begin to see the emergence of the true value of tacit knowledge and knowledge gained from experience with the product. This sets the industry up for the challenge it currently faces—how to capture, transfer, and use this emerging knowledge effectively; in other words getting knowledge to flow effectively.

Emphasizing the Capability to Capture and Transfer the Expanding Knowledge Base

Further evidence of the need to transform from a static toward a dynamic, responsive approach comes from the clear directive within ICH Q10 that the Pharmaceutical Quality System should assure that the body of knowledge is continually expanded during the *commercial manufacturing* phase;

> The pharmaceutical quality system should assure that the desired product quality is routinely met, suitable process performance is achieved, the set of controls are appropriate, improvement opportunities are identified and evaluated, *and the body of knowledge is continually expanded.*

This emerging nature of knowledge requires a significant phase-shift for an industry that has traditionally been focused on establishing and validating the *state of control,* and then maintaining that *status quo.* Indeed, with many of the traditional record and document management systems used in the industry, the requirements for strict audit control and data integrity led to design features that emphasized locking down control of the information and data held within; they focused little on the actual usability of that information as a source of knowledge.

Some recent pharmaceutical company knowledge management initiatives have addressed this issue of usability, through developing what are termed *connecting* and *converting* design features for information systems in conjunction with what is known as *collecting* functionalities.

In contrast to locking down control, what is required when dealing with emerging knowledge is a more *dynamic management of knowledge out of knowledge.*[11] Nonaka and Nishiguchi, in the introduction to their 2001 book titled *Knowledge Emergence,* noted that knowledge creation and knowledge transfer are delicate processes that must be *nurtured* rather than *managed,* and they necessitate particular forms of support and care. Another way to view this is that knowledge must be curated rather than managed.

Nonaka*, a leading academic in the area of knowledge, describes this link between knowledge creation and leadership in an article[12] published in 2000, assigning top management with the responsibility to "redefine the organization on the basis of the knowledge it owns," in order to articulate their knowledge vision and to communicate it throughout the company. Nonaka developed his renowned SECI model for knowledge conversion; this was in direct response to those traditional management models that focused on how to control the information flow within organizations. In contrast, the SECI process seeks to provide a "conceptual framework for the continuous and 'self transcending' process of knowledge creation."

This conversion concept hinges on the ease of access to existing explicit and tacit knowledge (*collecting*) and the ability to use, share, and transfer this knowledge with other colleagues in the organization (*connecting*). They in turn, bring their insights and experiences to enhance the understanding of a given element to create new knowledge, capture it, and embed it back within the organization (*converting*).

---

* Ikujiro Nonaka is a Japanese organizational theorist and Professor Emeritus at the Graduate School of International Corporate Strategy of the Hitotsubashi University, best known for his study of knowledge management.

This dynamic nature of knowledge, or knowledge flow, is central to Nonaka's description of knowledge management:

> What "knowledge management" should achieve is not a static management of information or existing knowledge, but a dynamic management of the process of creating knowledge out of knowledge.[11]

Nonaka's SECI model embodies this knowledge creation concept as a spiral, elevating the value of knowledge throughout the conversion process from tacit to explicit, and back again to new forms of tacit and explicit knowledge.

For completeness, let us include the definitions outlined by Nonaka et al. for tacit and explicit knowledge.

Explicit knowledge can be expressed in words and numbers and shared in the form of data, scientific formulae, specifications manuals, and the like. This kind of knowledge can be readily transmitted across individuals formally and systematically.

Tacit knowledge on the other hand, is highly personal and hard to formalize, making it difficult to communicate or share with others. Subjective insights, intuitions, and hunches fall into this category of knowledge. Difficult to verbalize, such tacit knowledge is deeply rooted in an individual's action and experience as well as in the ideals, values, or emotions he or she embraces.

The four stages of the SECI process show how knowledge is created through

- *Socialization* (from tacit-to-tacit), tacit knowledge can be shared or transferred between individuals and groups as peer-to-peer or expert-to-peer within an organization (i.e., *connecting*).
- *Externalization* (from tacit-to-explicit), the process of articulating tacit knowledge into explicit knowledge that facilitates the crystallization and translation of knowledge (i.e., *converting*) into readily available forms, which allows that knowledge to be shared by others and ultimately becomes the basis for new knowledge.
- *Combination* (from explicit-to-explicit), the process of converging existing explicit knowledge into more complex and systematic new explicit knowledge. Knowledge is collected (acquisition), exchanged (disseminated), and combined to create new knowledge and to make it more accessible.
- *Internalization* (from explicit-to-tacit), the process of embodying explicit knowledge. Closely related to *learning by doing*, knowledge that has been acquired or created is now shared cross-functionally throughout the organization (Figure 9.1).

The SECI process describes this dynamic spiral, where knowledge created is *organizationally amplified* as the conversion occurs, from the level of the individual employee right up through shared communities of practice (CoP), cross-functional teams, departmental, and divisional to organizational boundaries.

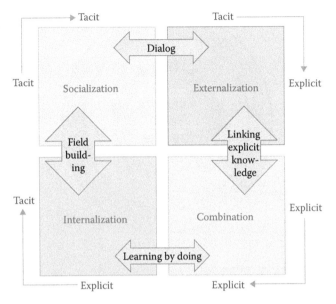

FIGURE 9.1
The SECI process. (Adapted from: Nonaka, I., Nishiguichi, T., *Knowledge Emergence*, 18, 2001; Nonaka, I., Von Krogh, G., *Organization Science*, 20, 635–652, 1994.)

This dynamic amplification process has real resonance for the pharmaceutical industry as new knowledge emerges *each batch, each day*. For example, through process deviations, daily operator experience when running a process, change control evaluations, in-process control and monitoring data, laboratory test results, and so on. All these provide information that, if captured and shared can be converted into new emerging knowledge about the process.

Understanding Knowledge Flow

Another useful way to view this concept of dynamic knowledge creation and conversion is to think about it in terms of enabling the *flow* of knowledge throughout an organization to enhance understanding. The importance of knowledge flow addresses very practically in the 2011 book *The New Edge in Knowledge* by O'Dell and Hubert.[14] This book outlines practices that organizations should employ to ensure they have the knowledge they need in the future and, more importantly, how to *connect the dots* and use that knowledge to succeed today. O'Dell and Hubert of APQC, a member-based nonprofit organization that engages in business benchmarking, best practices, and knowledge management research, have developed a KM framework and emphasize the importance of knowledge flow (Figure 9.2). O'Dell notes that

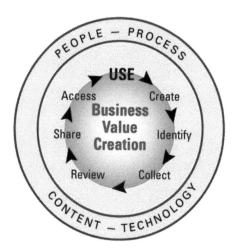

FIGURE 9.2
The APQC knowledge flow process. (copyright acknowledged).

"Knowledge is sticky. Without a systematic process and organizational environment it won't move."

Another critical aspect of this theory of knowledge emergence is the necessity of a nurturing environment (or an *enabling context* as Nonaka termed it), which must be provided through the support and *care* of the organization's leadership. Nonaka et al. describe this enabling context as *Ba* (based on a Japanese word roughly translated as *place*). Knowledge can only be shared, created, and amplified through personal interactions with others; *Ba* is the *knowledge space* where these personal interactions, that is, knowledge flow, can take place. This *knowledge space* may be physical, virtual, or mental, and will most likely be a combination of all three.

Establishing this enabling context and knowledge space requires leadership to encourage curiosity in the workforce, to reward proactive engagement by employees, and to build a culture of continuous improvement, which is knowledge-led, and which places the safety of the patient at its center.

More details on the power of utilization of tacit knowledge can be found in another recent article on this subject by this research team, titled *The 80/20 Rule of Knowledge*.[15]

Knowledge Activists as *Merchants of Foresight*

> Into every act of knowing there enters a passionate contribution of the person knowing what is being known, this coefficient is no mere imperfection but a vital component of his knowledge.

Michael Polyani

On the subject of tacit knowledge, Michael Polyani (1891–1976), a polymath, chemist, and philosopher, introduced the term *Tacit Knowing* in his 1958 book titled *Personal Knowledge*.[16] He based his theory on the fact that all knowing is personal. Since that time, the field of management science in relation to knowledge has increasingly recognized the critical role that *people* play in the knowledge process. It is the people and not the data that form the bedrock of an organization's knowledge base.

As an industry, the pharmaceutical sector and the GMP environment generally has long overvalued explicit knowledge at the expense of *know how*. It could be said that the sector has struggled with the *art* of unlocking and valuing tacit knowledge when the *scientific* approach favoured following the explicit specifications, forms, and procedures. Although ICH Q10 comprehensively addresses the enhanced roles and responsibilities of senior management, it does not specifically address the engagement of people as a route to building a proactive improvement culture. Many of the continual improvement recommendations included in Q10 for monitoring and review are more systems-based. If ICH Q10 were being written today, it is likely that it would contain more content on the influence that quality culture and employee engagement can have on delivering improvement. Indeed, the 2011 FDA guideline on *Process Validation*[17] goes some way toward this within the continued process verification (CPV) stage it describes, by recommending periodic engagement with production line operators and quality unit staff to, "... discuss possible trends or undesirable process variation, and coordinate any correction or follow-up actions by production."

One inspiring concept on the power and influence that *people* play in this knowledge challenge can be found in a book coauthored by Von Krogh in 2000, a Swiss academic, and Nonaka. It was titled, *Enabling Knowledge Creation*[18] and it addressed how to unlock the mystery of tacit knowledge and release the power of innovation. They introduced the role of *knowledge activists* as one of five key enablers to knowledge creation. These knowledge activists might be a business leader (e.g., Steve Jobs at Apple) or simply a passionate executive with specific responsibilities for promoting and curating knowledge-based work within their organization. Von Krogh and Nonaka identified three roles for knowledge activists as

1. Catalysts of knowledge creation
2. Coordinators of knowledge creation initiatives
3. Merchants of foresight

Linking back to the integration of the twin enablers of ICH Q10—QRM and KM—what a powerful image this projects. Consider a quality risk management activity being facilitated and executed by knowledge activists acting as *merchants of foresight* in assessing risks and protecting the patient!

Knowledge in Action—Smarter Decisions, Better Results[3]

Ultimately, what makes knowledge valuable is the capacity it provides us to take action. Davenport and Prusak[4] state (p. 6) that "knowledge can and should be evaluated by the decisions or actions to which it leads." Better knowledge enables smarter decisions, and this is the theme of a more recent book coauthored in 2010 by Davenport called *Analytics at Work: Smarter Decisions, Better Results*.[3] This book recommends strategies to improve our decision-making capabilities, by being more analytical in how we approach decisions. They define *analytical* as the use of *analysis, data,* and *systematic reasoning* to make decisions. The key, they say, is to "always be thinking about becoming more analytical and fact-based in our decision making."

They highlight (p. 9) the pitfalls known to many organizations of collecting and storing a lot of data but failing to *use* it effectively. "We have information and we make decisions, but we don't analyse the information to inform these decisions."

They ask the question—Are we simply working with known information or seeking to gain new insights? They present a useful framework for decision making that is based on two dimensions—time frame and innovation. They recommend that analysis that moves beyond purely information-based questions to queries involving insights are more likely to give a much better understanding of the dynamics of a given situation (Table 9.2).

Within our industry, when we reflect on the quality defects highlighted in recent regulatory actions or when we review the reasons given for many product recalls—what becomes evident is the poor quality of the GMP decision making that led directly to those defects and recalls. These cases demonstrate what happens when there is a failure to analyze the available knowledge to inform the necessary decision. Change control is an area of particular importance. Risks presented to patients by changes that have not been properly evaluated from a scientific perspective can be significant.

Poor decision making in relation to risk assessments conducted into quality defect issues can be another area of concern, as shown by the following case

TABLE 9.2

Key Questions Addressed by Analytics

	Past	Present	Future
Information	*What happened?* (reporting)	*What is happening now?* (alerts)	*What will happen?* (extrapolation)
Insights	*How and why did it happen?* (modeling, experimental design)	*What is the next best action?* (recommendation)	*What's the best/worst that can happen?* (Prediction, optimization, and simulation)

Source: Davenport, T. et al. *Analytics At Work—Smarter Decisions, Better results.* Boston, MA, Harvard Business School Publishing, p. 7, 2010.

identified during a GMP inspection. In this case, the manufacturer concerned performed a risk assessment in response to a number of complaints from the marketplace of potential quality defects related to the packaging of certain medicinal products. The risk assessment report identified certain situations in the packaging processes that could lead to the risk of reintroduction of rejected (defective) packaging components back onto the packaging line. The manufacturer rated two of these risks as *high* risk and risk-mitigating actions were identified. However, the risk assessment report indicated that those actions would not be completed on the line for a further eight months. During a GMP inspection of the facility an inspector reviewed this risk assessment report. It became evident that, in the meantime no interim risk control measures had been put in place for those two *high* risk issues. Therefore, between the time of the risk assessment and the implementation date of the permanent risk mitigations, eight months later, there was a lack of assurance that products with defective packaging components would not be released to the marketplace.[*] Better decision making in relation to that risk assessment might have ensured that those risks were managed in a more patient-focused manner.

This presents a clear case of where the information was available but was not used to inform the actual decision-making process. What we should be striving for is a situation where good people, working collaboratively, are enabled to evaluate, generate, and justify decisions that are beneficial to the patients who rely on those medicines. In the words of Davenport, "the best decision makers will be those that combine the science of quantitative analysis with the art of sound reasoning." This provides a strong endorsement of the many quality metrics initiatives, which are currently under exploration by the industry and regulators alike. It is the opinion of the authors that if adopted, quality metrics have the potential to present an additional tool in the armory of an effective Pharmaceutical Quality System.

Becoming Part of the Way We Work

> Knowledge creation is ... acquiring a new context, a new view of the world and new knowledge. In short, it is a journey "from being to becoming."
> Nonaka et al.[12]

The challenge; to embrace the journey from *being* statically-oriented to *becoming* dynamically-oriented in our quest to develop a quality culture that is based on continual improvement and excellence. To ensure this, knowledge

---

[*] Information provided by K. O'Donnell (HPRA), September 18th 2014, in relation to inspectional observations.

must become central to the way we work. It should be adopted firmly within the workflow and not perceived as an additional task to getting the job done.

There are good business reasons for striving toward this change. One immediate one is that it should improve performance in reactive activities such as investigation and CAPA work by getting to the true root cause of problems and enabling more effective preventative actions. This can help to reduce those internal costs typically incurred in issue resolution, that is, *the cost of poor quality*.

On the other hand, as knowledge begins to flow more effectively the fluency of the personnel in using the expanding knowledge base improves, this can facilitate proactively targeting the traditional cost structures of activities associated with the provision of *good quality*, for example, by improving supplier and materials quality management through effective knowledge transfers or by deploying process capability improvements through utilizing the available process knowledge.

Becoming more analytical in the decision-making process will also help to shift the focus from correction to prevention, resulting in a positive impact on the *cost of quality* and the overall business performance.

Another leading quality metric, which can be used as a means to demonstrate the elusive *effectiveness* of your Pharmaceutical Quality System is the percentage of preventative actions (PA) relative to corrective actions (CA) that are currently open within you quality management system. Presenting trended data on a *PA/CA ratio*, coupled with opportunities to correlate this performance with other measures, such as the *deviation occurrence rate* or the number of *active investigations*, offers opportunities to demonstrate the proactive nature of your quality system and the effectiveness of the actions taken, as required by ICH Q10.

Integrating the Twin Enablers—QRM and KM

One way of ensuring that knowledge is central to the way we work will be to truly integrate it with the quality risk management processes that underpin every activity within today's pharmaceutical manufacturing environment. This integration of ICH Q10's twin enablers can lead to a more effective Pharmaceutical Quality System, when QRM and KM operate in balance with one another, rather than as isolated entities.

In many organizations, QRM and KM operate, at best, in parallel and are neither well integrated nor well balanced, and may in fact be potentially disabling each other. It is important to reflect on the potential impact that *knowledge hoarding* may be exerting on your quality risk assessment (QRA) exercises. The practice of designating subject matter experts (SMEs) within many organizations may in fact be encouraging the practice of knowledge hoarding (holding onto knowledge as a means of power, influence, or reputation) and thereby actively disabling the KM enabler during these risk activities.

FIGURE 9.3
A balanced enabler-based representation of the ICH Q10 Pharmaceutical Quality System Model.

To promote a balanced integration of the two ICH Q10 enablers (QRM and KM), we present an alternative representation of the ICH Q10 PQS model. In this model, the four key elements of the ICH Q10 PQS, (product, performance, and product quality monitoring system, CAPA, change management and management review) operate in an environment in which the QRM and KM enablers are well integrated and in balance with one another. This is within the overall context of a GMP environment, which is encompassed by leadership oversight and management accountability, as shown in Figure 9.3.

This balanced approach offers many benefits and requires that all risk management activities are knowledge-led and scientifically sound. This rebalancing will require empowerment from the top-down, through enlightened leaders, and enablement from the bottom-up, through an engaged workforce.

Conclusions

The following key concepts and points are explored in this chapter:

- The current management of knowledge in the GMP environment is not leading to the ICH Q10 desired state of an effective pharmaceutical quality management system—this approach does not facilitate the management of current or emerging risks or the protection of the patient.

- Science and process understanding should be regarded as subsets of knowledge management (KM), with KM being the overarching concept within the desired state.
- Knowledge capture in itself is also not sufficient—we must find ways to transfer (or convert) our knowledge to knowledge that can be *used* to support patient-focused, risk-based decision making.
- Reaching a stage where knowledge is usable in an effective manner should lead to an environment of continual improvement and one in which current knowledge supports the development of new knowledge.

In this chapter we have presented and explored a range of ideas and concepts for how organizations can truly manage knowledge. Coupled with good *Quality Risk Management* processes, there is the potential for KM to play a leading role in delivering the desired effectiveness within your Pharmaceutical Quality System.

References

1. ICH, 2008. ICH harmonised tripartite guideline: Pharmaceutical quality system Q10. ICH, Ed.
2. Calnan, N., O'Donnell, K., Greene, A., 2013. Enabling ICH Q10 implementation—Part 1 striving for excellence by embracing ICH Q8 and ICH Q9. *PDA Journal of Science and Technology*, 67(6): 581–600.
3. Davenport, T., Harris, J., Morison, R., 2010. *Analytics At Work—Smarter Decisions, Better results*. Boston, MA: Harvard Business School Publishing.
4. Davenport, T., Preusak, L., 1998. *Working Knowledge—How Organisations Manage What They Know*. Boston, MA: Harvard Business School Press.
5. ICH, 2009. ICH harmonised tripartite guideline: Quality risk management Q9. ICH, Ed.
6. ICH, 2010. Q-IWG on Q8, Q9 and Q10 Questions & Answers (R4).
7. ICH Q-IWG, 2011. Points To Consider (PtC). ICH.
8. O'Donnell, K. 2016. Understanding the Context: Quality Defects and Recall Regulation in Europe—Role of the Irish Pilot Project, Regulatory Science Ireland meeting. Dublin, Ireland.
9. Kevin O'Donnell, K., Greene, A., 2014. From Science to Knowledge: An Overview on the Evolution of *Knowledge Management* in Regulatory Guidance. *Pharmaceutical Engineering: Knowledge Management e-supplement edition* [Online], 44–49.
10. FDA, 2004. *Pharmaceutical CGMPs for the 21st Century—A Risk Based Approach*.
11. Nonaka, I., Nishiguchi, T., 2001. *Knowledge Emergence*. Oxford University Press.
12. Nonaka, I., Toyama, R., Konno, N., 2000. SECI, Ba and Leadership: A Unified Model of Dynamic Knowledge Creation. *Long Range Planning*, 33.

13. Nonaka, I., Von Krogh, G., 1994. Tacit knowledge and knowledge conversion: Controversy and advancement in organizational knowledge creation theory. *Organization Science*, 20(3), 635–652.
14. O'Dell, C., Hubert, C., 2011. *The New Edge in Knowledge: How Knowledge Management Is Changing the Way We Do Business*. APQC.
15. Calnan, N., 2014. The 80/20 Rule of Knowledge *Pharmaceutical Engineering: Knowledge Management e-supplement edition* [Online], 54–58, May 2014.
16. Polyani, M., 1958. *Personal Knowledge*. Chicago, IL: University of Chicago Press.
17. FDA, 2011. Guidance for Industry, Process Validation: General Principles and Practices. Current Good Manufacturing Practices (CGMP) Revision 1 ed. FDA, Ed.
18. Von Krogh, G., Ichijo, K., Nonaka, I., 2000. *Enabling Knowledge Creation*. Oxford: Oxford University Press.

10

An Academic Perspective: Effective Knowledge Assessment

Mohamed A. F. Ragab and Amr Arisha

CONTENTS

> The restructuring of global markets toward a knowledge-centered economy has ignited a pervasive urge for better understanding of the dynamics of the knowledge assets in competing organizations. Knowledge assessment empowers organizations to identify the pockets of knowledge that create value to the business, and thus facilitates effective human capital planning according to an accurate assessment of the stock of knowledge assets within the organization. This contribution introduces an exciting new model (and Smart APP) that has been developed following rigorous academic research, to manage individual knowledge, known as the MinK Framework.
>
> **Editorial Team**

To know what you know and what you do not know, that is true knowledge.

Confucius

Introduction

A hallmark of the new business success is the ability of organizations to identify the economic value of knowledge assets. Although few scholars in the management literature question the validity of the popular quote by Lord Kelvin "If you cannot measure it, you cannot manage it," it is an established fact that the capacity to manage any organizational dimension becomes quite arduous without the ability to assess what is being managed. Despite being one of the most challenging and complex knowledge management (KM) endeavors, knowledge assessment has emerged as a key area of interest for both academics and managers and is often advocated for being an important prerequisite for the effective management of knowledge [1,2].

Successful KM entails knowledge assessment capability to enable the identification of *knowledge holders*, proper governance of an organization's value creation dynamics, alignment of strategic plans with available knowledge assets, and to support managerial judgment in the allocation of knowledge resources [3]. Discussions of knowledge assessment are often coupled with the related concept of *intellectual capital (IC)*—the compilation of organizational knowledge assets that drive organizational performance and value creation [4]. In the conceptualization where organizational knowledge is envisaged as a series of "stocks and flows" [5], IC represents the stocks of knowledge an organization holds at a certain point of time, comprising knowledge that has been integrated into the firm then used to create value. Knowledge flows, on the other hand, are the main concern of KM strategies, which focus on the acquisition and sharing of such knowledge [6].

Knowledge Assessment in Context

Organizations tend to assess knowledge for either internal monitoring purposes and/or external presentation. For an internal objective, managers often have little knowledge of the value of their own IC, nor where it exists within their organizations, despite it being their main source of competitive advantage [7]. Accordingly, IC measurement models attempt to discover *hidden* knowledge assets so they can be utilized more effectively to improve organizational performance [8]. After the firm's IC is unveiled, a measurement tool continues to be crucial to evaluate the impact of KM initiatives [9]. IC indicators are considered the gauge that justifies the enormous growth in KM expenditure—estimated in billions [10].

If it is for an external purpose, there is a widespread view that a company's value could only be assessed if intangible assets are taken into consideration. This view has emerged due to the widening gaps between companies' book and market values, where the ratio of the latter to the former has multiplied over the past decade [11]. An example is the acquisition of mobile instant messaging company *WhatsApp* by social media giant *Facebook* for an astounding $14 billion. Generally accepted accounting principles (GAAP), however, are only reporting on physical assets in the firm's balance sheet without considering any of the intangible ones. This "gap in the GAAP" conundrum [12] has led a number of researchers to propose alternative knowledge-based accounting methods that incorporate IC to reveal a company's true value.

Although the literature offers a plethora of knowledge assessment approaches, the main methods can be classified under the following: (1) financial methods, (2) IC scorecard methods, and (3) performance methods.

Financial Methods

This type uses financial models to estimate an overall value for a company's IC using information obtained from financial statements. The following are the most prevalent models:

Tobin's Q: Developed by Nobel laureate and economist James Tobin, Tobin's Quotient [13] is a technique that aims to evaluate investment decisions. It evaluates a company's market-to-book ratio and estimates the tangible assets of any organization using their replacement cost rather than their book values. Tobin theorizes that a Q that is higher than one, and is higher than that of rival companies indicates that the company possesses an *intangible advantage* with which it creates more value than its competitors. This advantage is assumed to be its IC.

Economic value added (EVA): EVA is a financial measure originally introduced as an indicator of shareholder value creation and involves applying 164 adjustments to the traditional balance sheet to account for intangibles, for example, by adding back research and development costs to assets [14,15]. It is then calculated by deducting the cost of capital from operating profit. Accordingly, an increase in EVA is an indicator of an efficient management of IC.

Human resource accounting (HRA): Originating in the 1960s, the objective of HRA is to use financial data to quantify the economic value of people as "human assets" [16,17]. To this end, researchers have suggested three types of HRA models: (1) cost models, (2) market models, and (3) income models. In *cost models*, human capital is valued as the cost of acquiring human assets (i.e., their recruitment and training cost), or alternatively the discounted value of employees' gross compensation. *Market models*, on the other hand, equate human value with the cost of buying an individual's services from the market, for example, via consultancy. Finally, *income models* use the present value of the revenues a person is expected to generate while working for a company to evaluate their knowledge.

IC Scorecard Methods

The second approach, IC Scorecard Models, splits a company's value into financial and intellectual capital, then breaks down the latter into different elements in the form of a scorecard, which are then evaluated individually [18]. When classifying IC, most authors agree with the tripartite classification proposed by Stewart [19] where IC is broken down into human capital (HC), structural capital (SC), and relational capital (RC). HC includes the combined knowledge, skills, and abilities that employees possess and is the core element of IC. Since HC cannot be *owned* by the organization, it is lost when employees leave [20]. SC is the supportive infrastructure the organization makes available for its employees including physical resources, information systems, and organizational processes. In contrast to HC, SC is owned by the firm, and so has been referred to as "knowledge that does not go home at night" [21]. RC refers to the combined value of a company's external relationships with stakeholders, such as suppliers and customers, who are valuable sources of both revenue and knowledge for the firm.

Following classification, sets of quantitative metrics are developed to measure each IC component in the scorecard. Metrics could be direct counts, monetary values, or ratios/percentages [22]. In cases where metrics measure a qualitative attribute (e.g., motivation), scale-based surveys are used to convert qualitative parameters into quantitative figures. The next step in a number of models is to aggregate all IC measures into a single quantum, using such methods as averages and weighted averages. In the final step, some frameworks then attempt a financial valuation of IC in monetary terms. Examples of the key models of IC measurement are the following:

> *Skandia navigator:* Cited more than 3000 times, The Skandia Navigator is the most prominent effort to measure IC of organizations. Led by Leif Edvinsson, the world's first corporate director of intellectual capital, the framework was developed by Skandia AFS, a Swedish insurance company and the first company to publish an IC supplement to its shareholders with its annual report. Skandia developed 112 metrics that cover four foci, in addition to a financial focus, where each focus relates to a component of IC. Consolidation in this model is achieved by combining all financial indicators into a single monetary value C, and converting all the remaining metrics into ratios and then aggregating them into an efficiency indicator I. In their seminal book, Edvinsson and Malone theorized that the overall financial value of IC is equal to I multiplied by C [23].
>
> *IC-index:* The IC-Index [24] is another method that, unlike the Skandia Navigator, does not propose specific metrics, but instead provides a framework by which every organization would set its own metrics in light of the company's strategy, characteristics, and the environment in which it operates. Organizations would start by determining

key success factors (KSF) in view of their mission and strategy. KSFs would then be converted into indicators, shortlisted, and categorized under HC (thinking part) and SC (nonthinking part). Indicators are then combined into a single index using a weighted average. The IC Index does not provide a financial valuation of IC, but instead indicates that there should be a positive correlation between a firm's index and its financial value if indicative and relevant knowledge measures are adopted.

Intellectual capital statements: As part of a project with the Danish Agency for Trade and Industry, Mouritsen and his team developed the intellectual capital statement framework adopting a more qualitative approach than most models [25–27]. Moving from a holistic view of knowledge resources, they propose the use of *knowledge narratives* that are a textual description of the firm's KM strategy based on its objectives and available resources. They are then used to define a list of associated KM challenges that the firm would have to overcome to be able to achieve the purpose of the narrative. The process of putting knowledge narratives into action is monitored through a set of indicators referred to as the *Intellectual Capital Accounting System*.

Other prominent IC models include the Technology Broker (IC Audit) [28], Intangible Asset Monitor [29], IC Rating [30], Value Chain Scoreboard [31], and the Human Capital Monitor [32].

Performance Methods

According to the literature, a group of researchers have adopted the view that knowledge *cannot* be measured and argued that the intangible and multifaceted nature of knowledge would thwart any effort toward its measurement. They recommended that efforts should be directed instead toward assessing the *impact* of knowledge, which in many cases is much more tangible and measurable than knowledge itself [1]. In another similar argument, few studies found that a number of firms still report limited improvement in post-KM organizational performance due to the lack of proper methods to evaluate KM performance [33], and hence they tried to address this research gap. Frameworks in this domain measure the performance of either KM *processes* or KM *outcomes*, or both [34,35]. Process performance measures are a type of leading measures that monitor the progress of a KM initiative and provide immediate feedback allowing management to take actions in real time [34]. Output performance measures, on the other hand, are lagging indicators that demonstrate the results of KM implementation in retrospect [34]. Their underlying logic is the comparison of corporate performance before and after the implementation of a KM project to examine its effect on the organization using a diversity of performance management (PM) approaches.

Methods that measure performance are often based on quantitative indicators that can be financial, such as stock price and profitability, or nonfinancial, like reductions in cycle time. Other studies use survey methods, such as questionnaires or interviews, to measure performance improvements based on respondents' judgments of the positive contributions KM has made toward their organizations. Moreover, the balanced scorecard (BSC) is a widely popular comprehensive approach that indirectly links PM with KM. Pioneered by Kaplan and Norton [36], the BSc offers a systematic methodology that uses strategy-linked key performance indicators (KPIs) to measure performance from four perspectives: financial, customer, internal business processes, and learning and growth.

Classification of Knowledge Measures

After an extensive review of existing models within the knowledge measurement literature, knowledge measures can be summarized in a five-fold taxonomy depicted in Figure 10.1.

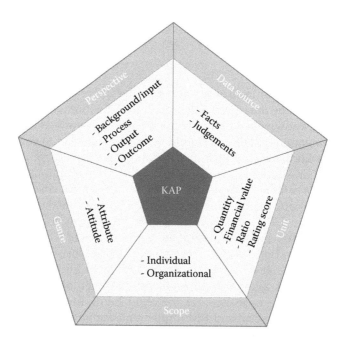

FIGURE 10.1
KAP classification framework.

This multidimensional framework is referred to as the *knowledge assessment pentagon (KAP)*. KAP suggests that knowledge metrics are categorized by the following:

- *Scope*: Level of assessment; knowledge can be measured on an individual, organizational, or national level.
- *Unit*: Knowledge measures are either quantitative (e.g., counts, costs), or qualitative, relying on rating scales.
- *Data source*: Data collected for measurement is either based on factual evidence or relies on *actor judgment* where an individual or group are asked to assess a certain factor based on their views [37].
- *Genre*: Distinguishes between HC measures that assess *attributes* (e.g., experience) and those who measure *attitudes*, such as ratings of employee motivation and satisfaction.
- *Perspective*: Denotes the time orientation of measures (Figure 10.2). It therefore includes the following:
  - *Background measures*: Assess inputs and enabling factors that empower the creation and exploitation of knowledge. Example measures include education levels and infrastructural resources.
  - *Process measures*: Monitor dynamic attributes of knowledge flows resulting from engagement in knowledge processes such as contribution and usage frequency of knowledge bases, or rates of social interaction [37].
  - *Output measures*: Evaluate the end results of knowledge creation and sharing processes. Examples include measures such as the number of patents produced.

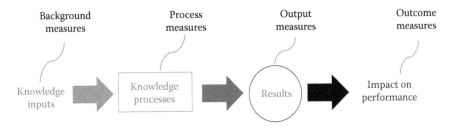

FIGURE 10.2
Knowledge measures perspectives. (From Malhotra, Y., Measuring knowledge assets of a nation: Knowledge systems for development, In *Invited Research Paper Sponsored by the United Nations Department of Economic and Social Affairs, Keynote Presentation at the Ad Hoc Group of Experts Meeting at the United Nations Headquarters*, New York City, NY, 2003; Bolisani, E. and Oltramari, A., Knowledge as a measurable object in business contexts: A stock-and-flow approach, *Knowl. Manage. Res. & Pract.*, 10, 275–286, 2012.)

- *Outcome measures*: Although KM outputs are the product of knowledge processes, KM outcomes are measures of the impact of such outputs on individual or organizational performance. Typical measures that fall into this class are increases in revenue or achievement of targets.

Individual Knowledge Assessment

Review of KM studies indicates that knowledge assessment models mostly adopt a holistic view of organizations by attempting to measure knowledge on the organizational level using the notion of IC. Limited efforts within the KM domain are dedicated to evaluating individual employees from a knowledge-based perspective. Current practices of individual assessment are mostly undertaken from other perspectives within the management landscape, namely performance appraisal and personality testing. Knowledge assessment differs from both by focusing on *what one knows* rather than *how they perform* or the *type of person* they are. The main characteristics of each perspective of individual assessment, in contrast with knowledge assessment, are highlighted in Table 10.1.

TABLE 10.1

Perspectives of Individual Assessment

	Performance Appraisal	Personality Tests	Knowledge Assessment
Description	Periodic assessment of job performance using certain criteria	Identification of the psychological traits of an individual	Evaluation of the knowledge held by an individual
Purposes	• Performance improvement • Motivation and reward • Succession planning	• Recruitment • Team building	Identification and allocation of knowledge resources
Types	• Results-based • Competency-based	• Personality types • Personality traits • Emotional intelligence	• Financial methods • IC scorecard methods • Performance methods
Assessment parameters	Company specific	Mostly standard tests For example, Myers–Briggs	Generic or firm specific
Method	Direct manager evaluates employee according to predefined criteria	Self-administered questionnaire	Diverse

Source: Ward, P., *360-Degree Feedback*, London, Institute of Personnel and Development, 1997; Fletcher, C., *Appraisal: Routes to Improved Performance*, London, Institute of Personnel and Development, 1997; Torrington, D. et al., *Human Resource Management*, 8th ed., Harlow, Pearson, 2011.

Among the recent research developments in this area is the *MinK Framework* [43–47], which is an integrated model to assess individual knowledge in organizational contexts. Evolving from a theoretical foundation that highlights the crucial role of individuals in firm knowledge dynamics, *MinK* is an assessment platform for decision makers that ensures that individual knowledge is accurately assessed from a number of perspectives using a well-defined set of theoretically grounded and industry validated indicators stemming from a multidimensional scorecard. Flexibility is embedded in the MinK framework, allowing managers to customize the key measures according to the firm's specific context. Adopting the 360-degree approach, the assessment process uses self evaluations and multisource knowledge appraisals to provide rich and insightful results. An *Individual Knowledge Index (IK-Index)* that denotes the overall knowledge rating of each employee is computed using a unique formula that combines a number of multi-criteria decision analysis (MCDA) techniques to consolidate assessment results into a single reflective numeral. The incorporation of technology through an advanced *MinK* app enables the complete automation of the assessment process and helps address parametric multiplicity and arithmetic complexity to guarantee the efficiency, security, and accuracy of the assessment process and the generation of indicative results.

Conclusion

The restructuring of global markets toward a knowledge-centered economy has ignited a pervasive urge for better understanding of the dynamics of knowledge assets in competing organizations. Knowledge assessment empowers organizations to identify the knowledge pockets that create value to business, and thus conduct effective human capital planning according to knowledge stocks and flows. Assessment of knowledge assets will also enable organizations to communicate their true worth to stakeholders. Research in the KM domain presents decent attempts of multitudinous frameworks that have objectives of assessing knowledge using each, or a combination, of financial methods, scorecard-based methods, and performance-oriented methods. The knowledge assessment pentagon (KAP), discussed in this chapter, presents a conclusive five-fold taxonomy of knowledge measures reported in the KM literature. Nevertheless, most of knowledge assessment models have tended to adopt a holistic view of organizational knowledge often overlooking the role of individuals in the creation of knowledge assets. Having said that, a newly developed framework (*MinK*) introduces an inclusive set of measures with an ultimate objective of identifying knowledge holders in organizational context [47]. This proposed innovative and applied solution places the individual knowledge holder at the core of KM while taking into account the unique setting and culture of the organization. MinK also suggests that effective KM strategies correlate with the proper management of knowledge workers, which can initially begin by helping organizations to know their knowers.

References

1. R. Ruggles, The state of the notion: Knowledge management in practice, *California Management Review*, 40, 80–89, 1999.
2. J. Liebowitz and C. Suen, Developing knowledge management metrics for measuring intellectual capital, *Journal of Intellectual Capital*, 1, 54–67, 2000.
3. D. Carlucci and G. Schiuma, Knowledge asset value spiral: Linking knowledge assets to company's performance, *Knowledge and Process Management*, 13, 35–46, 2006.
4. G. Schiuma, A. Lerro, and D. Sanitate, The intellectual capital dimensions of Ducati's turnaround: Exploring knowledge assets grounding a change management program, *International Journal of Innovation Management*, 12, 161–193, 2008.
5. F. Machlup, Stocks and flows of knowledge, *Kyklos*, 32, 400–411, 1979.
6. A. Al-Laham, D. Tzabbar, and T. L. Amburgey, The dynamics of knowledge stocks and knowledge flows: Innovation consequences of recruitment and collaboration in biotech, *Industrial and Corporate Change*, 20, 555–583, 2011.
7. N. Bontis, Managing organisational knowledge by diagnosing intellectual capital: Framing and advancing the state of the field, *International Journal of Technology Management*, 18, 433–462, 1999.
8. L. Edvinsson, Developing intellectual capital at Skandia, *Long Range Planning*, 30, 366–373, 1997.
9. G. Robinson and B. H. Kleiner, How to measure an organizations intellectual capital, *Managerial Auditing Journal*, 11, 36–39, 1996.
10. M. Khalifa, A. Yu, and K. Shen, Knowledge management systems success: A contingency perspective, *Journal of Knowledge Management*, 12, 119, 2008.
11. B. Lev, The old rules no longer apply, *Forbes ASAP*, 159, 34–35, 1997.
12. D. Skyrme, *Measuring Knowledge and Intellectual Capital*. Business Intelligence: UK, 2003.
13. J. Tobin, A general equilibrium approach to monetary theory, *Journal of Money, Credit and Banking*, 1, 15–29, 1969.
14. J. M. Stern, G. B. Stewart, and D. Chew, The EVA financial management system, *Journal of Applied Corporate Finance*, 8, 32–46, 1995.
15. B. Stewart, EVA: Fact and fantasy, *Journal of Applied Corporate Finance*, 7, 71–87, 1994.
16. R. H. Hermanson, *Accounting for Human Assets*. East Lansing, MI: Bureau of Business and Economic Research, Graduate School of Business Administration, Michigan State University, 1964.
17. E. G. Flamholtz, M. L. Bullen, and W. Hua, Human resource accounting: A historical perspective and future implications, *Management Decision*, 40, 947–954, 2002.

18. D. Luthy, Intellectual capital and its measurement, In *Proceedings of the Asian Pacific Interdisciplinary Research in Accounting Conference (APIRA)*, Osaka, Japan, 1998.

19. T. Stewart, *Intellectual Capital: The New Wealth of Organizations*. New York: Doubleday, 1998.

20. E. Carson, R. Ranzijn, A. Winefield, and H. Marsden, Intellectual capital: Mapping employee and work group attributes, *Journal of Intellectual Capital*, 5, 443–463, 2004.

21. T. Stewart and C. Ruckdeschel, Intellectual capital: The new wealth of organizations, *Performance Improvement*, 37, 56–59, 1998.

22. A. Lerro, F. A. Lacobone, and G. Schiuma, Knowledge assets assessment strategies: Organizational value, processes, approaches and evaluation architectures, *Journal of Knowledge Management*, 16, 563–575, 2012.

23. L. Edvinsson and M. Malone, *Intellectual Capital: Realizing Your Company's True Value by Finding Its Hidden Brainpower*, 2nd ed. New York: Harper Business, 1997.

24. J. Roos, L. Edvinsson, and G. Roos, *Intellectual Capital: Navigating in the New Business Landscape*. New York: New York University Press, 1998.

25. J. Mouritsen, H. T. Larsen, and P. Bukh, Intellectual capital and the "capable firm": Narrating, visualising and numbering for managing knowledge, *Accounting, Organizations and Society*, 26, 735–762, 2001.

26. J. Mouritsen, M. R. Johansen, H. Larsen, and P. Bukh, Reading an intellectual capital statement: Describing and prescribing knowledge management strategies, *Journal of Intellectual Capital*, 2, 359–383, 2001.

27. J. Mouritsen, P. N. Bukh, H. T. Larsen, and M. R. Johansen, Developing and managing knowledge through intellectual capital statements, *Journal of Intellectual Capital*, 3, 10–29, 2002.

28. A. Brooking, *Intellectual Capital*. London: Thomson Business Press, 1996.

29. K. E. Sveiby, The intangible assets monitor, *Journal of Human Resource Costing & Accounting*, 2, 73–97, 1993.

30. K. Jacobsen, P. Hofman-Bang, and R. Nordby Jr, The IC rating, *Journal of Intellectual Capital*, 6, 570–587, 2005.

31. B. Lev, *Intangibles: Management, Measurement, and Reporting*. Washington, DC: Brookings Institute Press, 2001.

32. A. Mayo, *The Human Value of the Enterprise: Valuing People as Assets: Monitoring, Measuring, Managing*. Naperville, IL: Nicholas Brealey Publishing, 2001.

33. I.-L. Wu and J.-L. Chen, Knowledge management driven firm performance: The roles of business process capabilities and organizational learning, *Journal of Knowledge Management*, 18, 1141–1164, 2014.

34. W. Vestal, *Measuring Knowledge Management*. USA: APQC (American Productivity & Quality Center), Houston, Tx, 2002.

35. V. Goldoni and M. Oliveira, Knowledge management metrics in software development companies in Brazil, *Journal of Knowledge Management*, 14, 301–313, 2010.

36. R. Kaplan and D. Norton, Putting the balanced scorecard to work, Douglas G. Shaw, Craig Eric Schneier, Richard W. Beatty, Lloyd S. Baird (Eds.), In *The Performance Measurement, Management, and Appraisal Sourcebook*, Human Resource Development Press, MA. pp. 66–79, 1995.

37. R. Mitchell and B. Boyle, Knowledge creation measurement methods, *Journal of Knowledge Management*, 14, 67–82, 2010.

38. Y. Malhotra, Measuring knowledge assets of a nation: Knowledge systems for development, In *Invited Research Paper Sponsored by the United Nations Department of Economic and Social Affairs. Keynote Presentation at the Ad Hoc Group of Experts Meeting at the United Nations Headquarters*, New York City, NY, 2003.
39. E. Bolisani and A. Oltramari, Knowledge as a measurable object in business contexts: A stock-and-flow approach, *Knowledge Management Research & Practice*, 10, 275–286, 2012.
40. P. Ward, *360-Degree Feedback*. London: Institute of Personnel and Development, 1997.
41. C. Fletcher, *Appraisal: Routes to Improved Performance*. London: Institute of Personnel and Development, 1997.
42. D. Torrington, L. Hall, S. Taylor, and C. Atkinson, *Human Resource Management*, 8th ed. Harlow: Pearson, 2011.
43. M. A. F. Ragab and A. Arisha, The MinK framework: Towards measuring individual knowledge, *Knowledge Management Research & Practice*, 13, 178–186, 2013.
44. M. A. F. Ragab and A. Arisha, The MinK framework: Investigating individual knowledge indicators, *Presented at the Ninth International Forum on Knowledge Assests Dynamics (IFKAD)*, Matera, Italy, 2014.
45. M. A. F. Ragab and A. Arisha, The MinK framework: Assessing individual knowledge, *Poster Presented in Knowledge Management Dublin Conference (KM Dublin)*, Ireland, 2015.
46. M. A. F. Ragab and A. Arisha. *MinK*, 2015. Available: http://www.minkindex.com.
47. M. A. F. Ragab, *The MinK framework: An integrated framework to assess individual knowledge in organisational context*, PhD, College of Business, Dublin Institute of Technology, 2015.

11

Knowledge Management and the Evolving Regulatory Process

Bill Paulson and Paige E. Kane

CONTENTS

Bill Paulson, Editor-in-Chief of the publication *International Pharmaceutical Quality* (IPQ) and Paige Kane, KM leader and regulatory researcher at Dublin Institute of Technology, Aungier St, Dublin 2, Ireland explore the background, challenges, and contradictions of KM, knowledge transparency, and the aging regulatory paradigm. The task of regulating the quality of medicines in the global environment in which they are made and distributed is a difficult one. In turn, navigating the regulatory shoals around the world is not an easy task for a global industry. In this chapter, the authors address how knowledge management intersects with these regulatory challenges and how the process could evolve to allow KM's potential to be realized for the benefit of industry, regulators, and most importantly, the patient.

Editorial Team

As pharmaceutical companies better understand the potential of knowledge management to enhance the way their processes are developed, controlled and improved across the product lifecycle, they are also seeing more clearly the *constraints* that an aging quality regulatory paradigm places in the way of realizing that potential.

In 2008, the International Conference on Harmonization (ICH) released a guideline addressing the current regulatory expectations for a modern Pharmaceutical Quality System (PQS), which more formally placed the concept of knowledge management in the regulatory equation. *ICH Q10* included a definition of knowledge management (KM), and highlighted KM along with quality risk management as PQS *enablers*.

Q10 [1] and the companion guidelines in the ICH Q series: Q8 on pharmaceutical development [2], Q9 on Quality Risk Management [3] and Q11 [4] on drug substance control were intended to bring more advanced scientific concepts of quality to bear on the regulatory process and its expectations for industry. The importance of the knowledge flow between suppliers, contractors, sponsors, and regulators is clearly indicated throughout the Q8-11 series.

Emerging in the current pharma/regulator dialog is a clearer picture of the foundational role that knowledge management plays in strengthening quality risk management and the control strategy, and in turn, facilitating process and product development, marketing clearance, and postmarketing process control and improvement. However, this role has yet to be well defined and institutionalized within companies, and more particularly in the context of how industry and regulators relate to each other through the quality regulatory process.

Knowledge management practices vary widely from company-to-company, and indeed even within a given company, across the product lifecycle. On the regulatory side, the expectations for KM - and just how much and what specific types of knowledge regulators need to receive or have access to - lack definition. This creates communication problems that make both the application review and inspection processes more difficult and can significantly complicate the picture for postapproval changes and continuous improvement. Therefore, running up against the constraints of the existing regulatory paradigm, the progressive Q8–10 principles have in some ways complicated rather than resolved the problems. Further heightening the dialog around these constraints is the *Q12* initiative that ICH has put in place to take a hard look at how the quality of medicines is regulated across their lifecycles and how the process could be transformed and harmonized to make it more efficient, transparent, flexible, science and risk based, and continuous improvement friendly.

A basic driver for the ICH Q12* effort is the belief that the regulatory process can be improved to better reflect the knowledge that has been gained over the lifecycle, resulting in

- Process improvements being made through a more flexible *do and tell* approach.
- More transparent and efficient industry/regulator communications.
- Enhanced inspector/reviewer interactions.

To help advance the dialog on KM, Regulatory Science Ireland (RSI) hosted a symposium in Dublin in March 2015. More than 150 industry and regulators gathered to share understanding of how knowledge management could be applied in the industry and intersect with the regulatory process.

During the symposium, ICH Q12 Expert Working Group (EWG) member Graham Cook[†] provided a clear explanation of the KM/Q12 relationship. In articulating the forces driving the Q12 effort and what it is hoping to achieve, he noted that the EWG's plan was to have a section in the guideline address the quality system aspects, which would include consideration of the risk-based change and knowledge management systems—the topics under discussion at the RSI symposium.

> If you are thinking about change management in the post-approval phase, Cook said, it is about linking the dossier to the quality system to make sure that we have consistent knowledge and then knowledge transfer from the companies to the regulators, between the assessors and the inspectors, and between development and commercial manufacturing. There is an underpinning of knowledge management that is crucial to understand and to get right in Q12.

Although Q10 defines KM and calls it out along with quality risk management (QRM) as key *enablers* of the quality system across the product lifecycle, little additional guidance is provided to better describe KM in the context of the development, technology transfer, manufacturing, and product discontinuation continuum. Although the more established nature of QRM concepts in the pharma regulatory arena did prompt ICH to develop a separate QRM guideline (Q9), ICH has not tried to do the same for KM.

However, the enabler of knowledge management has been addressed in one of the sections of a Q&A document developed by the Quality Implementation Working Group (IWG) discussing ICH Q8–10 issues in late 2010. The Q&A clarifies the basic relationship of knowledge management to the current quality regulatory process as defined through the ICH

---

[*] ICH Q12 is in development. The proposed title for the document is *Technical and Regulatory Considerations for Pharmaceutical Product Lifecycle Management*.

[†] Graham Cook is Senior Director, Process Knowledge/QbD, at Pfizer and represents the European Federation of Pharmaceutical Industries and Associations (EFPIA) on the ICH Q12 EWG.

guidelines. Addressed below are five issues that were felt by the IWG to warrant clarification in particular regarding KM:

- The impact of Q8–10 implementation
- If an ideal KM program is suggested in Q10
- The potential sources of information
- The mandate for a computerized system
- The inspection expectations

The following is the section on knowledge management in the ICH Q8–10 Implementation Working Group (IWG) Q&A.

ICH Q8–10 QUESTION AND ANSWER ON KNOWLEDGE MANAGEMENT

How has the implementation of ICH Q8, Q9, and Q10 changed the significance and use of knowledge management?

Q10 defines knowledge management as "Systematic approach to acquiring, analyzing, storing, and disseminating information related to products, manufacturing processes, and components."

Knowledge management is not a system; it enables the implementation of the concepts described in ICH Q8, Q9, and Q10.

Knowledge management is not a new concept. It is always important regardless of the development approach. Q10 highlights knowledge management because it is expected that more complex information generated by appropriate approaches (e.g., QbD, PAT, real-time data generation, and control monitoring systems) will need to be better captured, managed, and shared during product lifecycle.

In conjunction with quality risk management, knowledge management can facilitate the use of concepts such as prior knowledge (including from other similar products), development of design space, control strategy, technology transfer, and continual improvement across the product lifecycle.

Does Q10 suggest an ideal way to manage knowledge?

No. Q10 provides a framework and does not prescribe how to implement knowledge management. Each company decides how to manage knowledge, including the depth and extent of information assessment based on their specific needs.

What are potential sources of information for Knowledge Management?

Some examples of knowledge sources are

- Prior knowledge based on experience obtained from similar processes (internal knowledge, industry scientific, and technical publications) and published information (external knowledge: literature and peer-reviewed publications)
- Pharmaceutical development studies
- Mechanism of action
- Structure/function relationships
- Technology transfer activities
- Process validation studies
- Manufacturing experience, for example,
  - Internal and vendor audits
  - Raw material testing data
- Innovation
- Continual improvement
- Change management activities
- Stability reports
- Product quality reviews/annual product reviews
- Complaint reports
- Adverse event reports (Patient safety)
- Deviation reports, recall information
- Technical investigations and/or CAPA reports
- Suppliers and contractors
- Product history and/or manufacturing history
- Ongoing manufacturing processes information (e.g., trends)

Information from the above can be sourced and shared across a site or company, between companies and suppliers/contractors, products, and across different disciplines (e.g., development, manufacturing, engineering, and quality units).

Is a specific dedicated computerized information management system required for the implementation of knowledge management with respect to ICH Q8, Q9, and Q10?

No, but such computerized information management systems can be invaluable in capturing, managing, assessing, and sharing complex data and information.

Will regulatory agencies expect to see a formal knowledge management approach during inspections?

No. There is no added regulatory requirement for a formal knowledge management system. However, it is expected that knowledge from different processes and systems will be appropriately utilized.

Note: "Formal" means: It is a structured approach using a recognized methodology or (IT) tool, executing and documenting something in a transparent and detailed manner.

Knowledge Transparency at Issue

The issues around the relationship of KM to the quality regulatory process are both complex and highly significant.

Industry and regulators are wrestling with what the knowledge flow should look like across a product lifecycle—at pharma companies internally, and with suppliers, contractors, and regulatory agencies. But perhaps a more fundamental concern is that the *existing regulatory paradigm does not actively drive and reward that flow of knowledge*—a problem that becomes even more pressing when viewed at the international level. With pharma energies heavily responsive to regulatory influences, the effect has been a dampening one on KM's progress in the industry.

The expectation among both industry and regulators in developing the Q8–11 guidelines was that improved Quality by Design (QbD) understanding, risk management, control strategies, and quality systems would equate to greater operational flexibility over the product lifecycle. However, the benefits from the enhanced knowledge that were expected to accrue to industry, regulators, and patients from that flexibility at the commercial phase have not been fully realized.

Nevertheless, companies that are advancing down the Q8–10 pathway

- are certainly more aware of the knowledge gaps that need filling at the time they are applying to market their products
- learn more about their processes and products during commercial manufacturing, and
- understand better what needs to be done and can be done to improve them.

Unfortunately, this enhanced knowledge may only serve as a burden in the face of the unaccommodating nature of the global regulatory situation and the walls it continues to present in making any significant changes in a reasonable time frame. Ironically, as the injectable generics arena has sadly epitomized, the more needed the change the higher the walls become.

Regulators are sending a clear message to industry at public forums that the open flow of knowledge is essential to having a viable regulatory process that can facilitate product clearance and continuous process improvement. On the other hand, there are *disincentives rooted in the current paradigm* to this transparency, and the problem is further magnified as the diverging requirements of the different regulatory agencies come into play.

Contradictions of the Current Paradigm

> You know that there is a tremendous amount of time required for the regulatory go-ahead if you want to bring up a new line, a new facility, or whatever. And this holds back or blocks facility improvements, because people would rather not put that investment in and wait a long time and then hope they would get regulatory approval or actually have to wait for an extended number of years before they are actually able to utilize that investment. Many folks have told me this, and I completely understand it. Site changes and major upgrades are really formidable challenges for some of the companies.
>
> Manufacturers with robust quality systems actually should be able to manage such changes without regulatory oversight. Now I recognize that is a far distance from where we are right now. But really those are the fundamental principles of quality management and the quality revolution— that quality is built in by people who are committed to building in quality.
>
> **CDER Director Dr. Janet Woodcock, June 2013 [5]**

A review of the current quality lifecycle regulatory paradigm reveals inherent tensions and contradictions that will need to be addressed to open the knowledge pathways—within industry and between industry and the regulators—and allow KM and the Q8–10 principles to achieve their full potential in the pharmaceutical arena. Below are a few of the challenges that the current regulatory paradigm presents:

1. Sponsors are caught in a compliance no-man's land.
   a. On the one hand, the chemistry, manufacturing, and control (CMC) review system calls for manufacturers to submit an application and essentially commit to maintain the status quo. It is intended to make sure that firms do not vary from those application commitments without a fair amount of regulatory oversight.
   b. On the other hand, the advancing regulatory science principles as outlined by ICH Q8–10 call for control of risks by better knowledge management, advanced quality systems, and continuous improvement.
2. There is a lack of clarity in the current relationship between the CMC review and quality system (QS) inspection components of the quality regulatory process.
   a. The QS has traditionally been looked at as the domain of the inspection teams, but application submissions on the control strategy are crossing the borderline of assessor and inspection teams.

    b. Sponsors, particularly for more complex biotech products and processes, are wrestling with how much of the quality system information to provide, and agencies, in turn, with how much to require.

    c. Change management is a fundamental part of the quality system, but it also falls into the domain of the review process under the current paradigm. Reviewers therefore become the judge without the necessary knowledge base on the operations and change control processes inside the facility to ground their judgment.

3. The current review process discourages knowledge transparency.

    a. Similarly to a defendant/prosecutor relationship in a trial, applicants (defendants) are only incented to remove the doubt of reviewers (prosecutors) about their mastery of the process and product and not to reveal any information that might raise questions and slow the review or decrease the room for making adjustments and improvements postapproval.

    b. The process does not encourage transparency around either: the full knowledge base sponsors actually have at the time they are submitting a product for approval; or the limitations in that knowledge that will need shoring up as further experience with the process and product is gained.

    c. The knowledge sharing problem worsens as more agencies, particularly less sophisticated ones, come into play.

4. On the post-approval side, there is a disincentive to uncover problems.

    a. Finding problems undermines the firm's credibility on the claims made about the level of process knowledge in seeking application clearance. And the problems, in any case, may not be able to be meaningfully addressed given the high hurdles in the postapproval change (PAC) global regulatory arena.

    b. Increasing the disincentive is that quality system-minded agency inspectors may cite firms for not meeting the QS continuous improvement mandate if the problems become apparent in the inspection process—a "Catch 22" that the current regulatory paradigm puts pharma manufacturers right in the middle of.

5. Postapproval change (PAC) regulatory hurdles make solving problems and improving processes difficult.

    a. Under the current PAC paradigm, global companies have to wade into a quagmire of regulatory complexities and considerable delays in trying to make even minor, low-risk manufacturing changes that would clearly deliver control and efficiency benefits.

    b. The more significant the problems uncovered, the more significant the changes needed to address them, and consequently the higher the hurdles become.

c. There is a strong financial incentive for manufacturers to invest in their facilities and equipment in order to avoid expensive problems that can follow when they do not. And yet that investment is often not being made due, in part, to the difficulty of getting regulatory clearance for the necessary upgrades and changes from the various global regulatory agencies.

These tensions play out in many different dimensions of the quality regulatory arena—from process and product development, through application clearance, postapproval changes, and ongoing facility GMP compliance. Furthermore, the way they play out varies between innovator and generic drugs, biotech and combination products, and domestically versus globally.

This *status quo ante* orientation of the review process, and the hurdles it places in the way of using knowledge to drive change, acts as a significant dampener on industry's enthusiasm for investing more heavily in their KM efforts. *Knowledge management* in this context takes on a much less progressive meaning—something more akin to *knowledge manipulation*. The process becomes not really about how the knowledge is gained or stored, but how it is packaged and the level at which it is shared. This is an unfortunate result of the current regulatory pressures and the way industry is driven to respond to them, which is counterproductive to the actual intent and potentially disadvantageous to all.

Looked at as a whole, the limitations, constraints, and disconnects in the current CMC regulatory processes and postapproval requirements around the world are impacting the ability of manufacturers to make technology improvements and assure the supply of medicines to the patients that need them. A significant amount of valuable industry and agency resources are being consumed in this bureaucracy. Approval timelines are extended, varied, and unpredictable, with. Supply chains are hard to manage and are prone to breaking down while awaiting clearance from the various agencies involved. The result often is that manufacturing and control improvements are not made and the supply of medicines to patients is being jeopardized.

Meanwhile, applications are getting more detailed, particularly on the biotech side, reflecting the increasing complexity of the products and processes—further exacerbating the postapproval change problem. In turn, the learning curve is getting steeper, and technology moving faster—this enhances the need for regulatory flexibility around changes, on the one hand, and requires sophistication in a firm's ability to assess their significance and lower the risk, on the other. However, health authorities around the world are facing resource constraints that limit their ability to quickly review these increasingly complex submissions and conduct the related inspections. They need the benefit of full knowledge of how companies are dealing with the advancing technology and its control to regulate intelligently and to help them address the control challenges the technology is presenting. KM is key to the regulators' public health mandate.

Pondering on the *Blue Sky*

Regardless of how much the planned ICH Q12 guidance may accomplish, the initiative has been highly impactful in creating a better understanding of the limitations of the current paradigm and the transformation that will need to happen globally to address them. Also emerging into the common consciousness is a realization of just how big the stakes are for the patient and all the stakeholders involved.

The problems are pretty well delineated in the concept paper and mission statements that accompanied the launch of the Q12 initiative. However, what it will actually take to transform the current regulatory process into one that will achieve the envisioned goals is expanding in dimension and scope. Now under open debate is the question of how radical a departure from the current paradigm will be required, and how that transformation could be implemented on a global basis.

The following are some of the *blue sky* ideas that have surfaced in the industry/regulator dialog around Q12 on what a different, internationally harmonized approach might look like and the potential pathways to achieve it.

Transparency and Continuous Improvement across the Product Lifecycle

1. Use CMC applications as a template or roadmap for continuous improvement—informing the regulator on the control strategy and how it will guide the continuous improvement process—instead of using them to define the status quo.

2. Reward applicants for transparency and knowledge-rich submissions with fewer postapproval change (PAC) burdens. This would facilitate the power of technology, KM, and quality science to improve manufacturing and control processes rather than constraining it—recognizing that there may be more risk to the patient in *not* making changes than in making them.

3. Overcome the dichotomy between the current change-resistant CMC review process and the risk-based quality systems and continuous improvement mandates in ICH Q8–10 and the new process validation guidances.

4. Rationalize, better integrate, and internationally harmonize the roles that submission review and inspections play in assuring process control and fostering improvement. FDA has gone through a significant restructuring to create a more integrated quality oversight process, and can offer a model.

5. Call for firms to explain openly in their submissions where they are in their process and control learning curve and where the knowledge gaps lie that still need to be filled—creating space for a commitment

to fill them as that knowledge base is enhanced. The ability to explain what one does not know may be a more revealing regulatory risk barometer. This approach could actually simplify the filing process and potentially shorten applications, as well as accommodate the accelerated CMC development timelines and review challenges for *breakthrough* therapies and rapidly moving technologies.

6. Shift the focus from compliance to serving and protecting patients—not only in terms of product quality but expanding access and creating more efficient and reliable manufacturing processes to forestall drug shortages.

Driving toward International Convergence

1. Make common technical documents (CTDs) a stronger vehicle for driving knowledge communication between industry and agencies and across functions within companies—for example, by incorporating the learnings from FDA's question-based-review (QBR) experience. The goal would be to make CMC submissions fit into a more cohesive and harmonized QbD, science, risk, and improvement friendly regulatory process across a product lifecycle.

2. Create internationally recognized science and risk-based standards—for example, global comparability protocols—that would define the expectations for managing particular types of change for different product types. This would avoid the pitfalls and inconsistencies of working out the change protocol approach as a one-off between each sponsor and the individual regulatory agency and facilitate international harmonization. It would also better incorporate the knowledge and experience gained about the change management process more universally.

3. Along with a more defined and coherent submission approach, establish an international regulatory process or central authority to oversee the change management component of the regulatory process—for example, through the interagency Pharmaceutics Inspections Co-operation Scheme or World Health Organization. This authority would oversee the necessary standards development and the auditing process against the change management expectations, shifting PAC regulation away from paper review to a more quality systems orientation.

4. Potentially support this international change authorization process with user fees, which would represent only a small proportion of the savings that could ensue by avoiding the postapproval change bureaucracy. The industry-backed user fee system FDA has put in place for generic drugs—that provides the resources needed to restructure, better coordinate, and improve its review and inspection process—offers a potential model.

5. In line with the supra-agency approach, establish a direct electronic/cloud-based channel for sponsor/regulator communication and knowledge sharing regarding the management of quality system changes—for example, with the global change protocols. High-risk changes could be red flagged, necessitating further information to be provided.

6. Oversee the change process on a monitoring rather than preapproval basis through this living knowledge repository. The relevant regulatory authority could request additional information as needed. The electronic, cloud-based repository could also house deviation/performance metrics information thus closing the loop by monitoring the cause and effect of the changes.

The Patient Takes Center Stage

Ultimately, implementing some or all of these *blue sky* ideas could help to finally harmonize and centralize the entire manufacturing and control CMC/GMP oversight process internationally—resting it on science, risk, continuous improvement, and patient-based standards, with a seamless and transparent knowledge flow between the sponsor and the regulator over the process and product lifecycle.

Explaining the Q12 motivations and objectives at the 2015 Dublin KM symposium, EWG member Graham Cook noted that the ICH group has talked about the guideline as being "a hybrid document, having hopefully some high level principles, but at the same time some detail that delivers some immediate benefit." However, he cautioned, "we are all realistic that the constraints that we have, which are written into the concept paper—to work within the regulatory systems of the different countries and regions around the world. We are not supposed to be changing them. That means that we won't be able to deliver that straight away. But personally what I hope we can do by establishing some high level principles, maybe encourage appropriate evolution of the regulatory framework so that over time we do actually deliver the benefits that we need to deliver." The problem "is all of ours," Cook pointed out to the industry and regulatory representatives at the symposium, "and we need all of your help to get there."

What is clear is the international industry/regulator dialog is sharpening its focus on how the pharma quality regulatory process can more effectively protect and advance public health and serve the patient. The management and flow of knowledge undoubtedly has a key role to play in this new paradigm, and is finally taking its rightful place on center stage.

References

International Conference on Harmonization, "Quality Risk Management - Q9," 2007.

International Conference on Harmonization, "Pharmaceutical Quality System - Q10," 2008.

International Conference on Harmonization, "Pharmaceutical Development - Q8 (R2)," 2009.

International Conference on Harmonization, "Development and Manufacture of Drug Substances - Q11," 2012.

Woodcock, J. "Evolution in FDA's Approach to Pharmaceutical Quality," in *GPhA (Generic Pharmaceutical Association) Conference*, Bethesda, MD, 2013.

Section III

Practices, Pillars and Enablers: Foundations for Successful KM

This section introduces the *practices*, *pillars*, and *enablers* to establish a solid foundation to stage successful knowledge management (KM) implementation. In Chapter 12, readers are introduced to the *House of Knowledge Excellence*, a framework introduced by two-seasoned knowledge management program leaders (Paige Kane, who led KM for many years for Pfizer Global Supply, and Martin Lipa, who is responsible for KM in Merck's Manufacturing Division). This framework, further explained in detail in Chapter 12, is important not only in what it contains, but also in the relationship between how all of the elements work together.

Following the framework are two chapters: Chapter 13 from Pfizer Global Supply and Chapter 14 from Merck Manufacturing, which demonstrate this framework in action based on the experiences of the authors and the KM progress and realization of benefits within the respective organizations. Chapter 15 illustrates Lilly's progress in KM, and how KM is enabled through the quality management system.

With a highlight on the *enabler* of partnerships, Chapters 16 and 17 in this section focus on how KM and IT can work together to partner on the delivery of KM benefits. These two chapters complement each other, in that one chapter speaks from the role of how a KM professional can best engage their IT partner, whereas the other chapter shares how someone in the role of an IT professional can anticipate and best enable KM to succeed. Together, they offer leading perspectives on how this partnership can maximize its impact to the organization's KM goals.

12

The House of Knowledge Excellence—
A Framework for Success

Paige E. Kane and Martin J. Lipa

CONTENTS

In this chapter, the authors present a new model, *The House of Knowledge Excellence,* as a comprehensive framework for KM programs, which is based on benchmarking and experience of the biopharmaceutical industry and beyond. This model can be used to educate and build awareness of the benefits of a holistic approach, and also leveraged as a gap analysis tool for current KM programs to enable them to deliver additional value.

Editorial Team

Introduction

The case studies shared in Section IV attest to the great progress being made with knowledge management initiatives that are underway throughout several organizations, the most common pitfall for many KM programs (across multiple industries) remains the failure to consistently deliver on their intended outcomes. In the author's opinion, this comes about as a result of a lack of fundamental understanding about what organizations are really trying to accomplish with a knowledge management program.

Often, KM programs seek to deliver the outcomes of improved collaboration, or vibrant communities of practice, or the establishment of integrated knowledge repositories, or related goals. Although these aspirations may be valid *leading* indicators of KM success, we question are these really the meaningful outcomes that will garner sustained senior management support and ongoing investment?

Organizations are often tempted to pursue the *silver bullet* promised by software vendors (e.g., the next generation software tool that will solve all of the collaboration and connectivity problems faced by civilization) as their route to success. Although acknowledging these tools may well prove indispensable when utilized in the right context; deploying a collaboration, community, or connectivity tool risks being disconnected from what is most important to the business.

Even *ICH Q10 Pharmaceutical Quality System* (ICH Harmonised Tripartite Guideline 2008), which insightfully positioned KM front and center as a key enabler across the entire product lifecycle, has challenges from industry members and claim it is too vague or high level to establish any meaningful, sustainable focus for KM.

This lack of focus was highlighted in Knoco's 2014 Global Survey of Knowledge Management (Knoco 2014), which uncovered a huge diversity in organizational reporting lines for KM programs, is a marker that industry—including the biopharmaceutical industry—has not yet normalized on the delivery of KM and how value is derived. As discussed in the survey, the KM solid line was reported most frequently as being in to one of HR, operations, IT, strategy, learning and development, R&D, projects, or business improvement. However, the research shows more than 30 different areas of the business were cited as being responsible for KM programs, including a large percentage of KM programs that report directly to the senior management team. The authors do not suggest there is a single *right* answer, but this diversity is a clear signal there is some normalizing yet to occur on the *what* and the *how* of KM, and further it is the author's opinion that the *principal pitfall for KM lies in not inherently linking it to the Business Strategy.*

Without an effective KM program, organizations risk not achieving the objectives established by their overall business strategy, which is of course uppermost in the minds of most senior executives. Indeed, establishing that KM plays an active role in accelerating products to market, improving stability of supply, helping to identify risks to product quality, or reducing the threat of recalls—emphasizes to senior management the true potential for KM. Unveiling this line of sight to the business strategy is even more critical in the current environment of the biopharmaceutical industry as it continues to transform due to a variety of trends and drivers.

A *second common pitfall is from implementing that KM is not understanding how to establish successful, sustainable knowledge management solutions.*

Like the old parable about the blind men and an elephant (Sato 1927), each believing they know what it is they have discovered, some involved in KM solution roll out claim too quickly and too easily that a KM program or solution has been achieved without stepping back to understand the big picture and all associated interconnections.

It is crucial to understand this big picture, and create the right foundation on which to establish KM success. A holistic and comprehensive approach includes focus on people and process, in addition to technology and governance. Many have attempted KM, suggesting "if we build it [KM], they will come." However, experience across multiple industry sectors has clearly illustrated this is not the case.

A 2014 study by APQC (2014) on managing content and knowledge explored the link between content management and knowledge management. Content management (CM) is a close cousin of KM is sometimes even treated as a KM solution yet more often than not it is implemented as an IT solution without a larger understanding of KM principles and practices. The APQC study found that organizations who have incorporated content management as part of a formal KM strategy are *seven times more likely to report their overall content management is effective* than when content management is deployed on its own. This is just one example of how an effective KM strategy helps establish a solid foundation for business processes.

The purpose of this chapter is to present a new framework that describes how these key KM foundational elements of *practices, pillars and enablers* and overall *business strategy* relate to each other.

A Framework for Knowledge Excellence

Research has proven that knowledge management programs that are focused on delivering targeted business results or *outcomes* are more likely to be both successful and sustainable (Prusak 1999; Chua and Lam 2005). It is critically important to select KM outcomes that will help the organization deliver on its strategic objectives as these are more likely to sustain the business investment in the KM program, elicit sponsor commitment, and enhance employee engagement. Aligning the KM program with the strategic business objectives increases the chances that the KM initiative will withstand the inevitable transitions in leadership, portfolio prioritizations, and other unforeseen challenges likely to present over time. Thus, conveying the best possible chance that the KM program will remain relevant, and continue to build on successes for the long haul.

It is imperative to think about knowledge management not simply as a *solution*, or as a *tool*—these both have narrow, restrictive connotations—but rather as a knowledge-focused business *capability*. A capability is defined as *the ability to do something* and in this case that *something* is to manage knowledge. In this chapter, we will henceforth refer to these *knowledge-focused business capabilities* as KM *practices*, which encompass the holistic application of the concepts to be presented in the upcoming framework.

Before the framework is introduced one final key concept must be addressed. KM is generally understood to encompass caring for, curating, directing, and making decisions about organizational knowledge assets. More importantly, KM is about enabling *knowledge to flow* in order to achieve the desired business outcomes (higher quality, more stable supply, faster problem solving, increased employee engagement, etc.), which in turn, will help the business achieve its long-term strategic objectives.

Figure 12.1 introduces the *House of Knowledge Excellence*, which provides a framework to clearly link the KM program to the business strategy. By maintaining a clear line of sight from the business strategy to the supporting KM objectives, a thoughtful and relevant KM strategy and supporting KM program can be established. This alignment is critical to the KM program providing value in the eyes of senior leadership and other stakeholders within and outside the company.

The following sections will help to illustrate how the individual elements of the *house* relate to each other and will provide more insight and examples for each element.

Challenges Shaping the Face of KM in the Biopharmaceutical Industry

When considering KM, in particular as a key enabler in delivering an ICH Q10-based Pharmaceutical Quality System, there are several questions that arise regarding the impact of recent external trends and drivers (Figure 12.2):

FIGURE 12.1
House of Knowledge Excellence.

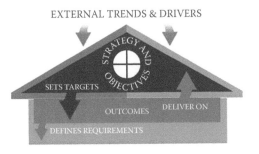

FIGURE 12.2
Understanding the impact of external trends and drivers.

- What are the opportunities for KM to have a meaningful impact in the biopharmaceutical industry?
- How can KM help biopharmaceutical companies deliver medicines and other therapies to patients more rapidly?
- How could an effective KM program support operational efficiencies for the company, improve employee engagement, and help address many of the other challenges that face the industry?

As discussed elsewhere in this book, in an environment of mergers and acquisitions, increasing regulatory expectations and baby boomers retiring (to name but a few), taking a fresh KM perspective of these challenges may be helpful.

Table 12.1 lists a subset of common challenges, trends, and drivers facing the industry today, coupled with elements of business strategies (and

TABLE 12.1

Challenges, Trends, and Drivers Facing the Biopharmaceutical Industry

(a) Challenge, Trend, or Driver	(b) Strategy or Objective to Address	(c) Illustrative Barriers/ Challenges to Knowledge Flow
Regulatory Driver(s)		
Regulatory expectation that knowledge is applied to improve patient outcome (e.g., ICH Q10)	More efficient postapproval changes product innovations	Difficult to surface prior knowledge from legacy products Difficult to understand rationale from past changes SMEs have left the company—losing knowledge of legacy products Design space not adequately defined
Regulatory expectation for improved understanding of risk	Improved risk assessment process and outcomes (standard process, routine frequency, etc.)	Difficult to understand rationale and decisions from prior assessments Lack of uniform assessment of risk across products, time, etc.
Business Environment Driver(s)		
Global competitiveness (pricing pressures, generic competition)	Operational Excellence (process capability, cost savings, etc.)	Inability to find historical knowledge to support process improvements Knowledge not flowing with the product throughout the lifecycle Past knowledge not easy to find/not findable Not knowing who the experts are Silo learning within groups, facilities, regions, etc.
Increased therapeutic area competition	Shorten time to market/ accelerate development timelines	Culture of not sharing Lack of processes to share Limited leverage/use of knowledge from prior products, modalities, etc. Inability to find knowledge efficiently to support development

(Continued)

TABLE 12.1 (*Continued*)

Challenges, Trends, and Drivers Facing the Biopharmaceutical Industry

(a) Challenge, Trend, or Driver	(b) Strategy or Objective to Address	(c) Illustrative Barriers/ Challenges to Knowledge Flow
Mergers and acquisitions	Increase technical capabilities, optimizing portfolio	Challenge to integrate new teams and capabilities Potential reduction in force Employees hording knowledge Often results in moving products from site-to-site—need for tacit knowledge that can be scarce (labor intensive and expensive) KM considerations not included up front—SMEs leave the company, knowledge transfer not planned proactively
Pressures to innovate to sustain growth	Operational Excellence (process capability, cost savings, etc.)	Inability to find historical knowledge for process improvements Past knowledge not easy to find/not findable Silo learning within groups, facilities, regions, etc.
Shift to outsourcing in multiple stages of the product lifecycle (e.g., clinical studies, product collaborations, contract manufacturing, and supply)	Leveraging external collaborations and third parties for competitive advantage	Contracts focus on regulatory needs not necessarily knowledge needs Culture of collaborators (third party or pharma organization) may not be conducive to sharing knowledge Concerns regarding intellectual property—reduction in learning Collaboration technology limitations
Emerging markets	Effectively and efficiently supplying products to emerging markets while satisfying evolving requirements in those markets	Location not conducive to collaboration Lack of resources Lack of internal capabilities in the markets Need to provide market specific products/packaging and labeling introduces great complexity to managing product knowledge

(Continued)

TABLE 12.1 (*Continued*)

Challenges, Trends, and Drivers Facing the Biopharmaceutical Industry

(a) Challenge, Trend, or Driver	(b) Strategy or Objective to Address	(c) Illustrative Barriers/ Challenges to Knowledge Flow
People/Talent Driver(s)		
Baby Boomer retirement	Business continuity	Retirees not replaced when they leave
		Lack of time/business process to transfer knowledge to colleagues prior to leaving
		Replacement is not in place/ knowledge transfer cannot happen
		Retirees not willing to share knowledge
Evolving workforce, (Millennials entering the workforce)	Innovation, attracting new and diverse talent	Technology platforms do not meet the expectations of the new workforce
		Different styles of working (discussion with peers vs. research alone, where content is stored, etc.)
		Notion that millennial will work at multiple employers over their career—more turnover and potential knowledge loss than in previous generations
Virtual/remote workers	Reduction of facility footprint	Reduction in people-to-people interaction—the *water cooler* and *coffee station* do not exist
		Risk of colleagues not developing an internal network—less connectedness and awareness of other peers in the organization

associated business objectives) that are typically invoked to address these challenges, with the potential barriers to knowledge flow identified that put that strategy or objectives at risk.

When examining challenges, trends, and drivers within the industry, it is important to emphasize there is no *one-size-fits-all* KM approach is available to address them. However, when challenges are described in terms of knowledge flow barriers, common themes begin to emerge. *Understanding these knowledge flow barriers is the first step in defining which KM practices will be required to achieve success.*

Furthermore, it is not just the KM practices themselves that will help to address the business drivers. In practice, a key factor lies in influencing *how* the work gets done. Successfully embedding KM practices within an organization to ensure knowledge flow requires changes in the behavior of the people doing the *day job*. This is especially the case in large, well-established organizations that can be slower to adopt newer, more agile ways of working in response to changes in technology and the incoming workforce.

In another recent study conducted by Knoco (2014) the top reason cited for *not doing KM* was that the culture was not yet ready for KM. For KM to truly succeed, employees must think and act differently in how they seek and share knowledge, and recognize the value and importance of knowledge *flow* rather than knowledge *hoarding*.

To address this *People* challenge, KM teams may opt to build skills and capabilities by leveraging tools from well-established change management methodologies. These methods can help to target the desired behaviors for knowledge seeking and sharing while identifying and addressing any risks to the successful realization of the KM program. Standardized processes and practices should be developed that embed knowledge seeking and sharing capabilities in the flow of the day-to-day work. These processes and practices should actively encourage employees to ask for help when solving a problem instead of using excessive time and resources to solve it through heroics.

It is also important to understand that as the millennial generation ascends to make up the majority of the workforce following the retirement of the *baby boomers*, work styles, norms, and company culture will begin to reflect this new generation of workers. Inevitably, the mechanisms and behaviors for sharing knowledge will also change and it is therefore imperative for the company to acknowledge and address this in their KM strategy if it is to succeed.

A good example of this change is evidenced in a recent internal focus group undertaken by Merck. The focus group discovered that millennials have a preference for *self-service* or the use of a trusted network for finding information. When asked how they gather information to solve a problem outside of work, their response typically was *Google and YouTube*, somewhat tellingly followed by *asking mom and dad*. This begs the question, is the biopharmaceutical industry ready for the expectations of this *new way of working*, or will companies quickly feel archaic for their bright new hires and become less attractive places to work?

An additional challenge worthy of note is the perception that the biopharmaceutical industry is somehow unique. Acknowledging that this largely appears to be an internal industry perception associated with concerns of protection of intellectual property and patents, the *first-to-market* race, the value of investment in new and novel drug development, and the extraordinary long regulatory approval timelines. However, this perception seems to overlook other industries (e.g., aerospace, nuclear, oil and gas, aviation) that face similar business and regulatory challenges as the biopharmaceutical sector that have successfully leveraged their knowledge to bring additional value to their business.

Learnings from publications, and knowledge networks such as APQC, indicate that strong leadership support is required to embrace and embed knowledge management within their organization and proactively build a culture that enables knowledge sharing. This will be discussed further later in this chapter.

MYTH BUSTING KM IN THE BIOPHARMACEUTICAL INDUSTRY (PHARMA)

MYTH: As Pharma is regulated it makes KM more complicated than for other industries.

FACT: Many other industries and organizations are highly regulated. It should be noted that all U.S. government agencies are highly regulated by multiple internal agencies. Many other industries also face regulatory oversight but have been successful in KM, including those in aerospace (FAA regulations).

MYTH: As Pharma is a regulated environment and we already spend a lot of time maintaining our records so KM is not needed.

FACT: Identifying and managing regulated records is only one component of a KM program and strategy. NASA (www.NASA.gov, similar to Pharma, has long product lifecycles—their missions may last 60 years or more. Business drivers such as an aging workforce, employees need for resources, and long product lifecycles can be addressed with a systematic approach to KM (Hoffman 2014).

MYTH: KM in Pharma is about managing "regulated content."

FACT: Regulated content is important, however process improvements, exploratory studies in the commercial manufacturing space as well as the transfer of technical and business processes are key in building new talent, transferring knowledge when employees leave, etc.

MYTH: KM is new in Pharma.

FACT: KM has been going on in Pharma for years, formally in small pockets of the industry but mostly on an informal basis. A formal approach requires a more holistic understanding of the knowledge flow challenges and ability and willingness to learn from others, whereas an informal approach can greatly reduce effectiveness.

MYTH: *KM in Pharma requires unique solutions.*

FACT: APQC has found that many industries share the same knowledge flow issues (see Chapter 4). Leveraging solutions such as expertise location, knowledge mapping, and after-action reviews can be applied regardless of industry.

Knowledge Management Strategy

If you don't know where you are going, any road will get you there

Louis Richardson,

KM World 2015.

How can you hope to deliver a meaningful, impactful KM program without a plan that defines what you are trying to accomplish? (Figure 12.3).

A *strategy* is defined as "a careful plan or method for achieving a particular goal usually over a long period of time" (Anon 2016). In the author's opinion, a strategy should also guide an organization to explore the potential risks that may threaten the realization of the proposed plan. This subject of strategy raises more questions to consider:

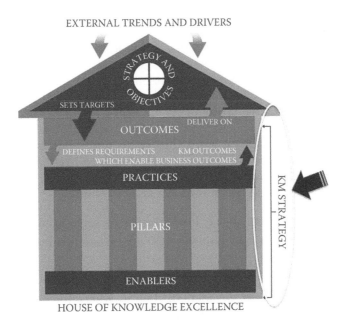

FIGURE 12.3
Examining the Knowledge Management strategy.

- Do you have a strategy for codifying and socializing the objectives for your KM program?
- Have you have defined what success looks like for your KM program, and how you will get there?
- Does your strategy *vertically integrate* the elements in the *House of Knowledge Excellence*, by establishing the key business outcomes KM will seek to enable, which KM practices will be required, and which associated foundational elements are currently in place?

Indeed, one could say that there are no knowledge management *projects*, per se. Rather, KM projects are really *business* projects that aim to address a business problem by *improving knowledge flow* (or inversely, by *eliminating waste associated with barriers to knowledge flow*) so that the business can achieve a desired outcome.

Therefore, a good KM strategy should begin with a clear understanding of the overarching business strategy it is supporting and which objectives will drive the achievement of the desired outcomes. The KM strategy should then work to achieve or enhance these outcomes. It is important to describe the benefits of the KM program *in the same language that senior leadership describes other business outcomes.*

A case in point—while it is highly unlikely that your overall business strategy has an outcome of *increased collaboration*, this remains a commonly referenced goal for many KM programs. Although enhancing employee collaboration may be a good installation (or *leading*) measure of progress, it is a step short of describing how KM can help achieve the broader business strategy. In other words, *the goal of the KM strategy should be to enable the business outcomes*, for example, shorter product development cycles, which is a realization (or *lagging*) measure, and increased collaboration is a lever to achieve this.

There are many ways to develop a strategy, but the most important feature is to *have* a strategy. One technique is to use a *Design for Six Sigma* methodology, which starts by understanding the customer needs. In the case of building a holistic KM program, this means understanding the internal customer or business needs and how the elements of KM strategy link to the desired outcomes (Lipa et al. 2014).

Other important attributes of your KM strategy should:

- Capture the current state of how knowledge is managed.
- Discuss why a change in the way the organization manages knowledge is necessary.
- Define the desired future state of KM for your organization.
- Set the direction to get there, including where the strategy will initially be targeted (e.g., pilot opportunities).
- Establish meaningful, reportable metrics or measures of progress toward realization.

- Define the known risks (and mitigation options).
- Define the guiding principles for KM at the organization.
- Align and concentrate resources for KM roll-out and support.
- Explore interdependencies with other work, groups, initiatives, etc.
- Define what you will not do (e.g., areas that are not a priority).

A note of caution when creating your strategy; avoid it being a set of glossy slides that sit in a binder on the shelf. Instead, focus on creating a strategy that you can use as a contract with those members of senior management sponsoring the effort and also to provide a tangible, practical guide for strategy execution.

To support the creation of a KM strategy, multiple KM maturity models exist that can be used to evaluate the current state of KM at your organization in a semiquantitative manner. For example, the APQC KM Capability Assessment Tool (APQC 2010) assesses KM maturity in terms of *Strategy, People, Process,* and *Technology.* By performing such an assessment, one can understand the current level of KM capability and identify what is required to achieve higher levels of KM maturity. This type of assessment will also highlight where peer benchmarking may add value. These steps can inform the focus of your strategy and identify the steps necessary to achieve it. By performing a maturity assessment on a periodic basis (e.g., annually), one can also measure progress toward realization of the strategy in an objective manner.

Knowledge Management Practices

When considering what it is that a KM program and the people that support it actually *deliver* to the broader business, it is helpful to think about the KM program as providing the *capability to enable knowledge flow.* This capability is enabled through one or more KM *practices* (sometimes referred to as *approaches, solutions, tools,* and *methodologies*) and includes the products and services provided by KM practitioners within the organization (Figure 12.4).

As an analogy, think about a Swiss Army Knife (O'Dell 2015). A Swiss Army Knife is a multipurpose pocket knife of individual tools, which when coupled with the right need and right enablers, give the user the *capability* to do something they could not easily do without the tool. For example, to cut a rope, open a bottle, or tighten a screw.

One can think about KM practices in a similar way. The right KM practice applied in the right manner at the right time *can enable the capability of the organization to do something they could not easily do otherwise.* The practice itself will typically promote knowledge flow, such as improved connectivity, access to experts, and sharing of lessons. This in turn, will enable specific business objectives such as speed to product launch, solving problems faster, and increased employee engagement. The premise being without, for example, a systematic *Lessons Learned* KM practice, the knowledge from a past

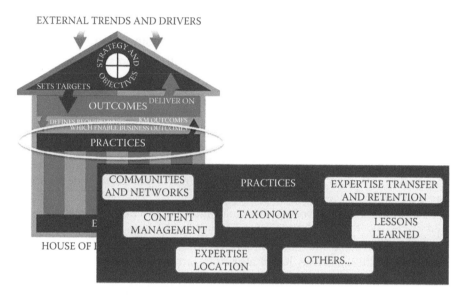

FIGURE 12.4
Common KM practices.

project might not have been shared, and the current product launch could take longer (or experience some other indication of decreased effectiveness).

Critical to the success of any KM program is that the right KM practices are selected. As per the theme already established in this chapter, this requires a clear understanding of the knowledge flow problems one is trying to solve. There are a variety of techniques to do this, from simple VOC (voice of the customer) and anecdotal stories from colleagues of past issues, to more structured and robust (and arguably more accurate) means such as *knowledge mapping*. It is important to understand the need to select the right practice to solve a given problem as opposed to just deploying a practice because *it is a good idea* or available.

The good news is the number of KM practices has grown substantially because KM emerged as a mainstream concept in the 1990s. There is now a rich array of KM practices to tackle the many knowledge flow problems. Several commonly used KM practices to improve knowledge flow include the following:

- Communities and networks
- Content management
- Taxonomies
- Lessons learned
- Expertise location

- Expertise transfer and retention
- Other practices, including transfer of best practices

In reality, these may be better described as groupings, or families of KM capabilities as there are many variants within each practice listed above and each addresses a different type of knowledge flow problem. More detail is provided below on these common KM practices.

However it is worth noting that currently, there are no clear trends or standard configurations across KM programs and essentially no two KM programs look exactly alike. This is demonstrated by the variety of case studies and other published KM literature that outline the different drivers prompting KM deployment, the influence of different functions *owning* KM, the impact of the culture of the organization, and indeed the size, spread, and complexity of the organization.

APQC has established a model to relate many of these KM practices titled, *Blended KM Approaches for Enabling Knowledge Flow* (see Figure 12.5), based on the following two criteria:

1. The level of explicit versus tacit knowledge being transferred
2. The degree of human interaction required

This is a useful reference model that organizes practices by these criteria and establishes a continuum for practices from the most basic self-service options (explicit knowledge requiring little human interaction) to the

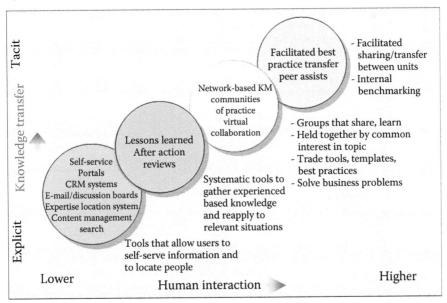

FIGURE 12.5
KM tools and processes. (Reference: APQC 2008).

most involved instances (high tacit knowledge requiring a high degree of human interaction to flow).

There follows a more detailed description of each of the common KM capability presented above, including a description and relevant considerations for each.

Communities and Networks

KM Practice	Communities and Networks
Description	Communities and networks (henceforth *communities*) are collections of people who share some level of interest and/or expertise on topic. A Community of Practice (CoP) is the most common type of community and can be defined as *Groups of people who share a concern or a passion for something they do and learn how to do it better as they interact regularly* (Wenger and Trayner-Wenger 2015). Communities may take many different forms (in-person, virtual, mobile, ...) and may exist for a variety of purposes, including best practice sharing, knowledge sharing, problem solving, and others. Communities are helpful to enable tacit knowledge flow, as the interaction between members is typically dynamic and context-sensitive. Communities may also be used to help codify tacit knowledge into explicit knowledge (e.g., capture of best practices) as well as to aggregate explicit knowledge around a topic via hosting a home page or work space, whether public or private. Communities can greatly vary in size and the types of interactions they have as being in person or virtual.
Considerations	Although communities have proven to be one of the most powerful practices to connect people and access the tacit knowledge that exists in the heads of people in the organization, the success rate for communities is relatively low. Proven practices for setting up a community for success include ensuring a value proposition for the individuals in the community as well as for the organization, having a community leader and/or steward, understanding that different communities exist for different reasons, and recognizing the motivations and participation of the individuals that make up the community. Communities may be informal or more formal. Informal communities tend to rely on the energy of a motivated few. Although more formal communities also rely on this energy to get started, the enabling structure (e.g., link to business outcomes) will help sustain them over time, especially as CoP leaders change.

Content Management

KM Practice	Content Management
Description	Content management is traditionally an IT-focused capability—a close cousin of document management—and generally describes the administration of digital content through its lifecycle, from creation through consumption, including editing, access administration, and publication. Additional features may include workflows, version tracking, coauthoring, and more.

(Continued)

KM Practice	Content Management
	Content management is a starting point for many KM programs, as the most visible, tangible issue many companies face is *not knowing* what content they have or where it all is. In the biopharmaceutical industry, *pain points* that illustrate this may include: the ability to quickly and confidently locate all relevant content for a technology transfer, research a problem. Given the long timeline and distributed resources often associated with product discovery and development locating content can often be challenging. Yet *SharePoint sprawl* has caused a proliferation of team collaboration sites, where in larger companies, on average, more than 100 new sites are created per month, typically with very little governance or stewardship associated, often with no standards behind them. This makes it exceptionally difficult for users to know how to store and more importantly, how to find content (Greenfield 2009).
Considerations	The success and relevance of content management can be greatly enhanced with application of the KM principles, pillars, and enablers, which is how content management becomes viewed as a KM capability. In addition to a holistic approach, creating an intuitive, user-centric taxonomy is a key success factor in people being able to find what they want, when and how they want to. Taxonomy also greatly enhances the effectiveness of searching the underlying content. There are many mature content management technology solutions in the marketplace, although be aware these typically focus only on the pillar of technology and perhaps some focus on process. A recent APQC study suggests that KM provides support and structure for content management, and that organizations that have content management as part of a formal KM strategy are seven times more likely to report their overall content management is effective (APQC 2014).

Taxonomy

KM Practice	Taxonomy
Description	*Taxonomy* is simply a way to group things together, typically by various characteristics associated with each entity. Taxonomies are often associated with IT systems, such as with content management and search but have much broader application. Taxonomies can enable standardization across areas of an organization, where different groups may use different terms to describe similar things. For example, it is likely that your organization has many virtual collaboration spaces across many different teams. But are they consistent and standardized in terms of what is stored there, how the folder structure is defined, and what the file names are? The answer is likely *no*, and this is normal. In reality, people are describing the content, for example, through the folder path and file name. However they are likely doing this at an individual or at best team or department level. Their view is limited to the work they do and how they describe it. Taxonomy can bring standardization and consistency across individuals, teams, functions, and even organizations, through defining a structure and common set of terminology to describe all relevant content.

(Continued)

KM Practice	Taxonomy
	Having a common taxonomy is a powerful enabler to search engines. Search engines help surface content, often unstructured, however, the content may still be lacking context. And when you search, you never know *what you do not find*. Taxonomies bring structure and this content can be very reliability surfaced through a search and be weighted with higher relevance. See Chapter 27 for more detail on key taxonomy terms and concepts.
Considerations	The business case for taxonomy may be more elusive than other KM practices but the anecdotal evidence is often a powerful motivator. Taxonomies must be designed with adequate input from the user base as they are—after all—to benefit the users and if they are not intuitive, they will not be as effective. Taxonomy must also adapt with changing needs of the business and evolution on how the content is viewed, so governance and a change control process are key considerations.

Lessons Learned

KM Practice	Lessons Learned
Description	Lessons learned refers to a collection of practices, including *lessons learned, after action review, postmortem*, and others that are typically associated with a reactive analysis or critique of a task or event. This analysis is intended to surface and describe the key *lessons* by the person or team involved in the task or event. The concept of lessons learned is often associated with things that did not go well, for example, *not repeating the same mistake over and over*, but is intended to capture all learning and insights, both *bad* (what did not go well to avoid doing again) as well as *good* (what went well to leverage in the future). Lessons learned is a key concept associated with learning organizations, which are able to be adaptive, and continually improve their work.
Considerations	The concept of lessons learned has been around for some time and is commonly practiced with varying levels of effectiveness in project management, most often after completion of a large and/or complex project. A common challenge associated with lessons learned include when to do them. For example, a *lessons learned* session after a three year capital project may not be ideal as early lessons may have faded from memory, and there will be a significant lag to extracting lessons to apply elsewhere. Embedding lessons learned into stage gates or other more routine checkpoints is a good practice to address this. It is often a challenge to ensure the lesson is *actually learned* by implementing the insight into work processes and practices, such that future work can benefit. It is great that the lesson may be identified, but the real value is ensuring it impacts future work through driving improvements to how work is done. Another consideration is the transparency with which lessons are surfaced, in particular *negative* lessons. It is common with the stress and pressure in the business environment to judge on what went wrong and to assign blame. These are barriers to effective lesson sharing. A *safe to speak up* culture must be nurtured to gain rich insights to how work is actually done and drive improvements.

Expertise Location

KM Practice	Expertise Location
Description	Expertise location is a general term that refers to a process or system to find a specific person with a specific skill or capability or a specific combination of skills and capabilities. Technology systems are typically referred to in a general sense as *ELS* or expertise location systems. Social tools (e.g., internal tools like Yammer or discussion boards) can also be used to locate experts to help, however, ELS have specific traits as described by an industry benchmarking study (APQC 2008) are • Formal • Brokered • Centrally managed • Expertise areas predefined In addition, some companies predefine *experts*, whereas others aggregate information from various sources and let the *seeker* determine if the person identified has the skills and experience to help. When experts are predefined, care must be taken to ensure there is a rigor in the selection process, whether it is testing or managerial decision point. When information is aggregated, it could be from the Human Resources database, training information, documents developed, self-declared skills and capabilities, or a combination thereof.
Considerations	Expertise location can be a very powerful KM practice; however, there should be a methodology to ensure that skills and capabilities are still relevant if the organization changes or employees have job changes. Regarding relevance: Skills—are they really specific to your company or is it a *Biopharmaceutical* skill? Organization names—with internal reorganizations, this could be a potential invalidating link, or causing a high amount of system mitigation. It is recommended to consider designing such systems in relation to the *work streams* within the organization rather than the organizational construct. Considerations should also be given to the change management activities associated with leveraging expertise location, as these types of systems only work if they expertise is maintained (update to date, high quality, relevant, etc.) and people leverage the process or system to seek out expertise.

Expertise transfer and retention

KM Practice	Expertise Transfer and Retention
Description	Expertise transfer and retention refers to a broad collection of practices that involve transferring tacit knowledge from one person to another. Perhaps the simplest example is on-boarding, whether an employee new to a company or new to a job functions. Arguably the most effective on-boarding occurs when the person leaving the position is involved so they can *show them the ropes* and explain how things really get done (tacit knowledge—hence the need to explain). On-boarding is typically not regarded as a KM activity but looking at on-boarding through a KM lens has the opportunity to greatly improve its effectiveness.

(Continued)

KM Practice	Expertise Transfer and Retention
	Other practices exist, such as knowledge retention interviews that seek to determine business critical areas of expertise by a certain individual and subsequently focus interviews on these topics. These interviews are often conducted per standard work to assess risk, determine topics, and capture results, and often by trained interviewers.
Considerations	This grouping covers a wide variety of practices. In business, expertise transfer is going on every single day when people interact and the vast majority of this transfer happens through the course of work. However, it is common for major issues to arise when the normal course of work does not ensure continuity of insights and expertise—for example, when a highly tenured expert leaves the company. These events may cause disruptions to business and as such require special attention—enter knowledge management. These practices vary greatly in effort and need to be tailored to the needs of the business. A very common type of expertise transfer in biopharmaceutical is a *technology transfer*. There is no specific practice associated with this per se, yet an established KM program will greatly support the success of doing technology transfers (e.g., existence of taxonomy for product knowledge).

Other KM practices

KM Practice	Other
Description	As presented previously, there are many different KM practices and variants of each to address many types of knowledge flow problems. Many other practices exist to solve niche issues or emerge as technology evolves. Examples of other often cited practices not explored here include, transfer of best practices, peer assists, before action reviews, knowledge mapping (see Chapter 26), and other practices in use and development across multiple industries.

How to Learn More

It was not the author's intent to fully define the KM practices listed above and there is no definitive answer for what might be considered a *best practice*. These and other practices are well documented in other KM sources not specific to the biopharmaceutical industry, often in the form of case studies. The case studies that follow in this book will highlight the use of many of these practices, so the reader may gain insight into the context in which they are applied. In addition, Section 4 contains a matrix to highlight which *practices* are evidenced in the respective case studies. Other selected sources of best practices and case studies on KM that the authors have found useful include *APQC, Knoco,* and *Straits Knowledge,* among others. In addition, several entities recognize leading KM practices through awards and industry recognition, including *KM World* and the *MAKE* (Most Admired Knowledge Enterprise) awards and worth a review for any prospective KM practitioner.

Pillars

Often overlooked in the design of KM practices is a comprehensive support-ing framework. A sound framework is critical to ensure that the practices move beyond a selection of optional tools or well-intentioned *good ideas.* Many experienced knowledge management practitioners have learned of the trilogy of knowledge management as *people, process,* and *technology.* Collison and Parcell included in their book *Learning to Fly* (Collison and Parcell 2004), that a successful KM program must have the right balance of these three elements. More recently, some knowledge management thought leaders have also included *governance,* as a key component of a success-ful KM program. Nick Milton, a prolific KM author, devotes considerable time and effort writing about the importance of governance. Based on the authors experiential learning in building KM programs, and witnessing both challenges and successes, governance had been selected as the fourth pillar supporting the *House of Knowledge Excellence* given its importance to suc-cessful realization and sustainability.

The pillars, as depicted in the *House of Knowledge Excellence* shown in Figure 12.6, are key structural aspects within the framework; it should be rec-ognized that applying KM practices alone would not guarantee a successful outcome. In fact, in a review of KM project failures two of the four main obser-vations noted in *Why KM Projects Fail* (Chua and Lam 2005), relate directly to the pillars of *People* and *Technology.* The remaining two observations are closely related to the other pillars of *Process* and *Governance,* and therefore help to validate their criticality. The top four observations for KM project fail-ures include the following:

1. Cultural factors are multilevel (related to people)
2. Technology issues are nontrivial (related to technology)
3. No content, No KM (related to process)
4. A KM project is nothing short of a project (related to governance)

The sections below will further discuss the four pillars and address how they provide a framework for the *House of Knowledge Excellence* (See Figure 12.6).

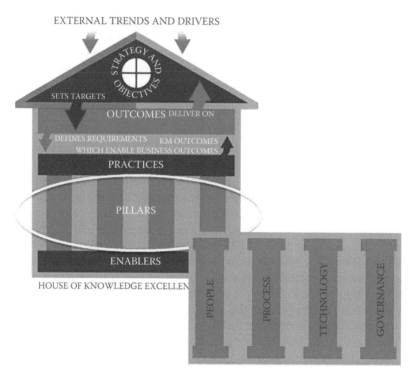

FIGURE 12.6
Pillars supporting successful KM.

People

People are at the center of knowledge creation and knowledge sharing. For each KM practice under development considerations should be given as to how people will engage and benefit from its implementation. When exploring the pillar of *People*, there are two primary components:

1. *People and culture* in regards to the creation, use, and value of knowledge and as leadership's role in developing and supporting a knowledge sharing culture.
2. *Dedicated KM roles* to design, deliver, and sustain a KM program (e.g., people that are needed to build and design the knowledge flow frameworks). Specific roles will be discussed further in this chapter.

People are at the center of knowledge creation and sharing. For any KM practice developed or implemented, considerations should be given how people will engage and benefit from implementation. The culture of the organization is an important influencer of norms and behaviors in how people create, share, value, and reuse knowledge. Ideally, the organizational culture should encourage and support investing the necessary time and resources to capture the available information, knowledge, and learning as a part of the day job—or in the flow of the work. Acknowledging, that with the current pace of the business, it can be a challenge to balance the *need to get the work done* with a longer-term view of conducting the work in such a way that the information and knowledge can be reused and leveraged. Listed below are some common challenges and barriers to capturing, seeking, and sharing knowledge *in the flow* of daily work and related opportunities:

Challenge	Opportunity
Lack of *clear expectation* from management or leadership to capture information/knowledge/learnings for future use	With senior leader sponsorship, expectations can be set that specifically relate to capturing and reusing information/knowledge/learnings—and associated accountability; consequences must apply, both positive to reinforce good behavior, or negative if expectations are not met
Lack of *defined business processes* to capture information/knowledge/learnings	Dedicated KM roles can enhance business processes to capitalize on knowledge capture and reuse, may need to design new processes to enable capture and reuse
Lack of *technical solutions* to aid in the capture of information/knowledge/learning	Senior leader sponsorship can help in attaining funding for technical solutions
Lack of *incentives—WIIFM* (what is in it for me)	A change management plan could be developed outlining incentives to engage colleagues in knowledge sharing and endorsed by senior leadership/KM sponsor

A combination of senior leader sponsorship and provision of dedicated KM roles can go a long way in addressing barriers that can be found relating to the impact of people and culture on the KM program. Additional information on KM roles will be further discussed in this chapter.

In Support of Dedicated KM Roles

To further enable the *people* pillar in the *House of Knowledge Excellence*, dedicated KM roles are key to addressing, not only the cultural challenges but also the business process challenges. When we consider the biopharmaceutical industry, it is safe to assume that every company has a dedicated quality unit. Yet, having a quality mindset is the job of each and every colleague every day. No one would challenge the expectation that everyone is responsible for quality; nevertheless there is still a dedicated group, whose primary role may be to release the product that has the responsibility to define, implement, and maintain the Pharmaceutical Quality System (PQS). Given the regulatory expectation in ICH Q10 that prior knowledge will be used and knowledge management is an enabler of an effective PQS, one could draw a corollary that dedicated roles are necessary to provide focus and expertise to develop the framework for a knowledge management program. Size and scope of a knowledge management team responsible for a dedicated knowledge management role is, of course, very specific to the business. It is, however, not uncommon in large oil and gas industry organizations to see a knowledge management team ranging from 10 to 25 people. Whereas, in the biopharmaceutical industry we tend to see slower adoption and smaller, less formal, less dedicated, and sometimes less integrated teams.

Three of the challenges identified above can be better addressed through the allocation of dedicated KM roles that provides a central/programmatic support to ensure a consistent approach and understanding:

Challenge	Opportunity
Lack of *defined business processes* to capture information/ knowledge/learning	KM role enhances business processes to capitalize on knowledge capture and reuse, may need to design new processes to enable capture and reuse
Lack of *technical solutions* to aid in the capture of information/ knowledge/learning	KM role collaborates with the IT functions—new IT technologies or approaches may be needed to enable knowledge flow (e.g., expertise location, search, discussion boards, and collaboration spaces)
Lack of *incentives*—WIIFM (what is in it for me)	KM role develops change management plan including incentives to engage colleagues knowledge sharing— to be endorsed by senior leadership/KM sponsor

Recognizing that dedicated roles are critical for KM success—having the *right roles* in the *right mix* is a key enabler that will enhance realization of the potential KM benefits. Further description of roles will be discussed in the *enablers* section later in this chapter.

Process

When considering the pillar of *process* in the KM space, there are two major components:

- *Business Processes* (e.g., new product introduction, technology transfer, and new employee orientation).
- *KM Processes* that enable knowledge flow (such as standardized processes for lessons learned, or communities of practice).

Business processes are a critical vehicle for KM program realization, as it is through these processes that the business actually operates. As discussed elsewhere in this chapter, a core tenant of KM is to improve the performance of key business processes. This can be achieved through analyzing the business process under review using knowledge mapping and other techniques in order to identify knowledge flow opportunities and the current pain points. Notably, it is in these very same places where the resulting KM practices can be embedded into the business process, or *operationalized*, as the new way of working. For example, during technology transfer (tech transfer), utilization of a document repository may be left to the discretion of the worker.

When KM is embedded into existing processes, such as the tech transfer example—use of a document repository may become a part of *standard work* for knowledge capture during tech transfer and as a result builds KM into the business process.

KM processes are also a major driver in the need for standard work in how KM is delivered in order to maximize its impact. For example, a standard work package should exist for how *lessons learned* are conducted—variable processes will only confuse the target audience and cause the practice to fail over time. Likewise, KM processes for community design and operations, taxonomy change management, metrics reporting, and others should be similarly standardized.

Technology

In this age of big data, information highways and social media, technology is an absolute necessity to support effective knowledge sharing and efficient knowledge flow. There is a multitude of articles and books written on how to successfully apply technology to business processes. One such success factor is a strong partnership with the information technology (IT) organization, which typically has long experience in helping their customers implement business solutions. However, a KM practitioner may look at technology through a different *lens*, hence the benefit for a strong partnership. When applying a *KM Lens* to a technology solution, the following items may want to be considered:

- Who is the target audience for this solution?
- Are there *secondary* customers that may only consume but not contribute? (Note: a knowledge mapping exercise of the business process may help to identify any secondary customers of the content proposed for the system).
- What type of metrics will the system generate to help inform the system steward or curator of progress and uptake?
- Who is using the system (from what region, department, etc.)? Details such as region and department can help determine where additional change management work could be applied. In addition, the ability to discern *content creators* and *content users* are helpful.
- What type of content or system functionality are users accessing the most (this would allow a steward to possibly encourage similar content to be added as it has high value).
- Can usage spikes be correlated to change management/user engagement activities?
- What other systems are out there that captures similar content? Can they be leveraged?

- Would this system benefit from any type of gamification* or providing visual acknowledgments or recognition for high volume contributors, etc.?
- Is there an easy to find *suggestion* box for users to provide feedback?
- Are success stories sought after via the system and are they visible from the system?

Many organizations start KM technology solutions with a pilot and build in learnings prior to expanding. Pilots can provide a cost-effective approach to test, and then rapidly embed key learnings, into the overall system design and roll out. In addition, leveraging cost effective COTS (commercial off the shelf software) may also be an option with a creative IT team.

Governance

There are two key aspects when considering *Governance* in the context of KM:

- Governance at the KM *program* level
- Governance at the KM *practice* level

It is important to take a holistic view, as implementing governance at program and practice will provide complementary and synergistic outcomes.

KM Program Governance

At the *KM program* level, governance requires that careful consideration have been given to establishing responsibility and accountability for the KM program, as well as for setting direction and program targets. It is recommended that program governance is formal, and sponsored by senior leadership such as through a sponsor or steering committee, with well-defined processes and procedures for approving business cases, changes to scope, funding, etc. As identified earlier, KM strategies should be linked to the business strategy. Leveraging formal KM governance is one way to help create such a link, and validates the importance of KM and helps to earn KM a *seat at the table*.

KM can also be effectively deployed through other business decision and direction setting processes such as *Hoshin Kanri* or IT portfolio prioritization, which help offer visibility as to where applying KM practices can provide additional business value. Effective governance at the program

---

* Gamification: 1. The application of typical elements of game playing (e.g., point scoring, competition with others, and rules of play) to other areas of activity, typically as an online marketing technique to encourage engagement with a product or service (Oxford Dictionaries).

level establishes a strong foundation for governance at the practice level (that could be less formal).

KM Practice Governance

At the *KM practice* level, governance refers to how KM practices are deployed, monitored, and sustained within a KM program. The intent of this governance is to set priorities for the practice (e.g., where it is deployed, who is trained, and key enhancements) and to monitor performance (e.g., metrics and success stories). In addition, governance is required to establish and operate the standard processes mentioned in the *process* pillar. One example could be a change management process, which requires that new skills must be input into an expertise location system. The level of formality may depend on the scope and complexity of the actual KM practice if one were to add skills or master data to an IT system, or modify a taxonomy, a more formal process of a steering committee may be utilized to ensure that all considerations are taken on board prior to making changes to standards.

NOTIONS ABOUT PEOPLE, PROCESS, TECHNOLOGY, AND GOVERNANCE

People: People are at the center of knowledge creation and sharing. For each KM practice developed or implemented, considerations should be given how people will engage and benefit from implementation. The success of changing business practices via KM practices to further enable knowledge flow is dependent on people accepting and embracing new practices.

Process: When thinking about process in the KM space, there are two major components: (1) business processes (e.g., new product introduction, technology transfer, and new employee orientation) and (2) the KM processes that enable knowledge flow (such as standard processes for lessons learned, or communities of practice).

Technology: Technology is an enabler to many KM practices; however, if technology is the primary focus, the technology tool could result in being the next *new thing* and not actually solve the underlying business process issue. Technology should be used to *enable* people and processes and thus be user friendly and attractive and accessible.

Governance: Governance is needed on two levels, first at the KM program level and second at the individual KM practice level. At the program level, governance means that thought has put into establishing responsibility and accountability, as well as for setting direction and establishing targets. Governance may be formal, such as through a sponsor or steering committee with well-defined processes and procedures for approving business cases, changes to scope, etc. At the KM practice level, it refers to how KM practices are monitored and curated within a KM program (e.g., a community steward is providing a type of governance through managing the community on a day-to-day basis).

Pillar Summary

The importance of embracing all four pillars cannot be understated. Their individual and combined roles have been demonstrated through the work of many KM programs, and are supported first hand by the experience of the authors. Table 12.2 attempts to illustrate the impact of omitting a pillar from a well-intentioned KM program.

TABLE 12.2

Impact Assessment of Omitted Pillars

People	Process	Technology	Governance	Impact
	✓	✓	✓	NO consideration for PEOPLE (culture or roles): KM FAILS. If knowledge seeking and sharing is not an expected and valued behavior, if people do not have the capability to engage in KM solutions, or if people are not motivated (by positive and negative consequences), then people will not engage in using KM practices, and knowledge will revert back to the old ways of working.
✓		✓	✓	NO consideration for PROCESS (business or KM processes): KM FAILS. Knowledge gets *"stuck."* Without a process, it does not flow (O'Dell 2011). If knowledge flow is not embedded into the business processes through which work gets done every day by every employee, it will forever be something extra to get done and not viewed as the same as core work. Further, there will be no structure to how KM practices are executed, leading to confusion and eventually abandonment of the change.

(Continued)

TABLE 12.2 (*Continued*)

Impact Assessment of Omitted Pillars

People	Process	Technology	Governance	Impact
✓	✓		✓	NO consideration for TECHNOLOGY: KM FAILS. Our work is intertwined with technology, and while some KM practices may have limited success independent of technology solution, the reality is that everyone uses technology every day to do their work—whether through e-mail, content management, search, etc. Leveraging technology helps embed KM where people are already working and it is through technology as a catalyst that KM can scale and enable virtual collaboration and knowledge flow on a global scale, as well as provide analytics and other technologies that can unlock new potential for KM.
✓	✓	✓		NO consideration for GOVERNANCE: KM FAILS. Without governance, there is no leadership, no control, no oversight, and perhaps no link to business priorities. Therefore, KM becomes disconnected from the core work and priorities of the business and in time, is likely to become irrelevant.

Enablers

All too often, organizations set off on a well-intentioned knowledge management effort but fail to recognize the importance of some key *enablers* and their role in delivering successful outcomes (Figure 12.7). Before we continue please consider these definitions of *enable* and *enabler*:

- *Enable*: To make (something) possible, practical, or easy
- *Enabler*: *One that enables another to achieve an end*

Key KM enablers include the following:

- Change management
- Change leadership
- Dedicated KM roles and skill sets
- Ownership and stewardship
- Partnerships

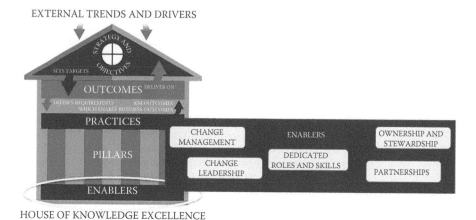

HOUSE OF KNOWLEDGE EXCELLENCE

FIGURE 12.7
Enablers for successful KM.

In fact, it is claimed that up to 70% of KM initiatives fail to meet their stated goals and objectives (Chua and Lam 2005; Knoco 2014). A variety of underlying reasons have been identified, which include lack of support and commitment, organizational culture, knowledge hoarding behaviors, change resistance, and others (Knoco 2014). Many of these failures can be traced to the exclusion, disregard, or superficial understanding of the key KM enablers explained in this section:

- Change management
- Change leadership
- Dedicated KM roles and skill sets
- Ownership and stewardship
- Partnerships

It is important to think about these enablers as critical to facilitating knowledge flow in a *practical, possible,* and *easy* manner. They are the foundations on which knowledge flow can be successfully established and sustained. The following sections will further discuss these key enablers and explore linkages as to why they are critical to KM success.

Change Management

Perhaps the enabler that has the most impact of all is effective *change management*. A word of caution, this is not to be confused with *change control*, which is a familiar compliance process. Rather, change management is about organizational change—sometimes called *transformational change*—and focuses on a structured methodology for leading an organization, and the individuals that make up that organization, through a specific change.

In the case of implementing knowledge management, there are often multiple changes in play, including changes to how people think and behave about sharing knowledge (e.g., fighting *knowledge is power*), changes to how people do their work on a daily basis (e.g., starting problem solving by understanding if similar problems have been successfully resolved), to changes in how people are incented and rewarded for seeking and leveraging the knowledge of others (e.g., fighting the *not invented here* syndrome).

Why Is Organizational Change Important?

Recognizing that organizational change is necessary and is absolutely essential to any successful KM program. If this is not recognized, then it is likely that your KM program is at high risk of not delivering on desired outcomes as essentially all KM practices require some change in the mindsets, behaviors, or actions of the target population. Change management provides methodologies to dissect a given desired change, so that it can be analyzed in a very fundamental manner.

For example, one popular change management methodology, illustrated below, helps to describe the change and the associated risks to successful realization in terms of:

- People (e.g., capacity and resistance)
- Intent (e.g., clarity and alignment)
- Delivery (e.g., resources and partnership)

This methodology also provides tools to analyze sponsorship continuity and techniques to secure leadership and sponsor support (Kotter International 2007). Other models for leading change exist, such as the *Kotter 8-Step Process for Leading Change* (Kotter International 2015).

Although these change methodologies and models are typically directed at large-scale transformational change within an organization they offer powerful approaches to better understand the changes required to ensure success of the KM program and therefore greatly increase the probability of success. It is recommended that all KM programs employ some skills in the art of change management, while adapting these methodologies to make them fit for purpose for your need.

Organizational culture change is closely related to change management. Some level of *culture change* will be required for the success of your KM program. However, it is better to understand this up front and address it in your KM strategy to ensure the appropriate level of sponsorship support from senior leaders. It is imperative to understand, at least at a high level, the culture of the organization going through such a change and the barriers that exist. If business leaders and KM leaders fail to recognize or understand what is changing and establish conditions for this to happen, how can they hope to be successful?

Change Leadership

Change leadership refers to the behaviors and actions of leaders in the organization that drive successful, sustainable change throughout the organization. *Change leadership* differs from *change management* in that it is an engine for change, not a methodology or set of tools and structures to keep a change effort under control (Kotter 2011).

Change management methodologies will often help to define which elements of change leadership will be required, typically described in the category of *sponsorship*. However, the concept of sponsorship in many places feels overused and underrealized. Leaders are typically responsible for sponsoring many change initiatives simultaneously—and may believe they are sponsoring them effectively—this is often not the case. Reasons for this might include the following:

- Sponsorship of too many simultaneous changes.
- Lack of commitment to or understanding of the change itself.
- Sponsors not modeling the desired behavior.
- Sponsors not holding people accountable for the new way of working.
- In many cases, sponsorship has become diluted and has lost some effectiveness (and often, they probably do not even recognize it).

Why Is Change Leadership Important?

One of the most common causes of failure cited for a KM effort is the lack of senior leader buy-in. Successfully engaging leaders in the organization—at all levels, starting with the leader who has the authority to initiate and legitimize a change—and leveraging these leaders to provide sponsorship for the change is a critical element of successful change. Ask yourself if you would rally behind a leader who says *do as I say but not as I do* (or who *talks the talk* but who does not *walk the walk*)? Leaders who are engaged, visible, modeling the change, and holding people accountable for the new way of working will bring more people through the change with them. Providing change leadership requires effort at all levels of the organization—it is not a spectator sport. According to Thien (2015), this includes taking time to learn, being committed to the change, creating the right environment, setting expectations, and remaining resolved. Other elements of change leadership may include the following:

- Effective, consistent communication.
- Adequate, regular attention to the change and how it is proceeding.
- Deployment of resources, including of people resources *and* financial resources.
- Support development of new skills.

Dedicated KM Roles and Skillsets

Roles, and associated defined responsibilities, are a *subset of the People pillar.* Having the right roles in the right mix is a key enabler that will enhance realization of KM benefits. This section presents further discussion on KM roles.

With many competing priorities, if a KM program is pursued it is important to maintain a focus to ensure tangible progress, whether the focus is defining a strategy, designing and deploying KM practices, or stewarding and sustaining a KM program. What roles are required and what skills and capabilities make up these roles?

There are a variety of well-documented KM roles, for example, by APQC (2015) list many KM roles including the KM leader, KM design team members, IT specialist, facilitator and supporting KM advisory group. These roles are central to getting a KM program off the ground. As a KM program matures and expands, other common roles may arise, such as:

- KM Champion (or KM team lead, or Chief Knowledge Officer [CKO])
- KM Specialists (or KM Analysts)
- KM domain specialists (e.g., deep expertise in lessons learned)
- Change Management and Communication specialists
- Taxonomists
- Program Manager and/or Project Manager

In addition, there are often other support roles required that may be akin to traditional IT roles such as:

- Business Analyst
- Business Architect
- User experience expert

Over time, other roles may emerge, for example, that of *knowledge stewards* who are responsible for the ongoing support of KM practices and processes.

Although some of the roles may be more clearly defined (e.g., taxonomist), many KM roles such as the KM Champion and KM Analyst are best realized through a combination of diverse skills. KM often sits at the intersection between disciplines—in many ways KM fills a gap between the functional areas it serves and information technology, human resources, learning and development, and operational excellence. Therefore, these diverse skills and competencies reflect the understanding and fluency in leading and facilitating a diverse team *while also* understanding the business context *and* having competency in knowledge management.

A recent KM survey across industries reported that KM team make-up varies across industries, but five core skills are common, including:

- facilitation
- change management
- organizational skills
- information technology
- information management

(Knoco 2014). Additional skills and attributes of KM leaders include the following (Leistner 2010):

- Service mentality
- Diverse experience across multiple fields
- Ability to inspire passion
- Multicultural experience

Why Are Dedicated Roles Important?

Committing resources to a KM effort is paramount. Understanding what roles and associated skills are necessary is fundamental in establishing a KM program for long-term success. This may require leveraging extensive benchmarking, outside help, or time to up-skill internal staff to catalyze the start of the KM effort.

Ownership and Stewardship

Ownership means ensuring someone is responsible and accountable for the various elements of a KM program. There must be a responsible party that *owns* the KM program, its definition and evolution, but is also responsible for the deployment and realization of KM into the business. This role is often referred to as the *KM Champion*. There needs to be clarity between the roles of KM Champion and the business sponsors on these expectations. A common model, linked to the broader culture change and associated change management topics discussed previously, is to think about the KM Champion as owning the KM program, the KM practices, and everything required to operate and support the program, whereas the business sponsor is ultimately responsible for driving the change into the business operations and realizing the benefits.

Another way of thinking about ownership is to ask—*who is responsible for knowledge management*, and a likely response is the *KM Champion* and then asks *who is responsible for managing knowledge* and this may prompt a much broader response. Ultimately, it is the staff of the organization who must manage their knowledge—for example, the scientists and engineers and others—who are *knowledge workers*, (Drucker 1999), that is, those doing the work every day where knowledge is created, synthesized, and is hopefully (by design) flowing.

This can be further illustrated by an analog to safety. Who is responsible in your organization for being safe? Is it the safety department? Clearly not—the safety department deploys the processes, builds capabilities, defines goals, and metrics to monitor progress, and keeps abreast of current best practice, legislative commitments, and requirements. But safety is the responsibility of every employee in how they approach and execute their work on a daily basis to ensure it is done in a safe manner and that everyone goes home safe at the end of their day.

Let us consider the role of stewardship next. Stewardship of a KM program is associated with the ongoing *care and feeding* of the operational aspects of the program. Stewardship can be considered conducting, supervising, or managing of something; the careful and responsible management of something entrusted to one's care.

For example, you may be familiar with stewards who are responsible for facilitating a KM community, or a lessons learned process. Think about these stewards as KM practice subject matter experts (SMEs) who are responsible for executing these processes on a repeatable and consistent basis while also helping to facilitate continual improvement. Furthermore, stewards allow a KM program to scale more easily into other areas of the business and have a stake in how a given KM practice is improving their *day job* on the shop floor.

Why Are Stewards Important?

Committing resources that have the right skills, competencies, and enthusiasm to support the KM effort is crucial. It is therefore important to clearly define the expectations and time commitments for these key ownership and stewardship roles. Unstated assumptions as to what each party does should be avoided.

Likewise, it is important to commit to stewarding to help ensure that KM processes do not fade from use as priorities change. Having stewards also helps to promote a community of practitioners as a KM program is more broadly deployed within an organization. This cohort can assist with new implementations, learning from each other, and helping to drive further improvements. Conversely, not having these stewards can cause processes to drift apart, eventually risking the intent of the KM program.

Partnership

There are many functions and disciplines that KM may need to interact with in order to establish a successful KM program. KM has the opportunity to build linkages to many *sister* functions that typically exist in organizations, including Learning and Development, Human Resources (HR), information technology (IT), and Operational Excellence (OpEx), sometimes also known as Lean Six

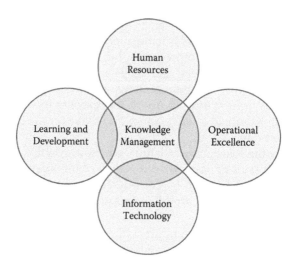

FIGURE 12.8
Linkages between KM and partners.

Sigma (LSS), as shown in Figure 12.8. Initially, KM practices may feel independent from these other disciplines, yet these disciplines are similar in many ways to KM in that they are also enabling *the business* to drive improvements and are often situated in some tier of a central organization and serve a broad user base.

Why Is Promoting Partnerships Important?

In many instances, what these sister organizations are trying to achieve is tightly linked to what KM is trying to achieve—and often their outcomes depend on improvements in knowledge flow. Furthermore, like KM, each of these functions is typically speaking to the business leaders, exploring what problems the leaders are trying to solve, and each, to some varying level of maturity and sophistication, has a strategy and plan to help the core business improve their outcomes.

One must recognize that business leaders may perceive these as a set of disconnected plans and strategies. Yet, there are synergies to be gained when understanding and partnering with these sister functions, for the mutual benefit of all. If the maturity—or desire—does not exist to ensure alignment, it is critically important to ensure there is not *misalignment*—as this will cause disruption to core work and may well render KM efforts ineffective.

In terms of partnerships, another interesting way to think about KM, is that KM often fills a gap between established functions and the business. KM programs can work in the *white space* between functions to address issues. As per the famous marketing slogan by BASF (Deutsch 2004) "We do not make a lot of the products you buy. We make a lot of the products you buy better." Likewise, KM can enable additional value from processes and tools that already exist in the organization. Table 12.3 lists some of the interdependency between these functions.

TABLE 12.3

KM Partnership Opportunities

Sister Function to KM	KM Partnership Opportunity
Learning and Development	At the core, learning and development is about building capability through competency development. In many ways, so is knowledge management. Knowledge management builds capability through the flow of knowledge, leveraging expertise to solve problems, learning from experts, collaborating as a community, and more. In fact, many learning models such as the 70:20:10 model (Lombardo and Eichinger 1996) of learning, depend on this. In this model, the *10* refers to the 10% of learning that occurs from structured courses and programs, the *20* refers to the 20% of learning that occurs by learning from others, and the *70* refers to the 70% of learning that occurs on the job. For many KM programs, the *20* and the *70* are squarely in the same space where KM is centered in trying to enable access to expertise, collaboration via communities, effective lessons learned, etc. So understanding and aligning with a learning strategy has clear benefit for both functions and can make the *learners* experience much more seamless and integrated (Lombardo and Eichinger 1996).
Human Resources	Human Resources, like L&D, also builds organizational capability through talent development programs, rotations, and other means, often including on-boarding processes. Also similar to the L&D relationship, KM can support capability development through improved access to experts and other tacit knowledge. There is also an opportunity to surface and set expectations for use of KM practices and knowledge sharing behaviors during the on-boarding process, and establishing a KM competency model for the organization. A common set of KM practices involving knowledge retention and transfer can be leveraged to create or enhance off-boarding processes in partnership with HR.
Information Technology	Perhaps the most common partnership with KM is with IT. This appears to be due to the many software tools that exist and the marketing that comes with them for how these tools solve the problems of sharing and collaboration. Many tools are directly marketed as KM solutions. Whereas many software solutions are key enablers to KM approaches, for example, collaboration spaces, search engines, social platforms, etc. Reality is that these are very rarely successful when deployed as an IT system without the benefit of many of the enablers and holistic approaches presented in this book. Therefore, the opportunity to partner with IT is one of great synergy where IT can bring tools that form the primary user experience to many of the KM practices, yet how these user experiences support knowledge sharing, and how the IT system is configured and established as part of the workflow can help IT realize benefit of an IT investment while enabling the broader goals of the KM strategy.
Operational Excellence	Opportunity for partnership should also be sought with OpEx functions in the organization, with a key opportunity being OpEx's quest for elimination of waste through the creation, deployment, and continuous improvement of standard work. KM's challenges is similar in many ways—to identify knowledge *waste* in a process through knowledge mapping or related means, and deploy practices to standardize this knowledge flow. Even more powerful is when KM practices can become fully embedded in robust standard work deployed by OpEx efforts.

TABLE 12.4

Key Barriers to KM Success and Enablers to Address

Barrier to KM Success (Knoco 2014)	Enabler(s) to Address
Lack of prioritization and support from leadership	Change leadership, KM strategy
Cultural issues	Change management
Lack of KM roles and accountabilities	Dedicated KM roles and skillsets Ownership and stewardship
Lack of KM incentives	Change management, change leadership
Lack of a defined KM approach	KM strategy
Incentives for the wrong behaviors	Change management, change leadership
Lack of support from departments such as IT, HR, etc.	Partnerships
Insufficient technology	Partnerships, change leadership

Enablers Summary

There are several enablers that can help drive success for your KM program. These enablers are intended to go beneath what lies on the surface when deploying KM practices (e.g., communities, taxonomies, and expertise location) in order to ensure proper organizational alignment and support, roles and skills, partnerships, and a deep understanding of what you are trying to accomplish. These are based not only on the experiences and observations of the authors, but are also broadly recognized by KM practitioners, as barriers to successful KM implementation. Table 12.4 reports the top eight barriers to KM, and contrasts these barriers to the KM enablers listed in this section.

Scale your efforts as appropriate but keep these enablers in mind as you design your KM approach and it will yield a more successful and sustainable KM program.

Putting It All Together

This chapter has presented a framework, entitled the *House of Knowledge Excellence* as a holistic model to reference when designing and delivering KM capabilities as summarized in Figure 12.9.

The power of this model lies not only in listing each element of the house, but also in the composite framework this creates, which is integrated from

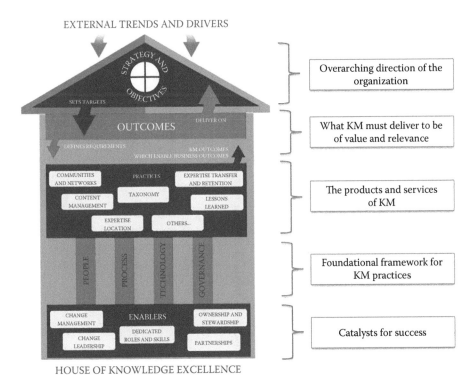

FIGURE 12.9
The House of Knowledge Excellence explained.

top to bottom. Often, organizations will jump into KM with good intentions but only focus on selected KM practices without fully appreciating the drivers that flow from above, and the foundations (pillars and enablers) on which those practices must sit. Based on the experience and beliefs of the full editorial team involved in this book, the best possible outcome can be achieved through an integrated approach.

In closing, as global regulatory health authorities continue to develop their understanding and concepts related to the ICH Q10 PQS-based enabler of *knowledge management*, there is a tremendous opportunity to continue to build industry maturity and share case studies in the knowledge management space.

To summarize, Table 12.5, which follows, presents some recommended *KM practices* to address the various barriers to knowledge flow identified previously in Table 12.1.

TABLE 12.5

Common Challenges, Trends, and Drivers Facing the Industry and Example KM Practices to Address

(a) Challenge, Trend, or Driver	(b) Strategy or Objective to Address	(c) Illustrative Barriers/Challenges to Knowledge Flow	(d) Example KM Practices to Address Barriers/ Challenges
Regulatory Driver(s)			
Regulatory expectation that knowledge is applied to improve patient outcome (e.g., ICH Q10)	More efficient postapproval changes product innovations	Difficult to surface prior knowledge from legacy products Difficult to understand rationale from past changes SMEs have left the company—losing knowledge of legacy products Design space not adequately defined	Knowledge mapping Content management Communities and networks Knowledge capture and retention program Taxonomy
Regulatory expectation for improved understanding of risk	Improved risk assessment process and outcomes (standard process, routine frequency, etc.)	Difficult to understand rationale and decisions from prior assessments Lack of uniform assessment of risk across products, time, etc..	Knowledge mapping Expertise location Communities and networks
Business Environment Driver(s)			
Global competitiveness (pricing pressures, generic competition)	Operational excellence (process capability, cost savings, etc.)	Inability to find historical knowledge to support process improvements Past knowledge not easy to find/not findable Not knowing who the experts are Siloed learning within groups, facilities, regions, etc..	Communities and networks Expertise location Content management Transfer of best practices Taxonomy
Increased therapeutic area competition	Shorten time to market/ accelerate development timelines	Culture of not sharing Lack of processes to share Limited leverage/use of knowledge from prior products, modalities, etc. Inability to find knowledge efficiently to support development	Communities and networks Expertise location Lessons learned Content management Taxonomy

(Continued)

TABLE 12.5 (*Continued*)

Common Challenges, Trends, and Drivers Facing the Industry and Example KM Practices to Address

(a) Challenge, Trend, or Driver	(b) Strategy or Objective to Address	(c) Illustrative Barriers/Challenges to Knowledge Flow	(d) Example KM Practices to Address Barriers/Challenges
Mergers and acquisitions	Increase technical capabilities, optimizing portfolio	Challenge to integrate new teams and capabilities Potential reduction in force Employees hording knowledge Often results in moving products from site-to-site—need for tacit knowledge that can be scarce (labor intensive and expensive) KM considerations not included up front—SMEs leave the company, knowledge transfer not planned proactively	Communities and networks Knowledge mapping Expertise location Expertise transfer and retention Transfer of best practices
Pressures to innovate to sustain growth	Operational excellence (process capability, cost savings, etc.)	Inability to find historical knowledge for process improvements Past knowledge not easy to find/not findable Siloed learning within groups, facilities, regions, etc.	Communities and networks Expertise location Content management Transfer of best practices Taxonomy
Shift to outsourcing in multiple stages of the product lifecycle (e.g., clinical studies, product collaborations, contract manufacturing and supply)	Leveraging external collaborations and third parties for competitive advantage	Contracts focus on regulatory needs not necessarily knowledge needs Culture of collaborators (third party or pharma organization) may not be conducive to sharing knowledge Concerns regarding intellectual property—reduction in learning Collaboration technology limitations	Collaboration spaces Communities and networks After action reviews Tacit knowledge transfer sessions
Emerging markets	Effectively and efficiently supplying products to emerging markets while satisfying evolving requirements in those markets	Location not conducive to collaboration Lack of resources Lack of internal capabilities in the markets Need to provide market specific products/packaging and labeling introduces great complexity to managing product knowledge	Communities and networks Expertise location Knowledge mapping

(*Continued*)

TABLE 12.5 (*Continued*)

Common Challenges, Trends, and Drivers Facing the Industry and Example KM Practices to Address

(a) Challenge, Trend, or Driver	(b) Strategy or Objective to Address	(c) Illustrative Barriers/Challenges to Knowledge Flow	(d) Example KM Practices to Address Barriers/ Challenges
People/Talent Driver(s)			
Baby Boomer retirement	Business continuity	Retirees not replaced when they leave Lack of time/business process to transfer knowledge to colleagues prior to leaving Replacement is not in place/knowledge transfer cannot happen Retirees not willing to share knowledge	Expertise retention Transfer of best practices Communities and networks
Evolving workforce (millennials entering the workforce)	Innovation, attracting new and diverse talent	Technology platforms do not meet the expectations of the new workforce Different styles of working (discussion with peers vs. research alone, where content is stored, etc.) Notion that millennials will work at multiple employers over their career (site reference)—more turnover and potential knowledge loss than in previous generations	Gamification Internal social media Communities and networks
Virtual/remote workers	Reduction of facility footprint	Reduction in people-to-people interaction—the *water cooler* and *coffee station* do not exist Risk of colleagues not developing an internal network—less connectedness and awareness of other peers in the organization	Communities and networks Internal social media Expertise location

References

Anon, 2016. Strategy. *merriam-webster.com*. Available at: http://www.merriam-webster.com/dictionary/strategy (accessed July 22, 2016).

APQC, 2008. *Using Knowledge: Advances in Expertise Location and Social Networking*. APQC Knowledge Base.

APQC, 2010. *Using APQC's Levels of KM Maturity*, Available at: www.apqc.org.

APQC, 2014. *Managing Content and Knowledge*. APQC Knowledge. Base.

APQC, 2015. *Resource Requirements for KM Roles and Responsibilities*. APQC Knowledge Base.

Chua, A. and Lam, W., 2005. Why KM projects fail: A multi-case analysis. *Journal of Knowledge Management*, 9(3), 6–17.

Collison, C. and Parcell, G., 2004. *Learning to Fly: Practical Knowledge Management from Leading and Learning Organizations*. West Sussex, UK: Capstone Publishing Limited.

Deutsch, C.H., 2004. A campaign for BASF. *New York Times*.

Drucker, P.F., 1999. Management challenges for the 21st century. *Harvard Business Review*, 86(3), 74–81. Available at: http://www.amazon.com/dp/0887309992.

Greenfield, D., 2009. SharePoint statistics: The real reason behind SharePoint's price tag (ZDNet). Available at: http://www.zdnet.com/article/sharepoint-statistics-the-real-reason-behind-sharepoints-price-tag/.

Hoffman, E., (May 2014). *Knowledge and Risk: Lessons from The Space Shuttle Experience NASA Chief Knowledge Officer A Tale of Two Shuttles. This was presented at the 2014 PDA/FDA Pharmaceutical Quality System Conference*, Bethesda MD.

International Conference on Harmonization, "Pharmaceutical Quality System - Q10," 2008. www.ich.org.

Knoco, (May 2014). *Knoco 2014 Global Survey of Knowledge Management*. Available at: www.knoco.com.

Kotter International, 2007. *Building Committment to Orginizational Change*. Atlanta, GA: Conner Partners.

Kotter International, 2015. *8 Steps to Accelerate Change in 2015*. Kotter International Available at: https://www.slideshare.net/haileycowan/kotter-eight-steps (accessed March 14, 2017).

Kotter, J.P., 2011. Change management vs. change leadership—What's the difference? - Forbes. *Forbes*, pp. 6–8. Available at: http://www.forbes.com/sites/johnkotter/2011/07/12/change-management-vs-change-leadership-whats-the-difference/.

Leistner, F., 2010. *Mastering Organizational Knowledge Flow: How to Make Knowledge Sharing Work*. Hoboken, NJ: Wiley.

Lipa, M., Bruno, S., Thien, M., and Guenard, R., 2014. A practical approach to managing knowledge—A case study of the evolution of KM at Merck. *ISPE Knowledge Management e-Journal*, 33(6), 8–18.

Lombardo, M. and Eichinger, R.W., 1996. *Career Architect Development Planner*, 1st ed. Minneapolis, MN: Lominger.

O'Dell, C., 2011. *Managing Knowledge in and Above the Flow of Business Defining KM in Relation to the Flow of Business*. APQC Knowledge Base.

O'Dell, C., 2015. Best practices in knowledge management. In *APQC's MENA Knowledge Management Conference 2015*. Dubai, Nov 18–19, 2015.

Prusak, L., 1999. *Action Review of Knowledge Management- Report and Recommendations.* Armonk, NY: IBM Institute for Knowledge Management.

Richardson, L., 2015. Facing our Creativity Crisis. In *KM World 2015 Conference,* Washington DC, Nov 2–5, 2015.

Sato, H., 1927. The blind men and the elephant. *Notes and Queries,* 153(DEC10), 425.

Thien, M.P., 2015. Creating a successful KM capability: A leaders responsibility. In *KM Dublin.* Dublin, Ireland. March 26–27, 2015.

Wenger, E. and Trayner-Wenger, B., 2015. Communities of practice: A brief introduction, pp. 1–3, April 2015. Available at: http://wenger-trayner.com/wp-content/uploads/2015/04/07-Brief-introduction-to-communities-of-practice.pdf.

13

A Holistic Approach to Knowledge Management: Pfizer Global Supply

Paige E. Kane and Alton Johnson

CONTENTS

> This contribution shares how Pfizer Global Supply leveraged a programmatic approach to implement Knowledge Management (KM) practices to address pain points in accessing and using internal knowledge.
>
> **Editorial Team**

Introduction

Pfizer Global Supply (PGS) is the commercial manufacturing and supply organization for Pfizer. Pfizer is a large international organization that is an aggregation of over 30 companies and PGS is composed of 64 international internal manufacturing sites, 134 logistics centers, 200+ supply partners, and 34,000 colleagues. There is tremendous complexity within the network with more than 850 major product groups 24,000 SKUs and more than 40 technical platforms. PGS serves 175+ markets in more than 50 languages.

The Business Case for Knowledge Management

Through the many years of acquisitions and divestures, there is been a high degree of change in the organization, not only from a company perspective, but also from a colleague perspective. There is a well-known dilemma in the knowledge management space—the baby boomers are retiring and valuable knowledge is leaving organizations. Along with movement of colleagues in and out of the business, there was also a desire to become more agile in decision making and product transfers. There was a belief that enabling the flow of knowledge would yield more effective use of our knowledge and lead to more efficient and effective operations.

A specific focus was placed on leveraging product and process knowledge to help enable speed and agility within PGS and leveraging the *know how* and *know what* of colleagues. In addition to the internal business drivers for enabling the flow of knowledge, regulatory guidance document ICH Q10 *Pharmaceutical Quality System*[*] highlights knowledge management (KM) as one of the two enablers to a Pharmaceutical Quality System. The PGS team needed to articulate how KM could further enable the business.

---

[*] Note: ICH Q10[1], which was released in 2008, describes a Pharmaceutical Quality System (PQS). The two enablers for a PQS are quality risk management (QRM) and knowledge management (KM).

Getting Started

In the early days (~2006), Pfizer had started looking into knowledge management. Investigative work was completed and a small group of passionate colleagues formed the process knowledge steering committee (Note: this was not related to Co-Development [Co-Dev] activities as later discussed). Around the same time, Wyeth, a company that was later acquired, was also implementing knowledge management in the context of Communities of Practice (CoP) and a Biotech KM Steering Committee. With the Wyeth acquisition by Pfizer in 2009, colleagues working in the KM space partnered and leveraged APQC[2] to conduct a baseline KM Maturity Assessment of PGS. The assessment was conducted in 2010 and provided invaluable insights material to understand opportunities for building and strengthening a KM program. Prior to 2012, KM approaches focused primarily on Co-Development[*] knowledge and Communities of Practice. In 2012, a comprehensive voice of the customer (VOC) was conducted in PGS and with key partners in pharmaceutical sciences (i.e., the development organization). Based on the learnings from the VOC and external benchmarking, the knowledge management framework (Figure 13.1) was developed to support the needs of the PGS business. Table 13.1 depicts the four KM framework elements and lists the focus as well business outcomes each element will drive.

One key learning from the VOC—a *one size fits all* approach is not appropriate to meet needs of the business. For purposes of articulating the VOC, the PGS network was sectioned into the following:

- PGS manufacturing sites
- PGS operational units
- PGS center/corporate functions
- PGS network (inclusive of all the above)

Common themes and needs were identified across the sections as a result. The knowledge management framework was developed by the knowledge management team residing within Global Technology Services (GTS), in order to champion KM on behalf of PGS. GTS is the technology and innovation arm of PGS.

The learnings from the VOC and external benchmarking also led the team to develop a holistic KM program for the business (i.e., inclusive of people, process, and technology elements). It is notable in the KM framework that people and retention of knowledge, not just data and information, are called

---

[*] Codevelopment is the term for the business process in which the development and the commercial manufacturing organizations collaborate to bring products to market.

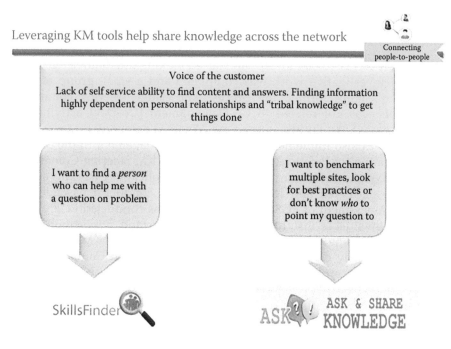

Leveraging KM tools help share knowledge across the network

Connecting
people-to-people

Voice of the customer
Lack of self service ability to find content and answers. Finding information highly dependent on personal relationships and "tribal knowledge" to get things done

I want to find a *person* who can help me with a question on problem

I want to benchmark multiple sites, look for best practices or don't know *who* to point my question to

SkillsFinder

ASK & SHARE KNOWLEDGE

FIGURE 13.1
Use cases for ASK and SkillsFinder. (From internal training materials.)

TABLE 13.1

Use Cases for ASK and SkillsFinder

KM Framework Element	Focus	That Drives
Retaining critical knowledge	Critical process and business knowledge	• Compliance (efficiency and performance) • Decision making/problem solving
Connecting people-to-people	Leveraging subject matter experts, global colleagues help solve problems	• Cost efficiencies • Product supply reliability • Colleague engagement and capabilities
Connecting people-to-content	Information findability, increase information, and knowledge reuse	• Cost efficiencies • Product supply reliability • Colleague engagement
Connecting and transforming data	Connecting systems to extract and transform data to enable new knowledge and insights	• Compliance • Decision making/problem solving • Agility • Product design • Product quality monitoring • Operational robustness

TABLE 13.2

Sample (Not Inclusive) of Practices and Tools within the KM Framework

		KM Framework Elements[a]		
		Retaining Critical Knowledge	**Connecting People-to-Content**	**Connecting People-to-People**	**Connecting and Transforming Data**
KM Practices and Tools[a]	Knowledge mapping	Enterprise search	ASK	Various pilots IbM University
	Process understanding plans (PUP)	PUP	SkillsFinder	Various data aggregation and statistical pilots
	Pharma Investigator™ (PI)	KM portal	Communities of practice (CoP)	
	Electronic lab notebooks (ELN)			
	Lessons learned			

[a] Neither the author or Pfizer is endorsing/recommending a specific software for KM processes.

out. Table 13.2 shows a sample of KM practices and tools within the KM framework that will be discussed in detail later.

Through this case study, the framework elements will be described. It should be noted the framework element of *Retaining Critical PGS Knowledge* will be discussed in the context of the product and process knowledge in the section below and also in the *Connecting People-to-Content* section.

Retention of Critical Knowledge and Connecting People-to-Content

Identifying Critical Knowledge: Knowledge Mapping

Working within the pharmaceutical industry, there are clear regulatory expectations for maintaining data and information for new products and marketed products. It is easy to focus on data and information required by predicate rules.* With such a focus on regulatory data and information, it is sometimes easy to forget about the other information and content created in the day-to-day business by colleagues across the entire lifecycle including support roles outside of the direct product impact roles. This could include things such as business processes studies, continuous improvement activities, and troubleshooting. It could be argued that the regulated content is the

---

* A *predicate rule* is any FDA regulation that requires companies to maintain certain records and submit information to the agency as part of compliance.

minimum requirement to be compliant and operate in a GxP environment and should be noted that the non-regulated knowledge content has the potential to add future value through innovation, creative ideas, and informative content and in some cases could be critical knowledge.

One approach to identifying critical knowledge (regulatory or business knowledge), is *knowledge mapping*. As Pfizer, and PGS, are an amalgamation of many companies, behaviors and methodologies for storing content varied across the business. Knowledge mapping is a KM business process that has been utilized to describe

- The explicit knowledge each area creates.
- The format (e.g., Word, Excel, PowerPoint, and e-mail).
- Who are the customers for the content (primary and possible secondary).
- The criticality of that content to the business in the event it could not be located.

Once the factors above are captured, it is then analyzed to determine if the content is being stored in a location that facilitates easy finding and retrieval. The outputs from the knowledge mapping exercise are the following:

- An Excel sheet *knowledge map.*
- Gap analysis report that identifies gaps, efficiencies, and possible remediation options (some examples are missing KM advocates and lack of change management plans).

The ultimate goal from knowledge mapping is to optimize findability for respective functional groups and customers. Knowledge mapping has proven to be an effective tool for understanding the work and the required structure for developing collaboration tools such as SharePoint collaboration sites and sorting content between internal customers and external customers.

Knowledge mapping is also used for specific business processes (e.g., investigations and workflows), in addition to mapping the explicit knowledge and content of organizational groups. More information and knowledge mapping can be found in this chapter.

Product and Process Knowledge

A Business Process for Product and Process Knowledge

Product and process knowledge is closely tied to the KM framework element *retention of critical knowledge*. A key component of the PGS knowledge management approach is the desire to retain knowledge throughout the product lifecycle. This is accomplished using common business processes and technology tools across the development and manufacturing lines.

The business process for how Pfizer develops new products is called *Co-Development* (Co-Dev). A cross functional Co-Dev team comprising development and manufacturing organization members spans many years of development. Their shared goal is a launched new product with a sustainable control strategy for commercial manufacture. The most visible outcome of the Co-Dev process is the regulatory filing, or common technical document (CTD). The CTD summarizes product and process understanding and provides the basis for the product submission to regulatory authorities.

Inherent within Co-Dev is the agreement that all products will use a Quality by Design (QbD) approach. QbD requires the use of quality risk management (QRM) to systematically build knowledge in a science and risk-based manner. The Pfizer QbD risk assessment approach was piloted in 2004, and the business process was eventually named right first time (RFT) for Co-Dev.

Prior to 2004, many of the tools used for risk assessment to guide prioritization of experiments (such as cause and effect matrices and process flow diagrams [PFDs]) did not have standards and were prone to documentation loss. Once the business processes were determined and pilots complete in 2005, only then were technology tools explored to more efficiently capture the knowledge in a way that would facilitate reuse.

At the conclusion of the pilot in 2006, in addition to the risk assessment standards in place, the PUP (process understanding plan) was established as the knowledge summary document. The PUP is the document that is built iteratively over the Co-Dev process and eventually handed over to the manufacturing organization upon readiness for commercial manufacture.

The PUP is currently developed as a Microsoft Word document that is controlled in a validated documentation system—it is considered a *roadmap* for product and process knowledge and contains links to the authoritative source for the product process knowledge. Key elements of the PUP are the following:

- It serves as the summary of product and process understanding and a principal reference for the project team and manufacturing sites.
- It includes an index of all technical reports and risk assessments that are accessible via hyperlinks.
- It serves as a platform at the end of the co-development process to inform continuous improvement by the commercial manufacturing team.
- It is maintained by the manufacturing organization after product approval.

Leveraging Technology for Process Understanding

Pharma Investigator™ (PI™)[*] is one of the technology tools used to enable the Pfizer knowledge management framework element of retention of critical knowledge. (*Note: Neither the author or Pfizer is endorsing/recommending a specific software for KM processes*). PI™ is a risk management and knowledge management software environment where development and manufacturing teams collaborate to generate process understanding and enhance application of manufacturing science and capability.

Cause and effect (C&E) risk assessments (part of the RFT for Co-Dev business process) are captured in PI™, which provides a visual for the C&E as well as a mechanism to capture the rationale and tacit knowledge from risk assessment participants. PI™ is also used to capture failure mode and effects analysis (FMEA). Like the C&E, the ability to capture rationale (often a blend of tacit and explicit knowledge) is a key component of the KM activity. The risk assessment studies (C&E, FMEA) are all based on a PFD within PI™ that spans the supply chain and allows the detailed capture of knowledge at the parameter level.

In order to build the control strategy, the risk assessment knowledge is summarized across the entire process flow diagrams as well as diving into the detail of the C&E matrices, FMEA's, and fishbone diagrams. Leveraging the functional relationship table (FRT), utilizing C&E scores, and connecting the experimental studies can provide a useful way to confirm a control strategy.

The combination of using a business process (i.e., RFT for Co-Dev) as well as leveraging technology tools has greatly assisted the PGS knowledge management journey in the following ways:

- Utilizing a technology tool (PI™) helps organize, structure, prioritize, and record prospective development via iterative risk assessments.
- Starting risk assessments with prepopulated, standardized C&E templates for unit operations that are common to many different products.
- The PUP and PI™ helps leverage experience and *institutional knowledge* gained over the past 100 years as a pharmaceutical company.

Note: PI™ has predominantly been used for recently launched new products and selected marketed products; it is not used for the entire commercial portfolio.

Pfizer is currently piloting a similar business process to RFT for Co-Dev that focuses on marketed product technology transfers. The knowledge management technology and business processes from Co-Dev is leveraged as much as possible for these similar business cases, and is focused on the needs of the distinct populations who generate and reuse knowledge.

---

[*] Pharma Investigator software developed by Light Pharma Inc., Cambridge, MA.

Enterprise Search and *Findability*

As mentioned in the previous section, knowledge mapping is important to understand content generated as well as the storage location. A common discussion topic with KM practitioners across multiple industries is how content is managed and found (e.g., enterprise search approaches). Feedback from users indicates that leveraging enterprise search is not as user friendly and provides a high number of results that are not relevant, leading to frustration. Differences between internal company searches and Internet searches with tools such as Google are due to the volume of content indexed as well as metadata associated with respective documents/content in addition to the search algorithms. Not all systems require input of metadata/content attributes; therefore, many search engines rely on full text indexing to return search results. Each business has different rules and different approaches for enterprise search. Colleagues have complained that it is like finding a needle in a haystack as the search engine returns too many results and the filters and refiners are not specific enough to narrow down hits to a reasonable and manageable number.

Knowing that colleagues found enterprise search difficult, the knowledge management team began collaborating with the Pfizer enterprise search team to clearly understand what document bases and web content bases are indexed by enterprise search. Based on learnings and input, the KM team develops fit for purpose guidance for PGS colleagues to guide respective groups within PGS to ensure that nonregulated content are stored in an area that is indexed by enterprise search. In addition, the KM team advocates for specific search applications that would benefit PGS colleagues. The team has found most colleagues would like their content to be found and used by others; however, there was a lack of awareness as to the best storage location for nonregulated content. With the KM team collaborating with the enterprise search team findability has increased, due to additional guidance of methodologies for storing nonregulated content and development of specific search applications for PGS.

Electronic Lab Notebooks

In an effort to bring efficiencies and findability into laboratory, electronic lab notebooks (ELNs) have been implemented in the GTS laboratories. GTS laboratories typically conduct process improvement research, investigate new technologies, as well as provide support to site technology groups. As there are multiple laboratories across several sites there was a desire to be able to share studies, thoughts, and research across multiple sites. Putting this information in electronic form and in a format that is findable has eliminated duplicate and repeat experiments across the laboratories. Implementation of electronic lab notebooks has been a key component of retention of critical knowledge in the laboratory space.

Connecting People-to-People

As discussed previously, PGS is a large ever changing network. In the past, colleagues would be bound by their personal networks to find who and what they needed. Today, it is more commonplace to have virtual teams that are global in nature; this, in turn, creates a new need for locating people and content. It was clear in the KM VOC that colleagues were looking for more effective ways to locate the people and the answers that could help them. The two KM tools that were developed in response to this need were ASK and SkillsFinder. Figure 13.1 describes use cases for ASK and SkillsFinder.

SkillsFinder and ASK are both applications built on a SharePoint platform. As such, they are cost effective to develop and deploy and the user interface is familiar. SkillsFinder would be considered an *expertise location system* (ELS), whereas ASK is a collection of discussion boards.

SkillsFinder

There was a desire to create community expertise locator. Previously such expertise listings were captured in Excel and very time consuming to collate and difficult to maintain. Pfizer was fortunate to participate in an industry best practice study focusing on expertise location systems led by APQC. Based on the observations and learnings from the APQC study, an expertise location system based in SharePoint was developed as a pilot. The structure of the SharePoint database was based around functional organizations and the skills and capabilities that reside within those organizations.

After much deliberation on how to future-proof skills and capabilities within the system it was decided to leverage the APQC process classification framework[*] (PCF) as the overall structure for activities that typically happen within a pharmaceutical manufacturing organization. As a result, 10 major knowledge domains were identified across the business, and serve as the framework for the skills and capabilities within SkillsFinder. One key critical success factor for the SkillsFinder is linking knowledge domains to capabilities to specific skills. This combination of information of knowledge domain capability and skills provides tremendous context of the capabilities and experiences of a colleague. One example would be technology transfer. Technology transfer is a very important skill—unfortunately, pinpointing the specific need within the large scope of what could be considered technology transfer, can be difficult. With the advanced search in SkillsFinder it is easy to locate a colleague that has experience working in a specific country

---

[*] APQC's process classification framework® (PCF) creates a common language for organizations to communicate and define work processes comprehensively and without redundancies. Organizations are using it to support benchmarking, manage content, and perform other important performance management activities.

on a specific piece of equipment and has experience with technology transfer. It is the combination of the skills and capabilities that help locate that *needle in a haystack*.

It is important to point out that not all skills and capabilities are populated within the SkillsFinder tool, by design. Rather, a framework was developed that was scalable and could be populated as new areas opted to engage with the tool. As groups and organizations decide to leverage SkillsFinder, additional skills and capabilities are identified and added in conjunction with a change management and communication plan for the respective area. More than 30,000 skills have been entered in the system since its release in 2014. The organization has had much success with this tool and as a result it won a Global PGS Vision award in 2015 (from more than 1800 submissions from across the PGS network). SkillsFinder continues to be a great addition to the KM toolkit connecting people-to-people when they need it most.

ASK

ASK is a collection of discussion boards that have been developed with the specific focus on capturing and retention/accessibility of the critical tacit knowledge that resides within our colleagues. Through the voice of the customer work it was clear that the PGS network is very willing to help anyone that asks. That being said, knowing *who* and *where* to *ask for help* was clearly a gap, based on VOC feedback.

Several groups had been using standard out-of-the-box discussion boards from various platforms and found that utilization was low. By applying a knowledge management *lens* to understand how colleagues would engage in helping others, specific functionality was identified and ASK was developed. Functionality considered was

- Ease-of-use of the system.
- Visual recognition that people were reading your request.
- Visual recognition of how many people have replied to the question.
- Curation/stewardship for each board.
- Easy to access metrics.
- Ability to link to an existing recognition system to thank colleagues for help.

Based on learnings from KM practitioners outside of Pfizer and the pharmaceutical industry and the needs of the PGS business, the KM team designed a custom SharePoint application to house the ASK discussion boards.

ASK was piloted with a few key topics such as engineering, active pharmaceutical ingredient (API), solid oral dosage (SOD), and operational excellence. The *engine* for many of these discussions boards were existing communities of practice that leveraged respective ASK board as their *virtual water cooler.* Examples of successes include a site that needed a piece of equipment to satisfy a regulatory observation—the equipment was on backorder from the vendor, however a quick shout to the ASK network and the piece of equipment was able to be shipped overnight to the site. Other examples include colleagues located in emerging markets who do not typically have extensive personal network; emerging market colleagues have leveraged ASK boards with great success in sourcing best practices and opportunities to replicate learnings.

ASK is built on a scalable platform that, at the time of writing, has 20 discussion boards. New boards are added when requested by the business. The KM team evaluates each request to ensure there is a sustaining topic and a large enough *customer base* to have robust discussions. In addition, the KM team works with each requestor to ensure that a moderator is identified and a change management/communication plan is in place to support the launch of a new board.

Communities of Practice

In 2007 (prior to Wyeth being acquired by Pfizer), the biotechnology business was looking for ways to connect the five biotechnology sites. Communities of Practice (CoP) were investigated as a potential solution to connect the sites and respective colleagues in key areas of the biotech manufacturing business. APQC was engaged to help the business build capabilities in the area of CoPs and helped design a Communities of Practice pilot. For the pilot, seven formal CoPs were designed and launched using the APQC CoP methodology. Each CoP had a named senior executive sponsor, a lead from a Biotech *corporate* function, such as training or engineering, a Co-Leader from a manufacturing site, and a *core team* that helped design, launch, and maintain the CoPs.

The pilot was endorsed by the head of the Biotech manufacturing business, who was a visible and charismatic change leader for Communities of Practice. The pilot was very successful with high engagement from the sites. In addition, a formal change management plan was implemented for the CoP pilot and any future CoPs. Results were focused in connecting people-to-people to build extended networks outside of the manufacturing sites and identification of processes/tools that could be replicated across the network to save time, money, and build operational efficiencies.

Communities of Practice have expanded across the Pfizer network, there is a small number (<10) of very formal CoPs in high impact and engagement areas, such as engineering (multiple focus areas) and training and many smaller less formal CoPs across the Pfizer wide enterprise. The KM team hosts a CoP leader meeting quarterly to develop connections between CoP leaders so that they can learn from each other sharing successes and challenges. In addition, the KM team serves as a resource for tools, template, and process assistance.

Formal CoPs maintain metrics and success stories and encourage the use of KM tools such as ASK and SkillsFinder to keep CoP members connected between CoP meetings. In addition, CoP core teams are encouraged to perform an annual CoP health check and ensure that recognition programs are leveraged to encourage sharing and reuse of community knowledge.

Partnerships

A key success of developing KM capabilities has been collaboration with partners within Pfizer (PGS and enterprise wide) as well as developing relationships with external KM practitioners.

In the early days, Pfizer joined APQC to benefit from the knowledge base on KM practices and case studies. Through more than 10 years of collaboration with APQC, Pfizer colleagues have met many KM practitioners across many industries. In addition, Pfizer has participated in APQC research studies in KM and as capability grew it also participated in KM advanced working groups (AWG). Relationships developed via APQC meetings and working groups have proved invaluable in learning how others approach business problems. Participation on an APQC expertise location systems (ELS) study in 2008 was a cornerstone that helped Pfizer learn about ELS from organizations such as NASA JPL, SUN Microsystems, and IBM. Learnings from this study became the foundation of SkillsFinder, the Pfizer version of an ELS system.

In addition to industry agnostic groups such as APQC, Pfizer also participates on the ISPE (International Society for Pharmaceutical Engineering) KM working team and the BPOG (BioPharma Operations Group) KM team. In these venues, business problems specific to the pharma industry are discussed and KM practitioners collaborate to share and learn from each other.

Internal collaborations with the PGS BT (Business Technology) organization have been pivotal to the progress of the KM journey. The PGS BT team that supports KM have assisted in development of funding requests, participated in APQC KM studies, and have provided resource for developing KM tools such as ASK, SkillsFinder, the KM Portal, and many other tools. The BT team also supports Pharma Investigator™ and serves as the BT link with the Pfizer enterprise search team. The relationship between the KM and BT team is seamless with both teams bringing ideas, learnings, and contributions to help address VOC feedback.

From a pharmaceutical perspective, collaborations with the development organization are compulsory as new products are handed off from the development organization (Pharmaceutical Sciences) to the commercial manufacturing organization (PGS). With the Co-Development process, this handoff is very well defined and a collaborative process. KM approaches are built into the process such as the Co-Development lessons learned process (See Chapter 23) and the Co-Dev PUP as previously discussed.

Other organizations in Pfizer have reached out to the KM team over the years to learn and share experiences. This has resulted in rich learnings across the business and has helped to further build KM capabilities across Pfizer.

Connecting and Transforming Data to Create New Knowledge and Insights

The fourth element of the KM framework is connecting and transforming data to create new knowledge and insights. This element is an emerging area for PGS and includes things such as the informatics strategy and pilots leveraging *big data* all under the umbrella of intelligence-based manufacturing (IbM). Intelligence-based manufacturing is a community of like-minded colleagues and leaders focused on harnessing the complementary power of data, models, engineering, and BT infrastructure to create a game changing paradigm based on the following: (1) transformation of data into knowledge and intelligence, (2) moving from reactive actions to proactive and preventative strategies, and (3) shifting from stand-alone and isolated unit operations toward integrated infrastructure at process, plant, and enterprise levels. One prominent activity that can be shared in this element is the IbM University (community), which collates and shares new technologies, and invites groups and colleagues across the business to share needs, experiences, and pilots with the larger organization. The IbM University *community* maintains an active collaboration space in addition to bimonthly meetings to encourage connections, sharing, and reuse of learnings.

How Colleagues Engage with Knowledge Management Tools and Practices

The PGS Global Technology Services (GTS) group not only champions the KM program, but also has colleagues in the organization that work directly with the manufacturing sites. These *site facing* colleagues understand site areas of need, pain points, and provide assistance with technical and engineering challenges.

To further engage the sites on the KM journey, GTS sponsored several positions at key manufacturing sites as a *KM lead*. Positions are sponsored for 12–18 months, with the option for the manufacturing site to continue to fund if the position added value to the site. This program has proved very successful with sites electing to continue the role to build internal KM capabilities that are allowing them to capture more efficiently product and process knowledge and leverage network wide KM tools. In addition to the site KM leads, GTS has nominated KM champions in each of the respective functional areas, with other parts of the business also naming KM champions. The champions have an understanding of KM processes and tools and not only help GTS colleagues engage but also help their customers (i.e., PGS sites) leverage KM tools for self-service help as well.

A component of self service includes the Knowledge Management portal. It was developed for as a resource to help the sites as well as provide a KM *one stop shop* for colleagues. The KM portal provides a pictorial diagram of the four elements of the KM framework. Each of the four KM framework elements is visually represented on the portal. Behind each icon for each respective element, is a page with a description of the element and a listing of related tools and business process to help capture, enable content, and knowledge flow. The portal also has prominent links to key KM tools such as ASK and SkillsFinder. Many organizations outside of PGS and within the larger Pfizer organization have also used the KM portal as a resource.

Summary

PGS has endeavored to leverage KM to accelerate knowledge flow across the business. Articulating KM for the business and developing the holistic framework have yielded a more connected network and provide a vehicle to address many of the elements identified in the VOC. The holistic knowledge management framework is described as

- Retention of critical knowledge
- Connecting people-to-content
- Connecting people-to-people
- Connecting and transforming data to create new knowledge and insights.

Into KM practices that have been discussed within the four KM framework element are the following:

- *Communities and networks*: ASK as leveraged by the business and Communities of Practice as implemented in PGS.
- *Content management*: Use of knowledge mapping as a business process, developing guidance for colleagues on how to store content and training materials on search best practices.

- *Search*: Collaboration with the enterprise search team, and development of fit for purpose training materials, and advocating for specific search applications for PGG.
- *Expertise location*: As discussed with the deployment and use of SkillsFinder.

The change management approach for implementing KM has been focused on the GTS organization; as GTS are technical consultants for PGS. With an organization of 64 internal manufacturing sites and more than 34,000 colleagues, starting a KM focus with GTS was practical and achievable as GTS problem solves and collaborates across the entire network.

GTS and PGS will continue the journey of KM; it is a marathon, not a sprint. The multipronged approach as articulated through the KM framework will continue to be developed and continually improved based on feedback from the business, keeping in mind, there will never be a one size fits all answer for the complex PGS network.

Acknowledgments

The KM journey was started many years ago in PGS. Over the years there have been many senior leader supporters: Sharon Timmis, Colin Seller, and Alton Johnson have all been responsible for KM to help enable the PGS business. Under the GTS leadership of Kevin Nepveux and Alton Johnson, the KM team has been able to deliver KM practices and tools that help PGS colleagues connect and get help faster, drawing from a vast network of experience. Holly Bonsignore and Mark Hand from the GTS leadership team have also been key advocates and sponsors for the GTS KM program.

- The GTS KM team, leading and advocating for KM approaches to assist the business consisting of Joe Brennan, Bernadette Tancini, and Paige Kane (author of this chapter).
- Our BT partners that have helped the team immensely: Vik Sharma, Rob Sim, Matt Gehrig, Bo Yang, and Jeff Moore.
- The Right First Time Program Office led by Vince McCurdy (Pharm Sci OpEx) that leverages Pharma investigator and PUP to build critical know during Co-Dev: Mariah Deguara, Ken Ryan, Joe Brennan, Dave Sharp, and Mike Rouns.
- The process knowledge steering committee led by Thomas Lantz and Joe Brennan.
- The PGS CoP leaders meeting led by Laura Eiler.

- Additional pharmaceutical sciences colleagues that have contributed to the KM Journey: Phil Levett and Rhea Bagnell.
- PGS KM site leads: Imelda O'Connor.
- Retired colleagues Dan Pilipauskas, Jay Campbell, Lou Pillai, and Cliff Sacks from the *early days*, endeavored to leverage KM for the PGS business.

As you can see, to develop and embed KM in an organization, it takes a *village*. We also acknowledge the thousands of colleagues that have subscribed to the ASK discussion boards to help their colleagues across the network and the colleagues that have shared their experiences in SkillsFinder. These colleagues are critical to enabling the flow of knowledge across the business so we can solve problems faster, become more efficient, and make it easier to get things done.

References

1. ICH Q10. *ICH Harmonised Tripartite Guideline: Pharmaceutical Quality System Q10*, June 2008.
2. APQC (American Productivity and Quality Center). www.APQC.org.

14

KM Evolution at Merck & Co., Inc.: Managing Knowledge in Merck Manufacturing Division

Martin J. Lipa and Jodi Schuttig

CONTENTS

Manufacturing at Merck (MMD at Merck & Co., Inc., Kenilworth, NJ, USA) created its KM strategy using the same methodology Toyota uses to design car doors—focused on the needs of the customer. Here we learn about how MMD has created a meaningful strategy for KM, founded on business outcomes, and has executed on that strategy to create an effective and far reaching KM program.

Editorial Team

Introduction

Over the past six years the knowledge management team Manufacturing Division at Merck & Co., Inc., Kenilworth, NJ, USA (MMD) have initiated a series of innovations to create channels through which knowledge can flow throughout the organization. By understanding the needs of the organization and bringing the right combinations of approaches, behaviors, and enablers together, knowledge management is enabling the organization to be more effective and efficient in many ways.

This section explores the journey in building knowledge management capability to enable MMD to achieve their strategic goals and objectives.

Case Study—The History of Knowledge Management in Merck & Co., Inc. Manufacturing Division

MMD has roughly 50 sites in more than 20 countries, with 200 external suppliers, and almost 18,000 employees. MMD produces more than 20 billion tablets annually, in addition to vaccines and biologics. Merck & Co., Inc. and MMD have historically been focused on small molecule products and vaccines but have a growing portfolio of biologics. Although the MMD KM CoE have established linkages to almost every division in Merck & Co., Inc. including research and development (R&D), information technology (IT), the Global Compliance organization, and others, MMD remains the primary area of support.

The KM team in MMD, known as the *MMD Knowledge Management Center of Excellence*, or MMD KM CoE, can trace its beginnings to a 2009 project to better define KM needs in global science technology & commercialization (GSTC), the technical organization within MMD. At the time, this organization was responsible for the late stage commercialization and lifecycle support of all small molecule products with knowledge as a key asset. Today the GSTC organization also commercializes and provides technical support for biologics and vaccines.

The impetus to begin the journey of adopting and integrating knowledge management within GSTC was the result of several factors, including the following:

- The *pharmaceutical industry was growing in complexity*, in terms of globalization, internalization, the acceleration of mergers and acquisitions, market pressures, and other external as well as internal drivers.

- Recognition that there was *insufficient capture of both tacit and explicit knowledge*. There were no standards for what knowledge to capture, how to capture it, and where to capture it.
- Emergence of *increased expectations and opportunities*, with the promise of leveraging prior knowledge as set out in ICH Q8 Quality by Design (ICH, 2009) and the need to translate that into practical use within the organization, as well as the recognition of KM as an enabler to the Pharmaceutical Quality System (ICH Q10) (ICH, 2008).
- Acknowledgment of past *knowledge management failures* such as the tendency to view knowledge management systems solely as IT systems, whereas a technology solution (IT) is only one of the pillars of the overall solution.

Over the past 5–10 years, Merck & Co., Inc. has not been alone in facing these challenges and pressures. Many pharmaceutical companies began various efforts in knowledge management. These efforts have been diverse in intent, scope, and approach, yet this characteristic is not unique to pharma. Larry Prusak, one of the pioneers of knowledge management and a guest speaker at Merck & Co., Inc. in January 2015, estimates that of the 250–300 knowledge management programs he has seen implemented across many industries, that only ~5% of those are the same. It appears that it is a matter of choosing which approach (or group of approaches) works best for an organization given the priorities and needs of that organization, taking into account variables such as the initiating organization, company culture, and initial drivers for action.

Taking the First Step

Given there was no established means to do this within Merck & Co., Inc.—no common strategy MMD could copy and adapt—the starting point seemed very ambiguous. There was an initial perception that the KM focus should be on content management to support more effective and efficient technology transfers—that is, primarily on explicit knowledge only. But was this right? What was missing? Bigger questions loomed. Should MMD:

- Focus on tacit or explicit knowledge? Or both?
- Focus on manufacturing-specific process-based platforms or on explicit products?
- Have a reference place for people to go to access knowledge and information, or enable the flow of knowledge throughout the organization's processes?
- Do something else?

A pivotal decision was then made. The senior leader of GSTC decided to invest in knowledge management for the long term by dedicating resources to define and execute a strategy for KM. MMD wanted to *learn by doing* in order to drive quicker, more relevant results rather than the strategy being a *nice set of glossy slides* that sit in a binder on the shelf. The team leveraged a Design for Six Sigma (DFSS) methodology, specifically DMADV (Graves, 2012) (Define–Measure–Analyze–Define–Verify) to define and ensure that the strategy was tightly linked to the needs of the organization. The strategy was intended to identify the *chess moves* they would make, the resources that would be necessary, to define the focus—and of equal importance—to define what they *would not* focus on.

Leveraging the DMADV framework, the team asked a series of questions while gathering voice of the customer (VOC) and associated data to support each question and through this process began socializing the anticipated impact of improved knowledge flow to build support for the changes that would be needed. These questions included the following:

- What are the goals of improved knowledge flow?
- How can these be linked to the business strategy?
- What knowledge is most important to the mission of Merck & Co., Inc. and MMD?
- How does this knowledge flow now, and what is the benefit to improving this flow?
- What is the desired future state and what steps are we going to take to get there?
- How will we know if our strategy is successful?
- How will we ensure we continue to learn and drive continual improvement?

This began with taking significant time to learn from other companies, primarily in other industries. Along with the strategy, the team devised a set of principles to guide how the strategy was to be executed to deliver rapid yet tangible value. Key principles included the following:

- Align with the business processes and associated business case
- Learn by doing (e.g., agile methodology vs. waterfall)
- Leverage common approaches, processes, and platforms
- Measure knowledge management approaches and associated business outcomes
- In addition, per Charlie Honke and colleagues (2008) while at IBM's Fishkill semiconductor facility, "think big, start small, but start."

Knowledge Management Strategy Comes in to Focus

The strategy set MMD on the course for a thoughtful KM program linked to business results. Among other outcomes, the strategy defined:

- The need for a small team of dedicated KM resources in GSTC.
- Roles and responsibilities of the dedicated KM resources, as well as support and stewardship roles that would be required by the functional areas.
- A general approach to developing a KM program.
- A framework for successful KM capabilities, including the pillars of KM.
- A general approach to conducting KM pilots.
- Current state and future state measures, including KM maturity benchmark.
- Initial pilot project opportunities and associated business cases.

As the team began to better understand what they were trying to accomplish through the knowledge management strategy, the team began to introduce and socialize the term *knowledge flow*. The analysis of the current state illustrated that there were many barriers to knowledge flowing freely—obstructions of many different types (not knowing where to look for documents, inability to find experts, lessons not captured, etc.).

One key lesson during the creation of the strategy was learning how to perform knowledge mapping, a process for which the KM team subsequently created standard work. Knowledge mapping proved to be a powerful diagnostic to look past ambiguous perceptions of knowledge management issues (e.g., *I cannot find my content*) to helping recognize underlying themes and trends and better characterize the knowledge flow barriers—that is, the *why I cannot find my content* (e.g., no common repository with appropriate access model, no standard taxonomy, and no stewards to oversee content). Further, it was through knowledge mapping that MMD really began to recognize the importance of tacit knowledge. Once initial knowledge mapping was complete, the KM team had mapped multiple process areas and had defined nearly 500 *knowledge assets* that were required to support business process execution. Analysis showed that nearly 80% of these knowledge assets were dependent on some way on tacit knowledge (e.g., required input from an expert). This was a key realization for MMD in recognizing the extensive tacit knowledge required in everyday work, and helped MMD begin to see past its initial focus on explicit knowledge.

Following the DMADV framework proved to be impactful for MMD. Through this process a strategy was produced that met the expectations of the initiating sponsor and established guiding principles and frameworks that guide the KM team to this day.

The Pillars of Knowledge Management in Merck & Co., Inc. Manufacturing Division

In MMD there are four pillars of knowledge management: *people, process, content,* and *technology*—in that order of criticality to success.

People

People comprise about 70% of the success factor for KM. There are two key contributors: (1) behaviors and mindsets and (2) KM roles.

> *Behaviors and mindsets*: MMD recognized that their knowledge management implementation was first and foremost a large, complex transformational change project that required robust change management. That is, at the end of the day, KM is about people having the information and knowledge when and where they need it to do their work. For this to happen, workers need to *seek* knowledge on a routine basis. This desire to seek knowledge must be a combination of (a) desired behaviors and mindsets (e.g., learning what worked or did not on the last project instead of planning to go it alone because *I am smarter than the last person*) and (b) clear expectations (e.g., the business process codifies knowledge seeking, perhaps through reviewing past lessons during project initiation).
>
> Similarly, people must *share* knowledge, and the same two conditions must exist—the desired behaviors and mindsets to *share* (e.g., this is what I learned) and the clear expectations (e.g., set milestones for capturing lessons). There were many cultural barriers to address in order for KM to gain momentum in the organization. It was recognized that change management methodologies could help address cultural barriers. MMD had previously subscribed to Conner Partners Change Execution Methodology (Conner Partners, 2007) and leveraged this framework for guiding the organization through the change. Not recognizing KM as requiring a focused change management effort is short-sighted, given that it is people at the center of knowledge management who must want to seek and share. This is perhaps why many KM efforts that focus solely on IT does not deliver on intended outcomes as they overlook the behavioral element. A key success for the team was their ability to link the KM change plan to a broader behavioral change occurring in MMD that enabled a new way of working for many (Guernard, Katz, Bruno, and Lipa, 2013).
>
> *KM roles*: The KM team also began by defining the knowledge management roles, both those as dedicated KM persons, and those in the business who are stewards, responsible in some way for sustaining a KM approach. In addition to the dedicated KM team, MMD has approximately 150 stewards across the various capabilities. Stewards

have a role to help knowledge flow within their area of capability, for example, through following up with SMEs to answer requests for help and ensuring documents are checked in and properly tagged. In certain areas, gaining stewardship required some negotiation, but logic prevailed. For example, historically any given product may have up to 50 SharePoint sites where product knowledge was stored. Each of these had an owner, who was the de facto steward. The starting point was therefore 50 people spending perhaps 1% of their time with exceptionally poor results in enabling knowledge flow. Contrast this with a single dedicated steward for 5%–10% of their time, providing highly effective stewardship over a single common repository. So the net result is a cost savings in stewardship effort, which is a small benefit compared to as well as the business efficiencies when people can actually find *what* they need *when* they need it. In addition, the stewards themselves realize the benefits for their careers, as they can see and influence the organization from a broader perspective in terms of hierarchy and connections.

Process

There are two facets to consider regarding process in MMD: (1) the business processes that ultimately deliver products to patients and (2) the KM processes that enable knowledge flow.

Business processes: The primary goals of improved knowledge flow are to enhance the outcomes of a business process (e.g., technical transfers done *right first time*, risk assessments done with *higher accuracy*, investigations closed out *more quickly*, supply disruptions *avoided*, …). Improved knowledge flow is not just for the sake of improved knowledge flow-it is to ensure that the business operates as efficiently and effectively as possible to achieve its mission. Therefore, business processes are the basis that defines requirements for knowledge flow through each respective process. In addition, it is in these processes that established KM capabilities are later embedded to ensure these outcomes are in fact realized.

KM processes: A second process facet is about having standard work to drive consistency and standards for KM, including well defined, repeatable, and measurable processes. For example, standard processes to capture and share lessons, to perform on- or off-boarding or whatever the relevant KM *subprocess* is—even how new communities are established and metrics are reported.

Content

Content is about the type and subset of what knowledge is key and must flow, and how that knowledge is described (taxonomy). The term *content* is a *logical* term only as the taxonomy MMD created also helps to describe tacit knowledge (e.g., metadata for identifying the subject matter experts [SME]). Taxonomy is seen as a common language to the business and can help unlock existing knowledge. Many examples of different groups doing similar work but calling this work and its outputs as different things were identified during knowledge mapping.

Technology

The KM team leveraged existing Merck & Co., Inc. enterprise technology to minimize upfront investment and lead time. As an unintended consequence, this allowed the team to commit more focus on the *people* components of KM. SharePoint was leveraged in several instances, including as part of the approach for managing product and process knowledge. Of note, the infusion of KM principles to SharePoint deployments resulted in a product that was often recognized across Merck & Co., Inc. as a *best practice* in how SharePoint should be leveraged as a collaboration platform. The team also designed the virtual technical network, or VTN, which today consists of 27 communities and 4000+ members. The VTN also leveraged existing enterprise technology (*NewsGator* at the time of VTN launch, now *Sitrion, www.sitrion.com*). VTN communities are technologically no different than the 300 communities across the company; however, the holistic framework of stewardship, rules of engagement, sponsorship, standard enablement processes, and other elements—set the VTN apart as the benchmark for successful communities across the company.

Knowledge Management Scope and Practice

As positive outcomes were realized and impact expanded, demand continued to grow for the KM team in GSTC. In time, this led to an increase in scope and responsibility for the KM team, expanding from focusing only on GSTC to supporting the entire MMD. The KM team was now recognized as a divisional Center of Excellence (CoE) for knowledge management. Adoption within the functions in MMD, including quality, supply chain, and the various lines of business varied based on a variety of factors but has generally been successful. The MMD KM CoE was able to solve similar knowledge flow issues across a much broader scope, extending their reach and impact, and ability to create value for the company.

The MMD KM CoE creates value through the collection of approaches to improve knowledge flow, often described internally as *KM capabilities*, as they provide MMD the ability to manage knowledge. As the CoE has evolved, solving new problems and serving new customers, these capabilities have expanded in scope and complexity. Each capability—product knowledge, process knowledge, connectivity, expertise, lessons learned, and knowledge mapping (Figure 14.1)—is a collection of one or more KM

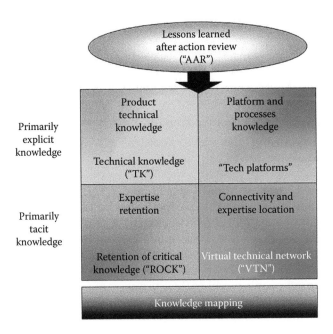

FIGURE 14.1
MMD KM capabilities.

products and services. Many of these have been described elsewhere in this book, but a brief description for the most common products and services follows.

Knowledge Management Capabilities (Products and Services)

Product Technical Knowledge

The MMD KM CoE established the standard for managing product technical knowledge in MMD. This primarily focuses on explicit knowledge but includes elements of tacit knowledge through improved ability to find subject matter expertise on various product-related topics. Drivers for the product knowledge capability included the poorly managed and distributed set of repositories and lack of standards for describing knowledge assets. The primary product, known as *TK* for *technical knowledge*, establishes the current standard, and includes several features such as the following:

- A common repository (SharePoint-based)
- Consistent look and feel across
  - Products
  - Drug substance and drug product
  - Modalities (e.g., small molecule, vaccines, and biologics)

- A standard taxonomy
- Well-defined stewardship roles
- Governance through a steering committee
- Appropriate processes to manage and control changes, for example, taxonomy change control
- Change management and communications processes
- Metrics reporting (analytics to prioritize areas of focus and health of the knowledge flow)

Closely related to TK is MMD's *content processing service* (CPS). CPS is an innovative solution developed in partnership with IT. Through a set of rules, CPS can analyze existing content in current repositories (*historical content*) and apply the TK taxonomy while moving these documents from several repositories into a single TK repository. CPS minimizes the effort necessary to manually migrate and tag thousands of older documents, and accelerates the change to get people working *the new way*.

Also, TK has provided a rich learning experience in taxonomy development and the resulting taxonomy forms a core of technical knowledge that the KM CoE expects to expand over time. This taxonomy is presented as a case study elsewhere in this book.

TK has delivered a variety of benefits, including eliminating significant wasted time and effort in storing and retrieving product-related knowledge. As a result, business processes such as tech transfer, problem solving, investigations, and process improvements have improved.

Process Knowledge

The MMD KM CoE established the standard for technology platforms and associated business processes. Like product technical knowledge, this primarily focuses on explicit knowledge but includes elements of tacit knowledge through identification of, and access to, the networks of persons responsible for the respective technologies. Drivers for the process knowledge capability included the lack of standards for how to maintain information on core technologies that are leveraged across many products (e.g., roller compaction), and standardization as critical to accelerating future product development. The primary product, known as *tech platforms*, establishes the current standard and includes several features such as

- A common repository (SharePoint-based).
- Consistent look and feel across platforms.
- Well-defined stewardship roles.
- Establishment of a knowledge stewarding CoE, responsible for sanctioning changes to the body of knowledge for a given platform.
- Change management and communications processes.

Tech platforms have delivered a variety of benefits including supporting accelerated development timelines and shortening time to competency for new employees.

Closely related to tech platforms are an emerging set of *business process platforms*. Leveraging the learning from tech platforms, several business process platforms have emerged with the intent to capture and share standard work and resources for topics important to a large audience, for example, problem solving.

Connectivity

The MMD KM CoE established the standard for communities in MMD. This primarily focuses on tacit knowledge but includes elements of explicit knowledge through the ability to search past discussions and other artifacts posted by the community. Drivers for the connectivity capability included the lack of connectivity across boundaries at MMD, including across functions, geographies, plant sites, modalities, and so on, and a known tendency for *reinventing the wheel*, as employees commonly did not know who to contact when looking for help. The primary product, known as the VTN establishes the current approach and standard for *helping* communities and includes several features such as the following:

- A common community model, based on enterprise technology (Sitrion + SharePoint).
- Consistent look and feel across communities.
- Standard business processes for community enablement, steward on-boarding, and capability building, and so on.
- Close linkage to divisional culture initiatives, including leveraging common language and creating assets such as *VTN rules of engagement*.
- Extensive change management plan, including communications, ongoing pulse checks, rewards, and more.

The most widely known of all of the KM offerings with nearly 4000 members, the VTN has delivered a variety of benefits, largely through problem solving and best-practice sharing. This has occurred throughout a multitude of processes associated with development and supply of the company's products.

Expertise Transfer and Retention

The MMD KM CoE has established the standard for reducing the risk of critical knowledge loss through a facilitated knowledge transfer process. This primarily focuses on tacit knowledge exchange but includes elements of explicit knowledge through the facilitation and capture of questions intended to help explore the thought and the thinking behind unique, critical expertise, and to enable focused knowledge flow to appropriate successor(s) and/or other channels (e.g., job aids and training) for retention and reuse.

Drivers for the expertise transfer and retention capability include minimizing the risk of losing critical knowledge and expertise from the organization, business continuity and sustained performance, and minimizing time-to-competency for successor(s). The primary service, known as ROCK (retention of critical knowledge), establishes the current standard and includes several features such as the following:

- Standard process for identifying, prioritizing, and transferring critical knowledge.
- Risk assessment for identifying and providing risk mitigations.
- Well-defined roles for knowledge transfer.
- Scalable *self-service* ROCK job aid for people managers and employees during routine transition of employees out of their current role or project.
- Delivery of structured facilitated ROCK for highly specialized business-critical knowledge.

ROCK has delivered a variety of benefits including identification of business-critical, unique knowledge, and expertise held by the employee, capture and transfer of key insights, thought models, and recommendations, and others. Moving forward the focus will be on increased visibility and usage of the ROCK risk assessment to proactively identify risk and further adoption of ROCK self-service.

Lessons Learned

The MMD KM CoE has established standard for lessons learned through an after action review (AAR) process. The focus includes both tacit knowledge exchanges through facilitated after action reviews and explicit lesson capture and sharing through knowledge transfer plans.

Drivers for the lessons learned capability included avoiding waste and rework associated with repeating problems and failing to leverage good practices. AAR was seen as a means to build continual improvement into standard work. The primary offering includes after action reviews that consist of several related products and services. AAR establishes the current approach and standard for identifying and learning lessons and includes several features such as the following:

- Standard repeatable process for after action reviews (Five standard questions).
- Scalable after action reviews based on size, complexity, and formality needed (simple, local five minute exercise to complex event or process requiring structured process).
- Focus on the *how* through inclusive mindsets and behaviors.

- Well-defined roles for facilitating after action reviews.
- Well-defined sponsorship for driving change.
- Change management plan, including risk identification and mitigation and communications.

After action review has delivered a variety of benefits including continual improvement of key processes, increased capability within the organization to identify and apply lessons, fostering a culture where it is safe to speak up on what did not go well, and recognition to build AARs into our daily work to become a more agile organization in a dynamic industry.

Moving forward the lessons learned capability will continue to grow moving to a more proactive model focusing on peer assist.

Knowledge Mapping

Knowledge mapping has emerged as a key service from the MMD KM CoE, based on the insights it can provide to successful business process execution. A key barrier for successful knowledge mapping remains the existence of a well-defined, standard business process. However, when a process is well defined, knowledge mapping is a powerful diagnostic to look at the process with a knowledge lens. Knowledge mapping has surfaced many barriers to successful, robust process execution, whether identifying *traditional* knowledge flow barriers or other prerequisites and enablers, such as training, job aids, and system reports. Standard work has been defined for knowledge mapping through the creation of a playbook.

Enablers to Effective Knowledge Management in Merck & Co., Inc. Manufacturing Division

There are many supporting enablers to the progress of the MMD KM CoE, and several of these have been described earlier in this chapter elsewhere in this book through the case studies of MMD colleagues. Following is further insight into several supporting enablers for effective knowledge management, including the following:

- Sponsorship
- Change management
- Success stories
- Partnerships
- Common approaches

Sponsorship

Sponsorship, or perhaps more appropriately *change leadership*, has proven to be a critical success factor in MMD. The KM program was fortunate to have the senior leader of GSTC as, not only a stakeholder, but also as a key visionary and thought partner during the KM journey. The concept of change leadership goes beyond sponsorship, in that it embodies the role of the senior leader to lead the organization through the change in a very holistic manner. The concept of sponsorship is still critical, but the typical execution is often more passive and hands off—more of a *do as I say not as I do*, whereas change leadership pulls in more *do as I do*. Key insights from the senior leader of GSTC as the initiating sponsor of effective KM in MMD include the following:

- Know what KM is and what it is not—for example, that KM is not just about documents, or about technology
- Be committed to learning
- Thoughtfully diagnose what problem you are trying to solve
- Take committed action
- Provide leadership
  - Learn
  - Commit
  - Create the right environment
  - Set expectations and hold people accountable, including positive and negative consequences
  - Remain resolved

Change Management

Much like inverting the concept of knowledge management to be about managing knowledge, the same applies for change management. Change management is really the process of managing change. But what does this mean in practice?

The change framework used in MMD is based on the Conner Partners Change Execution Methodology (Conner Partners, 2007). Change is seen as following a process that begins at the point where people may be aware that a certain change exists, but they do not really know what it is or how it affects them. For example, in MMD, a change may be using the VTN to ask for help on a problem instead of people solely relying on their own network. As people get more exposure and clarity to the change, along with the right expectations and reinforcement from leaders, people gradually move through change perception until a change is institutionalized—that is, the change is no longer a change but instead is seen as simply *how work gets done*. There are a variety of tools, templates, and techniques to help *manage change*.

In terms of behavior, the team is engaged in making sure people have the right conditions to influence their behaviors. This includes people having the right

- Direction (i.e., expectations)
- Competence (i.e., ability)
- Opportunity (i.e., the right chance to act in the right context)
- Motivation (reinforced by the right positive and negative consequences)

This framework leverages work by CLG's DCOM model that was deployed to change managers in MMD (Hillgren and Jacobs, 2010). When the team asks people about why they do not seek lessons on past work, for example, the feedback is telling. They may say they do not have time, or that it is not relevant to them, or that their work is different, or (perhaps implicitly) they want to figure it out on their own, or they do not know where to go, or that there is no incentive for taking time to look for old lessons, and more. All of these are linked to a failure in one or more of the conditions needed for driving behavior—direction, competence, opportunity, or motivation (DCOM).

With the combination of change management, linking to business outcomes and embedding it in the flow, people are not only realizing that their engagement in managing knowledge management systems is expected, but they can see that it is valued within the organization, and that is a key driver to business effectiveness. The speed of knowledge transfer is a key differentiator for companies (Miller, 2012).

Success Stories

Closely coupled with change management is an effective communications strategy, and perhaps the most effective element of the MMD KM communications strategy has been the *success story*. The team has accumulated a collection of success stories where improved knowledge flow has had significant positive business impact, whether saving time and money, to solving problems faster to ensure continuity of supply, to developing more robust products, to improved compliance posture, and so on. One such success story was when a piece of specialized equipment key to making tablets broke down. A replacement would have taken too long to acquire, approximately 20–24 weeks, due to warranty process issues and other factors. A colleague posed the question on the VTN to the 600-member powder processing community where another member in Asia sourced a spare in local inventory and delivered to another continent within a week. The person who posted the question had been hesitant to post on the VTN, unsure of what the reaction would be. However, once they did, they subsequently shared their realization that *People just want to help*. Due to the behavior change of asking the question to the global network the organization reached a quick and successful resolution. The person in this story went on to share "The VTN makes you feel like you are part of something larger."

Partnerships

Partnerships for the MMD KM team have been critical to its success. The most mature and effective partnership has been that with the IT organization, where KM is a program in the IT portfolio. Through this partnership, the KM team has effectively leveraged enterprise technologies, pursued select innovations (e.g., content processing), and influenced a much broader audience at Merck & Co., Inc. through its leading practices. Other partnerships exist, to varying levels of maturity and interdependence, including with the learning and development organization, human resources, and operational excellence. Each of these partnerships offers synergies for delivering solutions to the broader business unit if they can be unlocked.

Common Approaches

The impact of common KM approaches has also been a key enabler for the MMD KM program, for example, developing a standard for a community of practice, or a role definition for a steward, and applying that standard to multiple similar problems. The team found that once a problem was defined and a solution was identified that worked within a particular area, 90% of the success factors for that solution remained the same when applied to other areas. Said differently, the knowledge flow barriers the team encountered were largely common across areas—and a *helping community*, for example, is a helping community that could be applied in many different contexts where there was a need to connect people with the broader organization for advice and input. This discovery enabled a lot of rapid replication.

Challenges

There are many challenges along the journey, especially when one realizes the change effort that goes along with making managing knowledge as a way of working. Some of the challenges include the following:

- *Getting everyone in the organization engaged*: Part of being a scientist is to seek and share knowledge. It is fundamental to curiosity and the learning process. Yet we treat it like a discretionary activity, with no standard expectations for *what*, *when*, and *how*. Great progress has been made, and while some knowledge hoarding still exists (*knowledge is power* mindset) more and more people are being impacted by KM every day in MMD.

- *Change capacity*: People are overloaded with too many competing priorities—their capacity for change is full. Many people still see knowledge management as an initiative, rather than as a way of working. They are moving beyond that now in many areas, and leadership teams are recognizing more and more that they really need knowledge to flow in order to be successful in today's environment. Linking knowledge flow to a team's objectives, through strong sponsorship, is a key tactic to address this.

- *Articulating business value*: Asserting value can be difficult in terms of quantifying impact, as many of the larger benefits are circumstantial, indirect, or often never witnessed (e.g., problems avoided before they ever had a chance to happen). Success stories are a key lever here as they are relevant to the business, and employees can see their peers making an impact through demonstrating desired behaviors.

- *Knowledge workers of tomorrow*: The millennial cohort of MMD's workforce is growing in proportion every day, and has grown up with different expectations for their access to knowledge, information and people, due to the Internet (e.g., Google and YouTube), mobility, and a generally more connected world. Will they put up with how current systems work, when some of them have never had an e-mail account and are used to communicating through social networking channels? One goal of the team is to engage with the millennial population to ensure relevance of KM approaches while improving the attractiveness of Merck & Co., Inc. as an employer.

Summary

Currently, the knowledge management team in MMD estimates that they are about 20%–30% along the evolutionary journey of knowledge management institutionalization, with a long way to go. To date a solid foundation has been laid. Much of the work that remains is to scale more broadly to more users (more community members, more functions, etc.) and target content (more products, more processes, etc.), and adapt common approaches to other similar knowledge flow problems, all while continuing to demonstrate tangible impact and making knowledge flow easily through core work processes.

The key points to remember include the following:

- Aligning knowledge management capabilities with the organization's performance goals is important in order to ensure that knowledge management products and services are an integral part of the organization's processes rather than something separate and discretionary.
- The importance of change leadership and effective sponsorship cannot be understated, along with a robust and well-executed change plan.
- A holistic framework incorporating people, process, content, and technology supports the adoption of knowledge management across all areas of the organization.

It is impossible to provide a framework where every person in an organization feels safe to ask a question. However, the team has seen positive changes in how knowledge is shared over the past six years. It is now considered much safer to *make problems visible* and ask for help now than ever in the past, and this has been attested by people at all levels in the organization. This in and of itself is a major positive impact that KM has had on the organization.

Acknowledgments

The authors wish to recognize many leaders and key partners who have co-led or otherwise contributed to the success to date with KM in MMD, including Dr. Michael Thien, leader of GSTC and chief visionary; key partners in strategy development and early adoption in GSTC, including Dr. Jean Wyvratt, Anando Chowdhury, David Vossen, and the GSTC leadership team; the MMD Knowledge Management CoE team members, including Chris Tung and trusted partners Douglas Arnold and Yunnie Jenkins in MMD IT; Cindy Hubert and her colleagues at APQC for their steadfast guidance and connections to other practitioners; and the KM stewards and partners throughout MMD and beyond. And thank you to Katelyn Lipa for her support in graphics design.

References

Conner Partners. (2007). *Building Commitment to Organizational Change.* Retrieved May 27, 2016, from http://www.onthecourtcoaching.net/Documents/Building%20Commitment%20to%20Organizational%20Change.pdf

Graves, A. (2012, December 10). *What is DMADV?* Retrieved May 27, 2016, from Six Sigma Daily: http://www.sixsigmadaily.com/what-is-dmadv

Guernard, R., Katz, J., Bruno, S., and Lipa, M. (2013). Enabling a new way of working through inclusion and social media—A case study. *OD Practitioner, 45*(4), 9–16.

Hillgren, J. S. and Jacobs, S. (2010). *DCOM: A Proven Framework for Outstanding Execution.* Retrieved May 27, 2016, from http://www.clg.com/Science-Of-Success/CLG-Methodology/Organizational-Change-Tools/DCOM-Model.aspx

ICH. (2008, June 4). *ICH Q10 "Pharmaceutical Quality System."* Retrieved May 27, 2016, from International Conference on Harmonisation: www.ich.org

ICH. (2009, November). *ICH Q8 (R2) "Pharmaceutical Development."* Retrieved May 27, 2016, from International Conference on Harmonisation: www.ich.org

Miller, F. (2012). On-site training at Merck & Co., Inc. in West Point, PA, April, 2012 (related to Katz, J. and Miller, F., "Inclusion: The HOW for the Next Organizational Breakthrough," Practicing Organization Development—Third Edition, Jossey-Bass/Pfeiffer, 2009). Kaleel Jamison Consulting Group.

15

Integrating Knowledge Management with Quality Management Systems to Manufacture Pharmaceuticals Consistently and Reliably

Barbara Allen

CONTENTS

ICH Q10, the international guidance on the Pharmaceutical Quality System, published in 2008 was the first formal acknowledgment of the need to integrate and use knowledge effectively throughout an organization's quality management system. This chapter outlines how knowledge management has become invaluable at Eli Lilly and Company, and shares several examples of how KM is now enabled through the quality management system.

Editorial Team

Introduction

Over a decade ago, the author was involved in the development of the ICH Q10 *Pharmaceutical Quality System* guidance as a member of the expert working group representing Pharmaceutical Research and Manufacturers of America (*PhRMA*). During the development process for ICH Q10 it was discussed and agreed that knowledge management would be viewed as an enabler rather than as a system. There was a clear understanding that *knowledge* should be managed appropriately, but not as a separate system, rather knowledge should be integrated into how work is conducted. The purpose of this chapter is to give an overview of the ways in which knowledge management can be successfully integrated into the quality management systems of an organization, using Eli Lilly and Company (*Lilly*) as an example.

Over the past 10 years, the concepts of Quality by Design, of quality risk management (QRM), and of enhanced quality management systems have gradually become more integrated. Now, more organizations are systematically building knowledge management strategies in order to gain more advantage from this area.

Key Deliverables of Pharmaceutical Manufacturing

ICH Q10 reminds us that there are three key deliverables for pharmaceutical manufacturing. The first is *product realization*, or the achievement of the product with the quality attributes needed to meet the needs of patients.

The second is about establishing and *maintaining a state of control*, whereby effective and consistent control systems are developed and used for process performance and product quality, thereby providing assurance of continued suitability and capability of processes.

The third deliverable is to facilitate *continual improvement*. This involves identifying and implementing appropriate product quality improvements, variability reduction, innovations, and quality system enhancements, thereby increasing the ability to fulfil quality needs consistently.

Consistency and control are important to successful, robust processes, and operations, and to ensuring reliability of supply. It is important to facilitate continual improvement. Over the past 10 years advances in technology, data availability and infrastructure have improved, and it is possible and important to take advantage of this to drive continual improvement.

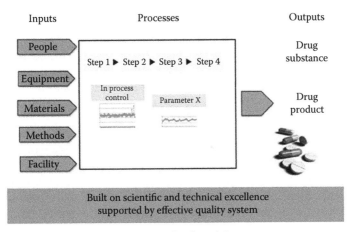

FIGURE 15.1
Representation of key deliverables of pharmaceutical manufacturing.

Pharmaceutical Manufacturing at Lilly

In the early 2000s Lilly sought to improve its manufacturing operation in a sustainable way. In large organizations it is important to establish an awareness at all levels of an organization of what the ultimate goal of that organization is. As a result, a clear mission statement was drawn up, which was expressed as "Safely and reliably manufacture quality medicines for our patients."

At a fundamental level, the primary purpose of manufacturing is to convert inputs to outputs, using carefully controlled processes, as represented in Figure 15.1 above. Achieving this consistently relies on scientific and technical excellence and on an effective quality management system.

Elements of Operational Excellence

Taking the mission of manufacturing further, Lilly drove toward a performance based on operational excellence. There are three overarching concepts within the approach, as represented in Figure 15.2.

- The first is scientific and technical excellence. This entails well-developed product and process control strategies, which are capable of controlling all aspects of the process.

- The second is well-defined management systems, with a focus on defined governance and using efficient and effective business processes, resulting in disciplined execution.
- The third is a continuous improvement mindset among employees, which entails broad ownership and accountability, visibility, a sense of urgency, and curiosity.

To enable this framework of operational excellence, Lilly revalidated each process and analytical method in order to establish a current state of optimal scientific excellence, with a focus on being in control and capable. Highlighting the patient as the key customer of the manufacturing output assisted in providing a solid frame of reference and focal point for risk management.

In addition, the quality management system was reengineered. This exercise involved redesign of quality standards, business processes, organization, and management controls. Knowledge management was not developed as a separate system, instead aspects of the management of knowledge was integrated within each of the management systems.

In conjunction with the reengineering of the quality management systems, Lilly promoted a continuous improvement mindset and culture. Members of the organization are encouraged to take what they see and make it better, expected to seek the answers to key questions, and are enabled to identify and act on needed changes. Improvements are implemented in a deliberate and controlled manner, with focus on reducing variability and reducing risk to the patient.

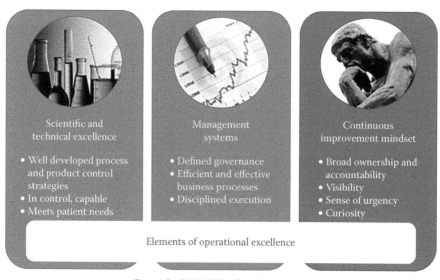

FIGURE 15.2
Elements of operational excellence for Lilly pharmaceutical manufacturing.

Building an Effective Quality Management System

The Lilly Quality Management System approach is represented in Figure 15.3 below and is based on four components. The initial development of the quality management system was within the manufacturing function of the organization, but the same model has since been deployed across the company in areas including marketing and clinical research.

The four components are the following:

- Integrated quality standards that provide clear requirements to meet compliance and patient needs.
- Business processes that define how work is to be conducted, both in terms of effective and efficient processes.
- Organization design that involves having the right people in the right roles with the right span of control and clear accountability.
- Governance/Management Oversight, including management involvement, escalation, decision making, auditing, and metrics programs.

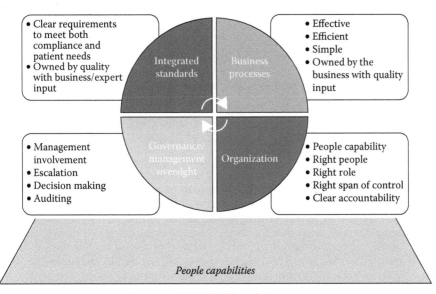

FIGURE 15.3
The Lilly Quality Management System.

Integrating and Using Knowledge Effectively through the Quality Management System

An organization may have lots of knowledge, information, and data on all the processes, including quality management processes; however, that knowledge is only useful if it is readily accessible and utilized. For Lilly, this is where knowledge management has become invaluable. Several examples of how knowledge management is enabled through the quality management system are shared below.

Example 1: Knowledge Management and the Quality Standard and Practices

Ease of Access to Codified Knowledge

Lilly designed a logical indexing tool within the quality management system to facilitate ease of access and use, with different codes for the various areas of the quality management system. For example, all facility, utility, maintenance, and equipment standards and processes are coded in a 200 series, whereas the laboratory related topics are coded in a 600 series. Each series includes the relevant global quality standards, regulatory references, business tools, and resources such as templates and papers. In this way, knowledge gained can be incorporated into quality practice and accessed easily by the organization.

Lilly maintains the quality system on an ongoing basis. This maintenance is based on both internal organizational needs and signals from the external environment, including any regulatory changes. Knowledge, analysis, and decisions are then reflected within the quality system and operations adjusted accordingly. Communication on emerging signals is also proactively provided, so that specific business areas have the context for priorities and decisions.

Example 2: Knowledge Management and the Organization

Providing People with Knowledge of Their Role

The organization component of the quality management system provides details on roles, responsibilities, and accountabilities. It specifically articulates responsibilities for all employees and additional responsibilities of individual management, as well as key technical and quality unit roles. Lilly Manufacturing has organized its operations with a site-centric model, such that the focus is on the production and testing of product. Everyone involved must have the appropriate knowledge necessary for their role. The design and detail of the quality management system facilitates this by providing people with the knowledge related to their roles. Responsibilities and accountability are continuously clarified, improved, and documented; for example, the role of the site head, the process chemist, or the engineer, and the quality unit personnel are all defined and documented to ensure clarity of purpose regarding their responsibilities for the production and testing of product.

Example 3: Knowledge Management and the Product and Process
Knowledge Generation and Utilization

Knowledge related to the product and process is key to enabling the fundamental role of the manufacturing function in running effective processes to produce quality products. The management of this knowledge is integrated into the quality management system in several ways.

- The quality management system requires a documented *process description* outlining the controls and supporting rationale. This process description captures the control strategy for the product and provides a link to the development history. It also serves as a key repository of the knowledge related to the necessary controls and continues to be modified in content and title as the concept of control strategy evolves within the industry and the regulations. The process description and associated knowledge are utilized as part of deviation and change management.
- The *annual product review* provides a key contribution to knowledge related to a product and process. Where warranted, any relevant knowledge obtained is incorporated into the control strategy and associated document as described above. The goal is to continuously build knowledge and then provide a mechanism that it can be easily referred to and maintained.
- Like many companies, some Lilly products are produced at multiple manufacturing sites. The quality management system includes a global product assessment that looks at how a product and process is performing globally. This global product assessment, which is reviewed by senior leaders from within various functions, focuses on the state of validation, and on both the process control and capability. Decisions are then made on where to make investments related to the on-going quality assurance and reliability of that particular product. Furthermore, the global product assessment captures and enhances the knowledge related to a product and its associated processes.

Example 4: Knowledge Management and Management Controls
Knowledge Generation and Utilization

Management controls at Lilly include formal reviews, metrics, and escalation processes. There are several processes associated with this component of the quality management system that contribute to the overall management and utilization of knowledge within the organization. For example, as outlined in ICH Q10, a review of the quality system and its effectiveness is used to identify areas of improvement and, where appropriate, leads to adjustments in standards and practices, thereby ensuring that knowledge gained is incorporated into day-to-day operations in a systematic and sustainable way. These reviews are conducted at both site and at global level and are reviewed by relevant level of management.

Knowledge Management and Information Technology

During the 2000s, Lilly made significant investment in IT platforms to manage operations consistently and enable data management. The investment was based on a model that identified which processes would be integrated across the manufacturing supply chain and which ones would be managed locally. This IT model has now been implemented across the entire Lilly Manufacturing platform and facilitates automated capturing of data. This has reduced the traditional burden of data compilation, enabling staff to focus on analysis, scientific evaluations, and knowledge generation and to make decisions accordingly. Although the scale of investment in IT capability may differ between companies, the approach to data and knowledge management should be deliberate.

Knowledge Management and Continuous Improvement

In the area of continuous improvement, Lilly has focused on variability reduction in its products and processes. The metric chosen to target this improvement is CpK; with an expectation that the CpK should routinely be above 1.33 and a goal of 2.0 (i.e., Six Sigma). By focusing on this aspect of performance all areas of the business seek to understand and control variability. This requires a greater knowledge and understanding of the product and processes. Through the use of the *annual product review* and the *global product assessments*, the collective knowledge of the organization is captured and utilized to drive this improvement. Furthermore, the engagement of senior leaders in this area ensures prioritization of effort and resources. This focus and level of variability control has resulted in greater reliability, predictability, and has lowered product losses and rejections.

Summary

Knowledge management at Lilly has been important in ensuring that the scientific information and expertise is captured and utilized. From experience, knowledge management is most effective when built into the day-to-day operations of an organization, rather than operated as a separate system. The examples above provide some insights to the approach at Lilly has utilized to integrated knowledge management within the quality management system. It is likely, and anticipated, that the approaches and tools will evolve over time; however, the goal remains same—the reliable manufacture of quality medicines for the patient.

16

Working with the IS/IT Function to Setup Your Knowledge Management Project for Success

Joseph Horvath

CONTENTS

> To be successful, KM champions need to take a pragmatic and informed approach to working with their partners in the IT function. This chapter shares one pharmaceutical company leader's experiences and insights on to how to leverage the KM–IT partnership to facilitate bringing the KM program to the next level.
>
> **Editorial Team**

Introduction

Knowledge management (KM) is not about technology but most KM solutions will not scale without the right technology having been well deployed. Although it can be smart to deemphasize IT at the outset of your KM effort, it is important to plan for success by considering what sort of technology you will need to turn a successful pilot project into a broader capability. The people in your company who can help you do this are in your information systems/information technology (IS/IT function). The IS/IT function is responsible for providing information technology solutions and services that meet the needs of the business—everything from network infrastructure to application architecture to help-desk support. They know what it takes to get systems designed, deployed, and supported in your company. Knowing how to work with them effectively will be a key to your success. KM ideas, visions, and project proposals tend to come bubbling up through different channels than do conventional IT projects. Often it is the folks who are intimate with the particular domain of knowledge who drive the initial conversation about managing it better, rather than the systems people who are the more typical *liaison* to the IT function. This can be great but it can also lead to disconnects and a relative late entry of the IT function into the conversation. In this chapter, we discuss why it is particularly important for KM advocates to work well with their IT partners and offer some basic advice for getting off on the right foot.

Unique Challenges

KM projects and resulting systems often have distinctive properties that are relevant to technology selection, design/configuration, and roll out, and these may present unique challenges. Your IT partner can help you to anticipate and work through these sorts of challenges, which may include the following:

Diversity of Users and Content

If KM is about getting knowledge to flow where it currently does not, we should expect that the content and user base for knowledge management projects will tend to be more diverse than that of the typical IT project. This brings a host of challenges, starting with the not-so-simple matter of making sure that everyone who needs to access and use the system can do so. Your intended users will likely be working at different sites and, depending on the state of your IT infrastructure, they may be on different networks, using machines with different desktop images, and/or supported by different IT groups. Depending on the project, they may even belong to subsidiary companies, suppliers, or other external partners. All this needs to be worked through with the help of your IT partners. They understand the current state of the company's network infrastructure, end-user computing environment, provisioning, and support processes. Engaging them early on will help to ensure that you do not find yourself going live with a system that excludes important user groups.

Beyond access, there are a host of challenges associated with the very diverse sources of content that KM systems seek to leverage. In order to make visible and usable all of the relevant, internal knowledge on a topic, it is often necessary to create linkages across multiple sources of information stored in different formats and file types. Frequently, these sources bear a definable relationship to one another and this relationship must also be represented (Horvath, 2014). For example, a search for information on the attributes of a particular compound may include the raw data from an experiment, a description of the method employed, information about the development history of the compound, the results of statistical analyses of the data, as well as presentations and reports that summarize and interpret the results. Ideally, the nature of the association among these information sources (i.e., they pertain to a single experiment on a single compound) should be apparent to the user. Conversely, the fact that they are encoded in different file types and may be stored in different repositories should ideally be invisible to the user. To further complicate matters, these sources of information are likely to participate in multiple webs of association (e.g., an experimental method employed across multiple projects or platforms) and need to be retrievable in those contexts as well.

Software vendors who target the KM market understand these challenges and have a range of technical approaches with which to address them. Some applications are optimized for specific knowledge domains and map data sources into a predefined semantic framework. Others rely on end users to tag content for retrieval, providing tools such as query-term expansion to make the task easier. Still others employ text mining algorithms to extract terms and concepts that associate diverse sources of content with one another. But to apply any of these technologies effectively in your company is as much about getting the semantics right as it is about getting the software right. Inevitably, the subject matter experts and content owners within the business need to agree on common ways of naming and classifying things if the desired integration across sources is to occur. Your IT partners can help you understand where such standards are required and can work with you to facilitate discussions within the business. They can also ensure that your system is designed to leverage current and evolving sources of company master data. These are the centrally managed and authoritative sources of information regarding key entities (e.g., products, people, and customers) that supply multiple systems and enforce a single version of the truth. Leveraging master data, designing integration strategies, driving appropriate standardization, and tracking dependencies between your project and related IT projects are all critical to the long-term success of the KM effort, and your IT partner should be able to connect you to the methods and the expertise you need.

Tacit and Voluntary Nature of Knowledge Sharing

Decades of research and practice have shown that the most valuable knowledge in an organization often has a richly contextual and/or tacit component (Horvath, 2014) (Nonaka & Takeuchi, 2008) (Pisano, 1996) (Sternberg & Horvath, 1999). This is knowledge that is not easily articulated and that cannot easily be separated from the situation of its use. For this reason, KM systems often need to be able to capture aspects of context through relatively sophisticated use of tagging, cross-linking, and annotation. They may also need to be able to point to people with relevant expertise in order to drive the types of conversations that convey tacit knowledge (Horvath, 2014). Those pointers need to be robust and well maintained as people's roles and interests shift over time and the *expertise location* functionality needs to be deployed in a way that is sensitive to the norms and culture of the organization.

For a variety of reasons—including the tacit nature or much of the organization's knowledge—it is very difficult to compel people to participate in knowledge sharing. It is just too easy to hold out and not share what we know with others. Although incentives and recognition can help, the use of KM systems is largely voluntary and this has implications for the technology needed to support it. First, usability needs to be high and barriers to usage low for both end users and for the content coordinators who are often needed to curate the contents of the knowledge base. Whatever the value of the content, it will be judged against the difficulty of using the system and it will take some really stellar and up-to-date content to get people to use a KM system with a poorly designed interface or a confusing repository structure. Similarly, performance support in the form of both documented job aids and *live* assistance needs to be present and effective in order to promote and sustain voluntary usage. In short, when people are not compelled to use a system, it needs to attract and hold them as users and this tends to set a high bar.

Ambiguity of Ownership

As described above, KM projects and systems often fall into the *white space* that is not fully covered by the company's organizational units and established processes. For example, who is responsible for getting the product and process knowledge acquired in commercial manufacturing back to the teams in process development who could benefit from it? Who is responsible for making sure that the knowledge acquired in early development is available to investigation teams in commercial operations? When a company develops systems to enable these knowledge flows, who are the business owners?

Part of the appeal of KM is that it has the potential to knit the organization together in ways that formal structures do not. But the boundary-spanning nature of KM systems can make it challenging to situate them within the company. Part of the challenge rests in the fact that the activities that KM systems are intended to support do not always lend themselves to a traditional process perspective. Indeed, the purpose of the KM system may be to enable types of knowledge exchange that cannot be anticipated. When an IT system automates or enables a standard process, the process owner is the logical business owner. For a KM system, though, it may not be that simple.

Questions of system ownership can usually be finessed in the early or proof-of-concept phase. The system can be owned by any sponsoring executive or his or her delegate without much consideration of whether this makes sense in the longer term. But when it moves from proof-of-concept to planning for full deployment, business and IT ownership need to be established so that resources can be allocated and accountability established. Some mechanism of cross-functional governance is also needed so that other system stakeholders will have visibility to and influence over the management and evolution of the system. These stakeholders may include not only business groups that will use the system but the owners of related IT systems that impact or are impacted by changes to the KM system. Although all of this can be tricky, it is something that your IT partner should be well equipped to navigate. Following their lead and their established practices will get you to a better answer, faster.

Working with Information Systems/Information Technology

A first step in working effectively with your IT colleagues is to get a basic handle on IS/IT organization, processes, and overall strategic direction. It is not your job to understand any of these things in detail, but if you take the time to ask questions and do some reading, you will gain a basic frame of reference and will be ready to partner effectively. For example:

- Who is your designated IT partner? What is their scope of responsibility? How do they prioritize their activities in support of the business?
- Who provides IT supports to the other business functions that may be in scope for your project?
- Who provides end-user support for IT applications (e.g., help desk services)? Is this activity outsourced? What about back-end support for applications?
- What is the structure of the typical IT project team? What roles and responsibilities are assigned to representatives of the business functions?

- What are the major activities and deliverables that make up the systems development lifecycle (SDLC) at your company?
- What is the calendar for planning and budgeting of IT projects and what is the process through which new IS/IT spending is reviewed and approved?
- How does the procurement function get involved? What are their policies and processes?
- What is the current application landscape and where is it heading as per the current IS/IT strategy? Has the company consolidated on global, shared platforms such as SAP, MS SharePoint, or IBM/Cognos? Are there plans to do so?
- What kinds of architectural or IT security standards might be relevant to your project? Does the company possess partner data that it is obliged to protect in particular ways? Does the IT function favor certain integration methods/tools? What about master data sources/processes?

Once you have gained a better understanding of how your IT function works and where it is headed, you will be ready to partner with them on your KM project. Every project is different, of course, but a few guiding principles are worth keeping in mind as you enter into this collaboration.

Do Not Start in the Middle

Like any professional offering consultation, your IT partner may ask you to take a step back from your initial ideas about a solution in order to make sure that the problem itself has been adequately defined. We all work with information technology in our jobs and we all have ideas about how it could be better, what should change. When we set out to actually invest, however, it is good practice to put those ideas aside, at least temporarily, to just focus on the problem we are trying to solve. That is, what gap or opportunity is the business facing that an investment in IT can help to address? What is the scope of the problem? Who are the stakeholders? What part will the new or improved IT system play in solving the problem? What parts of the problem will need to be addressed through other means? Asking these questions can help to identify gaps in our thinking and surface differences of opinion or interpretation.

Being willing to start at the beginning—at problem definition—is a key element in working effectively with IT. So is a willingness to keep the initial technology design simple. Many IT managers have had the experience of working with an internal customer whose business requirements for a system, already in hand, are based on the full set of features identified in a vendor brochure or demo. This tends not to be a productive approach. It can be difficult to accurately predict which features the users of a new KM system will actually use and it is very easy to overwhelm them with complexity. As a rule, new ways of working should be supported by enabling technologies that are as simple as possible. Over time the absence of needed features will make itself clear and can be addressed. Often however, project champions worry that they will only get *one bite at the apple* and try to cram as much functionality as possible into the first iteration of a system. Remember, if you launch a system that adds value to the business, you will have the opportunity to enhance it over time. By starting simple, you allow those enhancements to be grounded in experience rather than speculation.

Understand and Play Your Part

As you embark together on your KM project, your IT partner will be looking to you, as the lead representative of the business function, to play a defined role. In general, that role will revolve around bringing the business to the table whenever and however needed to make the project a success. Bringing the business to the table will mean different things as the project progresses. Early on, it will mean securing executive sponsorship, contributing to the development of a business case, and identifying project stakeholders and resources. Later, it may mean driving efforts to harmonize processes, practices, or terminology across business areas. During vendor evaluation, your IT partner will need your help to ensure that the right people are present in demos and vendor evaluation discussions. You may also need to referee and/or escalate disagreements. Similarly, when design work commences, it will be up to you to ensure the right representation, to set expectations for the design phase, and to shape the discussion in the direction of a pragmatic, workable design. As system implementation and roll-out approach, you may be called on to secure business resources for software testing as well as the development and delivery of training and communication materials. Finally, once the system goes live, you should expect to participate in the on-going governance of the system including the collection and reporting of outcome measures and the identification of needed course corrections or enhancements. In summary, you will be a full and equal partner with your IT function in making the KM system you envision a reality and you, and your management, should understand the required commitment at the outset.

Understand the Role of IT in Your Business

Often we use the term *internal customer* to refer to the relationship between the IT function and the business units it supports, but this is not quite right. In the marketplace, customers have little or no obligation to their suppliers beyond paying on time and suppliers are considered to be justified in charging all that the market will bear. By contrast, you and your IT function have a strong, mutual obligation to do what makes the most sense for your company. This means that IT's responsibility is not only to your business case but also to the company's overall need for a manageable, secure, and cost-effective IT environment. To meet this need, a mature IT function will have established a range of standards and processes that your project will need to accommodate. Maybe your company has made a recent commitment to a single business intelligence and reporting platform. Assuming that platform meets your projects core requirements, you will be encouraged and expected to use it, even if your stakeholders prefer another product. Almost certainly, your company will have established an SDLC to ensure the quality and compliance of the systems it deploys. Your project will follow that process, even if some of the steps and deliverables seem unnecessary to you. You can rest assured that your IT partners feel a strong accountability to help you meet the needs of your business unit as expeditiously as possible—it is their reason for being, after all—but it is worth remembering that your own obligations go beyond those of the typical customer.

Conclusion

Technology is only one part of realizing the value of better knowledge management but it is an important one. As the imperative to do a better job managing critical knowledge becomes clearer to both health authorities and business leaders, so does willingness to invest. When we pair this with the rapid and continuing evolution of KM software products and supporting infrastructure there is ample reason for KM champions within a company to plan for success by thinking early on about how their efforts will scale. This requires forming a strong partnership with the IT function. That partnership will be critical to navigating a range of challenges including but not limited to (1) the broad diversity of users and content that tend to characterize KM systems, (2) the tacit and voluntary nature of knowledge sharing, and (3) the difficulty of positioning such broadly multidisciplinary systems within the organization. These can be daunting challenges but your IT partners are ready to help and, by preparing for them together, you can take your KM project from a great idea to a valuable new capability for your company.

References

Horvath, J.A. Implementing a successful knowledge management program. *Pharmaceutical Engineering: Knowledge Management Supplemental Digital Edition,* 26–30, 2014.

Horvath, J.A. *Working with Tacit Knowledge.* In Cortada, J.W. and Woods, J.A. (Eds.) *Knowledge Management Yearbook 2000–2001.* Boston, MA: Butterworth-Heinemann, 2000.

Lipa, M., S. Bruno, M. Thien, and R. Guenard. A case study of the evolution of KM at Merck. *Pharmaceutical Engineering,* 45(4), 2013: 94–104.

Nonaka, I. and H. Takeuchi. *The Knowledge Creating Company: How Japanese Companies Create the Dynamics of Innovation.* New York: Harvard Business Review Press 2008.

Pisano, G. *The Development Factory: Unlocking the Potential of Process Innovation.* Boston: Harvard Business Review Press, 1996.

Sternberg, R. and J. Horvath (Eds.). *Tacit Knowledge in Professional Practice.* Hillsdale, NJ: Lawrence Erlbaum Associates, 1999.

17

Knowledge Management Implementation: A Guide to Driving Successful Technology Realization

Doug Redden, DBA

CONTENTS

> IT Professionals have the opportunity to become key partners and cata-
> lysts to create successful KM programs. Following through from the
> last chapter, this contribution shares the perspectives of an IT strategy
> leader and offers frameworks that the IT function can leverage to accel-
> erate and enhance the KM value proposition.
>
> **Editorial Team**

Introduction

Organizations that are on a journey toward efficiency and innovation may
view knowledge sharing and reuse as one lever in realizing this goal. The IT
function can play a central role in a business-driven KM journey (Sharma,
2013); however, it is left to the IT professional to navigate the plethora of technologies available and work toward implementation options that capture,
connect, and reuse organizational knowledge. Complicating things further,

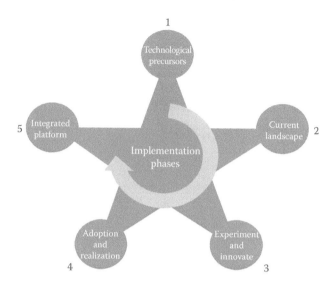

FIGURE 17.1
KM implementation star model.

each organization has different processes, tools, individuals, and goals, making each KM program unique to that organization, emphasizing a complex problem that each IT professional faces when partnering with the business. For example, the purpose of a KM effort might focus on the help desk function, cross-functional learning modules, driving community discussions among manufacturing sites, or even finding expertise from a global network. Your business colleagues should be considered as your subject matter experts in understanding the required capabilities of a KM solution, and in-turn will look to you as the IT expert to help navigate technology options, repositories, and integration components, leading to an appropriate deployment for your organization. This chapter offers an implementation star model with five focal areas (Figure 17.1) intended to help provide guidance as to *how* one might conduct a technology implementation for an organization. The model presented above is based on experiences from IT professionals that have gone through large KM efforts, offering a suggested IT-centric approach applicable to KM. An in-depth look at each point in the star is provided below.

Technological Precursors

Initial conversations with the business should be void of technology talk, and rather focus on foundational elements such as the KM programs' intended value offering, initial processes that are in scope, and the role the business sees IT playing. This helps to appropriately set the stage for follow-on technology discussions. It is important to understand the underlying motivation around a KM effort; therefore, the IT professional should seek full clarity around the following areas:

- Executive sponsorship
- A KM definition from the business (including benefits)
- Business benefits expected to obtain
- Processes that the business considers critical that the technology will support

Commitment toward a KM initiative will be evident in the strength of sponsorship. Sponsors effectively communicate the organizations vision, goals, and expectations to the team throughout a program (PMI, 2010). As the IT partner, having a clear understanding of key sponsors and advocates is a necessary step. Sponsors are your biggest allies, as they will provide support in *use* of the technologies being invested in. In turn, having a well-articulated definition for KM and its intended use helps to ring-fence the scope of the initiative. The following questions may offer a starting point:

1. What problem is the business looking to solve?
2. What are the critical success factors for the KM effort?
3. What is the outcome the business seeks to achieve?

If the KM effort is not clearly defined, this can be the first place IT offers help. The IT professional with requirements gathering, user experience documentation, architecture, and business case development, are well equipped to ask the right questions related to scope and boundaries of the effort. Doing this can foster a stronger relationship with the business up front and demonstrate the effectiveness of IT outside of only technology delivery.

Current Landscape

How KM is defined will determine the boundary conditions for the IT effort. Once equipped with this, consider these three critical questions:

- What technologies does the organization currently have that provide KM capabilities?
- What additional technologies are required to meet the goals of the KM effort?
- What systems need to be considered to provide integration for seamless user experiences?

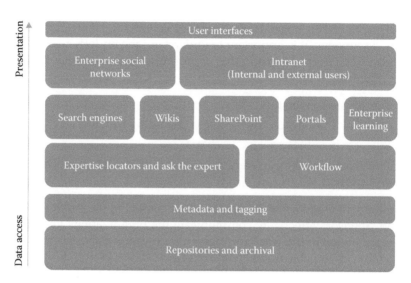

FIGURE 17.2
KM IT capability model.

Classifying each technology by capability will help. A KM capability model (Figure 17.2) can help with mapping current systems for your KM effort. This model has been developed based on KM initiatives at large organizations. Depending on the size of the organization, this exercise may not take much time (small organization with few technologies), but in a larger, more mature environment, you may have several content management repositories, workflow tools, Wikis, portals, search engines, and more. In addition, documenting the interdependencies between systems is important as you think about creating a holistic picture of the KM landscape.

This current state map demonstrates, through reuse of existing technologies, proper fiduciary responsibility by the IT professional. It will also demonstrate accelerated implementation and reduced risk given the systems are already in your current landscape. Look to understand the user community, scaling required, and footprint of any new user groups to ensure appropriate service level expected by the business.

Once you have an appropriate handle on the current state of technologies, there is an opportunity to now see what is missing and begin to identify what additional tools may be needed. For example, e-mail may be used as the primary discussion thread tool, but for future state, moving toward intranet discussion threads will allow for broad consumption and response. Having these type of *trade-off* discussions where you discuss the benefits of one technology capability over another (e-mail vs. discussion thread) with your business partner will be critical as you look to develop and implement the required future state.

KM tools that are bundled and integrated provide far more value to the end user than a set of distinct and independent technologies. It is the concept of one-stop-shopping for KM, instead of users navigating separate system logins, screen orientations, and missing data integrations between capabilities. Developing toward integrated platform architecture should be considered a long-term goal, with a clear understanding that it may take several years to get there.

Experiment and Innovate

During your KM journey, demonstrating value through the use of current technologies provides partnership trust between IT and the business offering benefits such as faster speed to market and higher return on investment as two examples. In conjunction, pushing current technology boundaries and considering new innovations may prove valuable in advancing the organizations KM agenda. Spend time researching and connecting with external organizations to find emerging capabilities that may be suitable. Offer experimentation ideas to your business colleagues emphasizing the learnings that will come out of any experiment, successful or unsuccessful. Experiments also demonstrate to the business what may be possible and lead to follow-on experimentation.

Experimentation can prove beneficial when clear goals have been set. If the experiment(s) is seen as a loosely structured activity that burns through funding, the business will not sign up. Here is a list of the following activities that should be included for an experiment:

- Develop a short-term project plan (roughly 8–12 weeks depending on effort).
- Articulate the goals for the experiment. This could be as simple as saying it is a *discovery exercise* to learn more about internal process, or may be very specific such as *integration of three content sources with search included.*
- Provide continual updates back to the business as to the learnings to date and budget projections.

Content integration, personalization, and mobile are the three possible areas to explore that have seen vast improvement in technology alternatives and opportunity to provide efficiency gains. As you begin to frame your experiment(s), bounce them off of your business partner, discuss vendor options, timing, budget, and most importantly discuss the advantages of learning from the experiment.

Adoption and Realization

A common misconception within IT is that success is defined as *technology installation* and that IT has little to any role in accountability for realization or user uptake. This is simply not true and successful KM programs have accountability from IT to partner with the business to achieve realization. Avoid the temptation to follow this historical pattern. The business will not see any value from the system(s) put in place without user uptake, and IT has an opportunity to drive user adoption through several change mechanisms.

Training, support, user value sharing, and active communication are four areas that should be considered:

- Develop training materials and have go-live training sessions that help gain momentum in the right direction. Continue reinforced training sessions to *make it stick*. Try hosting lunch-n-learn style reinforcement sessions allowing for users to continue education of the system, asking questions, and offering possible learning for advancing the next version of the system.

- Create a small core team that is focused on offering one-on-one assistance on system functionality and use. In addition, this team may even be called a *realization team* focused on achieving the outlined business outcomes. Consider a direct contact number with 24/7 support for a few weeks during go-live. Successful *realization teams* go beyond system functionality, rather they relate to the business challenges and necessary steps that an IT system can offer in being more effective and efficient.

- Users need to understand how this system makes it easier or faster or less burdensome for them. Emphasize this in training and communications campaigns. Provide executive sponsors with speaking points and testimonial accounts that offer benefits for each individual.

- Develop a communication plan that accounts for sharing features, functionality, and benefits currently offered by the system, but also look to share the future improvements that users will see. Without showing the progression, users may see initial roll out as limited in function, but this can be avoided by sharing the big-picture view for what is to come. The communication plan should be tightly aligned to the overall program's change management plan.

Integrated Platform

At this point, you have identified your current technology landscape, the required new technologies necessary, and innovation efforts that are intended to advance your KM agenda, and have an adoption or realization plan. Individual solutions can serve as the foundation for KM capability development, while considering how to integrate and provide a more seamless interaction. It is important to help the organization see *what is possible* by implementing and realizing this future state. Be able to articulate a future state technology platform (collection of capabilities or technologies) that will serve as your IT *north star* as you develop a plan of activities to get there. An example of a conceptual platform is provided in Figure 17.3. This figure is based on KM program architecture implemented within a large organization.

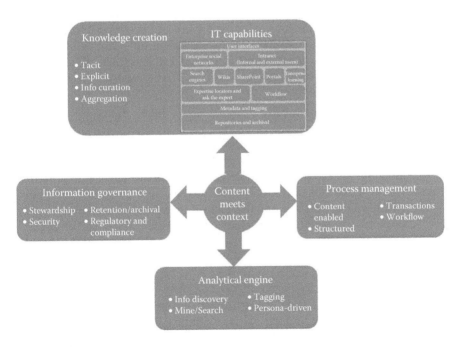

FIGURE 17.3
KM integrated platform (conceptual).

Working from the top box and moving clockwise:

- With *knowledge creation*, platforms may consist of a collection of KM capabilities and tools for capturing explicit as well as tacit knowledge. This could be in the form of document retention to project after action reviews, or even critical knowledge retention repositories.

- Knowledge is created and leveraged throughout daily business process activities. Through transactional activities or even workflow driven, *process management* offers ties to both technology and manual knowledge creation. Any technology solution should be *in the flow* of existing processes, meaning they should be synchronized with the way a person works and not add additional burden. In fact, they should streamline activities for a user.

- The bottom of the diagram, *analytical engine* depicts the analytical support engine that can begin to provide information discovery, rich text search, advanced tagging, and offer knowledge back that is personalized based on user type.

- On the left-hand side is *information governance*. This is critical as you think about information stewardship and user rights to certain information. The lifespan of information and any possible regulatory obligations to retaining or even archiving information falls under governance as well. This is not a technology capability, rather the connection to functions/groups that drive proper use of the KM platform.

- As these four come together *content meets context*, meaning, content that is captured can now be offered back to users based on contextual aspects. An example of this is finding an expert within the organization. By having the appropriate business process (process management) to identify expertise, information can be captured about an individual's skills, current projects, and so on. Knowledge capture, managed through the proper information taxonomy and guidelines (information governance), and mined (analytical engine), for users to find.

This diagram, although conceptual, begins to offer a glimpse into what it might be like taking it to the next level and showing the technologies that support and interconnect to deliver not only content, but also relevant contextual knowledge back to users. That is something your business partners can get excited about and feel is worth the journey to get there, knowing there is enormous opportunity in moving the needle toward a rich set of organizational KM capabilities.

Closing

Offered in this section is a recommended path toward supporting your KM partner in technology implementation. KM technology delivery nirvana may be a fully integrated set of technologies that capture, connect, and reuse organizational knowledge as needed. As a leader within the IT group you must set an appropriate plan in place that takes steps toward the intended future state, which appropriately binds technology implementation with user adoption and realization. This is easier said than done, however, and by incorporating the KM implementation star model (Figure 17.1), you will be better equipped to tackle such a momentous effort. Go for IT!

References

Sharma K. 2013. Knowledge management and its impact on organisations: An assessment of initiatives in the industry. *Journal of Technology Management for Growing Economies* 4(2): 51–66.

Project Management Institute. 2010. Executive Engagement: The Role of the Sponsor. Accessed February 17, 2016. http://www.pmi.org/Business-Solutions/~/media/PDF/Business-Solutions/Executive%20Engagement_FINAL.ashx.

Section IV

Practices and Case Studies in Enabling Knowledge Flow

This section contains a collection of case studies from across the biopharmaceutical industry which profiles a variety of current knowledge management (KM) efforts. Some of the stories shared are broad and far-reaching KM programs, whereas others are localized. Some are sponsored by various functional leaders, whereas some are grassroots initiatives. To aid review, each of these efforts has been coarsely categorized by the editors to the *product lifecycle phase* in which it is focused (per ICH Q10). Furthermore, each case study has also been mapped to indicate which specific *practices* have played a role. The following matrix and associated mapping should prove to be a useful reference guide for KM practitioners to see how various companies have approached KM.

It is however important to note that the case studies that follow are not necessarily a *how to* for a given *product lifecycle phase* or *practice*. Rather, these case studies represent real and tangible KM efforts, often in the context of a business problem, that show how the author and their company leveraged knowledge management to address a specific business problem. These studies also serve to highlight various organizational drivers for KM, definitions of KM, and targeted outcomes.

These case studies do not necessarily represent the entirety of a firm's KM efforts, nor what functions or lifecycle phases they focus on, nor where in the lifecycle they focus. These are included as real-life examples to illustrate and teach. Finally, keep in mind that not all language and descriptors used in the case studies may be consistent with how the editors have described *practices*, *pillars*, and *enablers* in Section III.

TABLE 1

Mapping of Case Studies to *Product Lifecycle Phase*

Chapter #	Author	Case Study Title	Pharmaceutical Dev't	Technology Transfer	Commercial Mfg	Discontinuation
18	Sandra Bush and Matthew Neal	Let us Talk About Knowledge Management—Learning from the Library of Alexandria Disaster				X
19	Anando Chowdhury	Rapid and Robust Product Development Powered by Knowledge Management Capability	X	X		
20	Eda Ross Montgomery, Mani Sundararajan, David Lowndes, and Gabriele Ricci	Knowledge Management Case Study: Using Near Real-Time Data Analytics and Performance Metrics to Ensure a Robust and Resilient Supply Chain			X	
21	Beth Junker	Knowledge Management Elements in Support of Generation of Chemistry, Manufacturing, and Controls Regulatory Documentation	X			
22	Sian Slade and Catherine Shen	A People Approach to Managing Knowledge: Who Are You Working For?			X	
23	Phil Levett	Developing a Lesson-Learned Process—Where Lessons Are Learned: A Case Study of Pfizer Pharmaceutical Sciences	X	X		
24	Dave Reifsnyder, Kayhan Guceli, and Kate Waters	Capturing Critical Process and Product Knowledge: The Development of a Product History File			X	

(Continued)

TABLE 1 (*Continued*)

Mapping of Case Studies to *Product Lifecycle Phase*

Chapter #	Author	Case Study Title	Pharmaceutical Dev't	Technology Transfer	Commercial Mfg	Discontinuation
					Product Lifecycle Phase	
25	Renee Vogt, Joseph Schaller, and Ronan Murphy	Communities of Practice: A Story about the VTN and the Value of Community		X	X	
26	Paige Kane and Chris Smalley	Identification of Critical Knowledge: Demystifying Knowledge Mapping		X	X	
27	Adam Duckworth, Vince Capodanno, and Thomas Loughlin	The Practical Application of a User-Facing Taxonomy to Improve Knowledge Sharing and Reuse across the Biopharmaceutical Product Lifecycle: A Case Study		X	X	
28	Marco Strohmeier, Christelle Pradines, Francisca F. Gouveia, and Jose C. Menezes	Knowledge-based Product and Process Lifecycle Management for Legacy Products			X	
Other Chapters						
13	Paige E. Kane and Alton Johnson	A Holistic Approach to Knowledge Management: Pfizer Global Supply		X	X	
14	Martin J. Lipa and Jodi Schuttig	KM Evolution at Merck & Co., Inc. Managing Knowledge in Merck Manufacturing Division		X	X	

Note: X indicates an area of focus.

TABLE 2

Mapping of Case Studies to *KM Practices*

Chapter #	Author	Title	Community and Network	Content Management	Taxonomy	Lessons Learned	Expertise Location	Exp Transfer and Retention	Other KM Practices
			Knowledge Management Practices						
18	Sandra Bush and Matthew Neal	Let us Talk about Knowledge Management: Learning from the Library of Alexandria Disaster	+	+	+	+			+
19	Anando Chowdhury	Rapid and Robust Product Development Powered by Knowledge Management Capability	+				+		+
20	Eda Ross Montgomery, Mani Sundararajan, David Lowndes, and Gabriele Ricci	Knowledge Management Case Study: Using Near Real-Time Data Analytics and Performance Metrics to Ensure a Robust and Resilient Supply Chain							++
21	Beth Junker	Knowledge Management Elements in Support of Generation of Chemistry, Manufacturing, and Controls Regulatory Documentation		++	+				++
22	Sian Slade and Catherine Shen	A People Approach to Managing Knowledge: Who Are You Working For?	+	+			+		+

(Continued)

TABLE 2 (*Continued*)

Mapping of Case Studies to KM *Practices*

Chapter #	Author	Title	Community and Network	Content Management	Taxonomy	Lessons Learned	Expertise Location	Exp Transfer and Retention	Other KM Practices
		Case Study	**Knowledge Management Practices**						
23	Phil Levett	Developing a Lesson-Learned Process—Where Lessons Are Learned: A Case Study of Pfizer Pharmaceutical Sciences				++			+
24	Dave Reifsnyder, Kayhan Guceli, and Kate Waters	Capturing Critical Process and Product Knowledge: The Development of a Product History File		++					+
25	Renee Vogt, Joseph Schaller, and Ronan Murphy	Communities of Practice: A Story about the VTN and the Value of Community	++				+		
26	Paige Kane and Chris Smalley	Identification of Critical Knowledge: Demystifying Knowledge Mapping		+					++
27	Adam Duckworth, Vince Capodanno, and Thomas Loughlin	The Practical Application of a User-Facing Taxonomy to Improve Knowledge Sharing and Reuse across the Biopharmaceutical Product Lifecycle: A Case Study		+	++				

(Continued)

TABLE 2 *(Continued)*

Mapping of Case Studies to *KM Practices*

Chapter #	Case Study		Knowledge Management Practices						
	Author	Title	Community and Network	Content Management	Taxonomy	Lessons Learned	Expertise Location	Exp Transfer and Retention	Other KM Practices
28	Marco Strohmeier, Christelle Pradines, Francisca F. Gouveia, and Jose C. Menezes	Knowledge-Based Product and Process Lifecycle Management for Legacy Products	++	++				++	++
Other Chapters									
13	Paige E. Kane and Alton Johnson	A Holistic Approach to Knowledge Management: Pfizer Global Supply	++	++			++		++
14	Martin J. Lipa and Jodi Schuttig	KM Evolution at Merck & Co., Inc. Managing Knowledge in Merck Manufacturing Division	++	++	++	++	+	++	+

Note: "+" indicates degree relevance (none, + low, ++ high).

18

Let's Talk about Knowledge Management—Learning from the Library of Alexandria Disaster

Matthew Neal and Sandra Bush

CONTENTS

> History is meant to teach, yet how should organizations enable themselves to be ready to learn from their past experiences? This case study challenges the preconceptions around the value of social media and other societal norms of how we share and use what is currently known by the complex collection of individual contributors within a corporation. It discusses how these individual contributions can be harnessed to establish a collective corporate *memory* by outlining the key elements of an effective learning organization are.
>
> **Editorial Team**

Yes, the Library of Alexandria (El-Abbadi 2016) (we'll get to this). But first...

The typical approach to knowledge management in the pharmaceutical industry is focused on a regulatory view, big data, or product lifecycle management as outlined in ICH Q10 Pharmaceutical Quality System (ICH Harmonised Tripartite Guideline, 2008). This is a valuable way to think about knowledge management, but lacks the important detail that work is done through and with *people*. Knowledge is exchanged through and with people. To that end, it was important to Global Manufacturing Operations within a large biotech to focus knowledge management efforts on the exchange of experiential information to build organizational knowledge and improve the overall operation.

The highly regulated nature of the pharmaceutical industry ensures development of robust processes that mandate content is maintained, transferred, and accessible with a focus on regulatory compliance, while turning much less focus on learning in the organization. In this context, learning is not training. Any pharmaceutical company has highly trained staff, but how are companies sharing their experiences and learning as an organization to ensure mistakes are not repeated? How can we protect that knowledge and embed it in the way we operate so we do not lose the knowledge?

Our journey into the development of a knowledge management capability began with a simple question: *How can we avoid repeating the same mistakes?* To go one step further, how can knowledge management enable an *effective learning organization*? Through external research and benchmarking outside of the pharmaceutical industry, we defined an effective learning organization by the following characteristics. It is through these characteristics that an organization can fully leverage the knowledge it gains from experiences to improve its performance. We will explore each of the following characteristic in further detail:

- Thinks systematically and solves problems with analytical thinking
- Learns from its successes and failures and from those of others
- Takes initiative to capture and share knowledge
- A willingness to change based on experience
- Leaders are teachers and improve their teams through coaching
- Commitment to personal mastery (Senge 1990)

It seems intuitive that an organization would find it valuable to be an effective learning organization (ELO), but it would seem, that unless they can rally that desire around a specific problem, they will not achieve ELO performance. ELOs benefit from reduced error rates, have integrated their experiences into their processes, and spend less time *relearning* from mistakes. We determined that they did not have an effective knowledge management infrastructure. This was evidenced by a lack of high quality knowledge objects, a lack of a searchable knowledge repository, and no formal learning group structure or communities of practice. As a consequence the company was not effectively drawing on, distributing, or consuming experiential knowledge from either internal or external sources and was not supporting the flow of knowledge within and across their sites and networks. To correct this gap, our organization committed to delivering a knowledge management capability that enabled purposeful learning and sharing.

What Does an Effective Learning Organization Look like?

1. *Thinks systematically and solves problems with analytical thinking.* The most important aspect of solving a problem is truly understanding the problem. A variety of methods enable a methodical approach to root cause identification; however, the approach taken should be of sufficient rigor to ensure that the team or investigator does not jump to solutions before truly understanding the problem they are trying to solve in the first place. An effective learning organization takes understanding why a problem occurred as seriously as solving a problem and is *willing to get uncomfortable so they can assure open communication.*

2. *Learns from its successes and failures and from those of others.* Once an organization understands why a problem occurred, it is their mission to learn from the experience so they do not repeat the actions that led to the problem. You can relate that to children touching a hot stove. They experience the pain of a burn as a result of touching the stove and learn not to touch the stove. In an organization it is not enough to be the only one that has learned not to touch the stove because in large complex organizations, many people face similar problems, and we cannot afford to learn these hard lessons one by one (or have everyone running around with burned fingers). Organizations must harvest learnings and ensure that they share them, so that many people across a variety of boundaries will benefit. They must also actively seek out peer, cross-functional, and site experiences. This active seeking of knowledge helps the organization minimize repeat failures across the organization.

3. *Takes initiative to capture and share knowledge.* Knowledge sharing is not a passive activity. It takes time for an organization to hone the skills to share and communicate what they have learned in a meaningful way. Rewarding people for knowledge sharing activities is more effective than setting knowledge sharing goals. This typically results in higher quality knowledge—at least—qualitatively, considering breadth, depth, and the frequency of update. We have also learned that simple knowledge-sharing templates coupled with a knowledge-content repository enable sharing and consumption of knowledge across functions and sites.

4. *A willingness to change based on experience.* As intuitive as this characteristic may sound, manifesting willingness to change is more difficult than it might seem. Often we have people, technology, and processes that are impacted by a change that needs to be made as a result of learning. Ensuring active change as a result of learning from failures is a key imperative for any effective learning organization. This level of commitment to change and improvement must be embraced at all levels of the organization. It requires a level of commitment and energy throughout the organization that should not be marginalized.

5. *Leaders are teachers and improve their teams through coaching.* We depend on our leaders to help us lead the change. That means they need to support their staff in sharing and utilizing knowledge. They need to coach their staff to get to the true cause of problems and encourage them (strongly encouraging or even *requiring* participation is strongly encouraged) that they seek first to *learn what the organization knows about similar problems before seeking to solve a problem*. This also means that they support leveraging organizational knowledge before they make changes and give their staff enough time to proactively seek learning opportunities. They should also emphasize the importance in learning with their actions.

6. *Commitment to personal mastery (Senge 1990).* The concept of a learning organization is dependent on the individual striving to improve. Having a commitment to personal mastery of their discipline and striving for that mastery ensures that individuals continue learning and seeking opportunities to gain more knowledge and experience.

What Does a *Good* Knowledge Management Capability Look like?

Fun Fact: It is speculated that the entire world's knowledge was at one time stored at The Library of Alexandria, Alexandria, Egypt. It is also speculated that it was destroyed several times at different points in history. Sometimes that destruction was unwittingly self-destruction. It is unknown what information we had to relearn multiple times or, worse yet, may have yet to rediscover.

How Do You Start from Zero?

Our company did what any other company would do; they engaged experts in knowledge management and asked to hear more about their knowledge management capability. Key executives and key staff visited companies outside the pharmaceutical industry. They also conferred with a leading consulting firm that runs their business through effective knowledge management. In the end, they were able to identify areas of focus. There were three major areas to develop:

- The Marketplace: an accessible portal
- The Customers: user communities and learning groups
- The Products: knowledge objects

The MarketPlace

The first activity was to develop an easy to use, searchable content repository that allowed for explicit knowledge to be documented with supporting evidence, keywords, and metadata based on a simple taxonomy. This was accomplished through the use of Microsoft SharePoint, workflows, and enterprise search tools. The initial system was developed as a pilot and was used to illustrate the possibilities and benefits a more robust system could provide. This system was used as a template to engage the global information systems team and gain their support for the development of a more robust system that was aligned with the company's information systems roadmap and could be supported with *in house* talent. This way of iterating the development of the repository also allowed the team to make adjustments and design improvements to the system as the content was being developed.

The Customers

Even with the strong endorsement of a group leader with a targeted audience—the need to campaign, educate, and inform that audience should not be marginalized. Knowing the audience and paying special attention to the development of a reward and recognition program that promotes knowledge capture and use will build excitement and momentum. The expansion of the capability to a wider organizational construct is still a daunting task. Our organization invested a considerable amount of effort into supporting the customer groups and nurturing the correct knowledge-sharing behaviors.

The Products

Our products, like One Point Lessons and Alerts (our knowledge assets), were artifacts of this endeavor. (We cover both in much more detail later ... keep reading).

Negative Events Are Learning Opportunities

Complicating factors and inciting events impact KM. A significant negative event will drive a learning agenda within all other facilities once a root cause has been identified and the event has been fully investigated. Dynamic learning agendas are developed when leaders and staff actively seek content from the knowledge management system (KMS). Employees will also review shared practices and disseminate information across their network. Knowledge captured in the flow of business events feeds continuous improvement. An effective and robust KMS infrastructure and support system will take the process described above and not only put it in practice for isolated events within departments, but share knowledge with other departments in the enterprise and create *synergy* (Ward 1918).

Lifecycle

An interesting word when it comes to knowledge and a key focus for ICH Q10. We could certainly agree that information flows out of the entire lifecycle. Information is a building block of knowledge. But just how long is this lifecycle? In the pharmaceutical industry, product lifecycle outlasts most careers. The process of building toward a licensed product can take 15 years—and that is just the beginning. Then you follow the product, and follow the product, and follow the product, and new people answer questions about it because the lifecycle of a product is much longer than the average individual's career span with a single company. Passing on knowledge over a long period of time is essential and yet, it seems the pharmaceutical industry is just starting to think about keeping track of it.

Whether or not the known world's knowledge was contained in the Library of Alexandria or not, the fact remains that for as long as human beings have been able to pass along information, they have done so as a means of survival. Knowledge retention is essential for survival. How to store that information, however, has really only been an issue as new favorite mediums have developed. One can imagine that before written and oral tradition were possible (because humans did not share the ability to speak, write, or draw), perhaps tradition had origin in illustrating to offspring through example and hoping that after enough repetition, they learn how to survive. Shared communication, language, and written tradition certainly made this easier.

In the pharmaceutical industry, in particular, standard operating procedures (SOPs) are a known quantity and essential to many of the processes that drive the business. However, the quantity of information about SOPs, their origins, their importance, their nuance, and their shortcomings are rarely shared or understood. In addition, these are the documents that are held up for compliance and adherence to practices. Staff departures, departmental moves, promotions, new jobs, and new assignments—the things that happen to the humans who create the products and procedures—are rarely, if ever, captured in a way that is extended to those that follow. Without a comprehensive KM process and mindset, we have minidestructions going on all the time and only have a chance to realize it when a question is raised for which no answer is readily available. Although we can never anticipate everything, KM gives organizations a chance to put as much in place as possible to avoid such disaster.

Focusing on *in the Flow* Delivery of Knowledge

How does one put knowledge in front of staff in an effective way? Before social media came around, that would have been a difficult question to answer. But now, in your personal life, you almost cannot avoid information about yourself and everyone you know, every day, all the time. Why do we settle for less when it comes to the information we *actually* need to get our job done and how do we do it in a compelling way?

Social Media for Business

Social Media (Wikipedia n.d.) has become its own entity. Whether it is Twitter, Facebook, Instagram, or SnapChat, (and any other platforms that rise and fall in the time it takes to publish this book), the category seems to be defined as everyone's platform. Social media has become a platform where even regular people have a voice. Social media implies a collective. Their rise to prominence in any age group or demographic is an indicator of attention, user preference, and social prominence and that landscape is shifting rapidly.

When it comes to business, the prevailing mindset (perhaps this is generational, current workforce, etc.) is that social media is only for *fun* and no real work can come from it. Dispelling this notion is very difficult even though it is only been in our vernacular for a decade or so and only ubiquitous for less than five years.

The fact is, collective work is extremely valuable in business. It is collective work, defined here as more than one organizational unit collaborating with another, which enables our organizations to move information quickly, not suffer the same mistakes repeatedly (there is substantial return on investment in this alone), and allow everyone who participates to share in the victory and contribute in a meaningful way and gain appreciation for contributions that matter (Morgan 2014).

Webster defines social media very simply as "forms of electronic communication (as websites for social networking and microblogging) through which users create online communities to share information, ideas, personal messages, and other content (as videos)" (Webster n.d.). So, to label social media good or bad seems misguided. This is the uphill battle that KM practitioners face. It may be that the best sponsors for a knowledge management effort, and specifically those that involve social paradigms, are those who use social media at home to solve problems and *just get it*. They come from a place of acceptance of the power of the tool, not proving return on investment (ROI). They know that ROI for knowledge exchange is soft and difficult to quantify.

ON ROI OF NEW STUFF YOU DO NOT UNDERSTAND

A newly appointed CEO of a prominent social media company gave a lecture. His opening story was about his experience using e-mail within the first few weeks of taking over. He described how the new e-mail software he was using did not function as he was expecting, because he had nowhere to organize his e-mails with folders. After a few days of frustration with this, he reached out to his team to find out why they had chosen to make the system function this way. He could not very well continue being the lead champion for software he himself did not find useful. Once he discovered (after a logical explanation from developers) that his e-mail box was using archiving and search features that were much more powerful than his own memory or folder filing system, he was hooked. Folder free and boldly going where he had not gone before, he learned that he was saving time he had not previously thought he was wasting. ROI for that CEO? Tough to calculate because he never really gave much thought to how much time he was wasting. Organizations moving forward without KM will never realize how much they are missing until they need to discover it all over again.

When upper management dismisses social media as something their kids are doing and it does not apply to them, you have a tremendous obstacle to overcome. Now, you are faced with the same challenge that start-up CEOs have to endure when attempting to get venture capital or private funds. The difference is, you are already aware of success—but you have to prove it to the nonbelievers that have not yet bought into the benefits of social media enough to justify the investment relative to the balance of the project portfolio.

Let us start with an example that occurred in our organization. On one occasion, an employee was packing a column of resin and was not sure if the column was packed tight enough to be effectively used in filtration operations. The column was clear and the compaction of the resin could be visually seen through the side of the column. The staff member took a picture of the column and posted it on the social media networking site to see if the column pack looked as dense as it should. Collaborating with network experts through visually verifying the effectiveness of the column pack prior to use would help keep production on time and save product and materials. Within a few minutes of posting it, and well within the threshold of time to have a *go or no-go* decision on using the column and keeping the batch on time to serve patients, several responses came in reassuring them that the resin pack was visually normal. The column was used without production delay and in the end was within all parameters. A brief five-minute exchange—question posted, image posted, experts respond, crisis averted—saved the company time, effort, meetings, and ultimately, money. The result was education not only for the worker who posted the question, but everyone who would work in that area again. Photos to show what a correctly packed column looks like were added to the SOP and created a working *knowledge* that would only have come with experience.

It is examples like this one that lead others to join and participate. Critical mass is the lifeblood of social engagement and interaction. If Facebook only had a few thousand members, it would not stay interesting for very long (see box below). It is imperative for social networks to grow and continue to include more and more participants in order to be self-sustaining, self-correcting, interesting, and useful. Leaders play a key role in *making it okay* or encouraging the type of behavior they want to see from their organization. Simply setting up a social network and turning it on will not create the critical mass needed to make it useful. Working out loud needs to become the norm.

> There are graveyards of abandoned social media sites that lost attention and became useless dating back almost to the start of the Internet. AOL, Prodigy, Chat Rooms, MySpace … the list goes on and on. So, though history will certainly have a place for Facebook, it may not stay that way for long as different generations start to decide how and where they want to connect. But there will always be a need for critical mass to make it effective and stay a part of daily life. Think of coming into your office every morning and literally "plugging in" to the rest of your organization. It might be overwhelming at first, but imagine how quickly problems could be solved!

Behavior change is at the heart of bringing social into business. Historically, the mode of operating in companies was, in many ways, a selfish behavior. People generally operated under the assumption that sharing the knowledge that only they knew made them obsolete—or at least giving it away somehow diminished their value or contribution. Keeping information close to the vest made them valuable. Well, they were only half right. The knowledge was valuable. What they did not know (and maybe in some cases, still do not realize), is that when they shared it, the knowledge itself gained value. The sharer also gains value—even if it is just a *like* on their comment. That small gesture can trigger the pleasure center in the human brain (Ritvo 2012). The risk is low. The benefit for the company is huge. Knowledge gains value when it is written and shared. It gains even more value when it is consumed and results in changes that help the company.

This behavior does not need to be taught to people who never knew the world without the Internet. The *Millennials* that are entering the workforce today will find it laughable that people worked in isolation. The sharing economy (*The Economist* 2013) (Kessler 2015) and the value of knowledge in a collective environment, without borders, without time zones, and without restrictions is something that comes naturally to the children of the iPad era. Not only can a two-year-old operate an iPad, they do operate an iPad. As soon as they are able to read, the expectations of interaction with others will be established. Working in an environment of open sharing will come naturally. That is not the case for current purveyors of knowledge management and sharing. It is a short term (though very steep) learning curve for many of the information workers that currently hold positions in corporate institutions. Using a variety of techniques, these information workers can be influenced by the value knowledge sharing with the processes, people, and technology can bring. What follows are a few examples of such techniques.

Let us explore what happens after a major marketing application. In most cases, the large team that came together to create the submission would gather notes and discuss how things went down. This is sometimes called a *postmortem* analysis. Quite a grim name and equally grim experience if you have been through it. It used to entail an endless stream of PowerPoint slides that detail the happenings of the filing. One of the biggest *attractions* of a postmortem slide deck was the *lessons learned* section. Inevitably, someone would declare at some point that there were things we really will *never do again*. Whew! Are you not glad that nightmare is over? Here is the problem—that declaration took place in a conference room and that slide deck was put in a file share or a document management system (DMS) somewhere never to be seen or heard from again. And alas, if you were not present at the meeting, you missed the knowledge exchange, as well.

With knowledge management in place and functioning well, those lessons can reach the entire community, they are searchable, and they should enable change ensuring that history does not repeat itself. In theory, your postmortem becomes more like a celebration because you did not have to make the same mistakes your predecessors made in the same situation. Learning takes place and collective wisdom grows exponentially. We should not only stand on the shoulders of giants, but read over their shoulders to save time.

Here is a brief overview of some tools we used to lay that bread crumb trail for future travelers on the same path.

One-Point Lessons

There was once a freezer that held multiple finished lots of drug product. The manufacturer of that freezer added a very secure door mechanism. In order to close the door properly, one had to lean on the door and hear an audible click sound. This was probably in the instructions or in the owner's manual that came with the freezer. However, after the installation of the freezer, the manual was not seen by floor staff again. As a result, this particular freezer was not closed properly on several occasions. The only explanation for that is the instructions on how to properly close that freezer were not shared.

That is where the one-point lessons come into play. The beauty of the one-point lesson was that they only contained, you guessed it, ONE problem and one solution. In the case of the freezer, a one-point lesson was developed and hung on the door of the freezer that very clearly stated that in order to properly close the door and avoid destruction of the lots contained in it, one had to use two hands, close the door, and listen for an audible click to make sure that the door was properly sealed. There were even pictures of a freezer door that was not properly closed and had not sealed and one that was properly closed.

Would this work if you hung a thirty five page owner's manual on the door? Even if you put a tab on the page that described the proper closing procedure? Probably not. One problem, one solution = one-point lesson.

Once established, these simple lesson artifacts start to become a very good way to distill and share information that is essential for getting the job done. The brand itself starts to catch on, and soon people who are working come to expect that more tasks have established one-point lessons. And then, a strange thing starts to happen. People start to make them for themselves. Content starts to proliferate and good ideas start to spread.

One-point lessons were the knowledge sharing work horse. The simplicity of the document allowed it to be shared easily, but also challenged the authors to focus on the problem and condense it to a point of crystal clarity. As a result, this type of knowledge object was developed and consumed more than any other type (approximately 60% of the assets in the repository were one-point lessons, and one-point lessons and alerts were consistently the top assets queried). The sharing of knowledge was done through the content repository, e-mail, and social media, at the point of use, one-on-one exchanges, learning groups, and shift change meetings.

Content Repositories, Taxonomies, and Search

Once you start creating content and putting it somewhere, the attention needs to turn to being able to quickly surface that information and present it in a meaningful way. Here we hit another major behavior change challenge.

It is strange how people's habits become so pervasive. The first time we met a computer visual filing system—for most of us, Windows 2.0 at the earliest (Microsoft 2015)—was sometime in the late 1980s. This meant that our *stuff* was in folders somewhere. We had to know what the folders were to find things. Since we named the folders, well, of course we would be able to find them later. How many of these early adopters named all of their documents *mydocument1*, *mydocument2*, and so on. The concept of folder naming and file naming became crucial to ever finding anything ever again. As a result of that urgency, behaviors were ingrained and for many, many years after that—it was just how things worked. Especially in institutions of learning and businesses where Microsoft Windows became the most used operating system, we just followed the way people did it because computers first entered the workplace. Not much has changed even today—almost 30 years later.

That is where a new paradigm comes into play. Search enables users to find their stuff. The first few times organizations launched search, it involved new terminology like *Boolean* (Elmer E. Rasmuson Library 2014). It almost requires you to be a computer expert to find your stuff. You had to learn a different language to navigate this new world. So, needless to say, people still relied on their old favorite—file and folder naming. The problem with that is, you had to know what your organization methods were, and you had to tell the next person that needed to find your stuff how you stored them. This perpetuated inconsistency at a level that generated an absolute mess.

In comes natural word searching, Google. The algorithms and Boolean logic happened in the background and changed the landscape of search from a mathematical pursuit to a natural experience. People do not naturally run queries—they ask questions. This is an important shift, and knowledge management can benefit greatly in any organization once they accept this shift and think about employees as a vast resource.

As many companies are slow to adopt new technology, we are stuck between the old world reality and the new world expectations. Imagine that you had to actually conduct a Boolean search of an obscure actors name when you are having a bar stool discussion about a random movie you watched a few weeks ago. Insane, right? You just Google it (ok, maybe if you are really into it, you have the IMDB—Internet Movie Data Base app on your phone).

One of the keys to a good knowledge management system is good search. Natural, relevant search results drive adoption and avoid redundant work. Good search should not be a slave to taxonomy. Taxonomies change and search tool must be dynamic enough not to rely on a static taxonomy that is never going to work for your whole organization.

Global Networks

A pervasive problem in large organizations is the naturally occurring phenomenon of silos. Functional silos are the sworn enemy of cross-functional collaboration and communication. A solution for such issues is global networks of interest. If you are a member of one organization that relies on information supplied from a different organization, paying attention to what that other organization is up to becomes an essential part of your job. A global network in the context of a knowledge management system enables that behavior by creating connections to those other facets of the organization through collaboration. The same concept of *following* something you are interested in—something that has become a standard part of participation in something like Facebook, for example, once a novelty, is now a way of life. So, in business, being connected to the complex facets of the organization and staying in constant communication—on the pulse, without a meeting—will become part of the way we work. It is already the way many industries, from small start-ups to industry giants like oil and gas have found success in collaboration and blowing up those silos.

Alerts

If your organization has ever had repeated mistakes, an alert process might be worth considering.

Our team developed the notification and alert process to ensure that events were well communicated, action was taken, and learning from the event would not be lost. This process required that events meeting specified criteria be communicated with a formal notification. Notifications are distributed within a specified time frame, so that other facilities have an awareness that an event has occurred. Notifications are distributed prior to a root cause being identified, so that awareness can occur without delay. Alerts are then distributed to specified groups of individuals after the root cause is identified and include specific actions and accountable personnel to correct similar conditions and prevent similar events. The actions are tracked to closure.

Often a significant negative event will trigger a timely assessment within all other facilities. To support this activity, an event critique process was developed. The critique process was developed to ensure that within 24 hours of a significant event all involved parties and leadership will go to the scene of the event to investigate the facts. These activities will start with a review of the event timeline and the pre and postevent activities. This helps establish the root cause of an event, so that appropriate actions can be taken to prevent similar events from recurring.

Time and attention are the rarest of commodities in the workplace. Therefore, making sure that people understand when something they are *or should be* interested is happening is absolutely essential. By setting up a technology whereby users can be subscribed to important things or can self-select items of interest, alerts are able to direct attention where needed. Imagine the speed and power of an organization with that much agility!

So What Did We Learn?

Step outside of everything you know about the world for a second and consider the fact that there once was a time where if you did not know the answer to a question, you just did not know the answer to that question. That was the case for a really long time. Perhaps you happened to actually research it or ask the right person, or stumble across the right set of circumstances to give you an answer. Even then, that answer may or may not have had anything to do with reality. It was just the closest approximation you could come up with for a relevant answer to your question. How many of those answers actually were rooted not in fact, but in *confirmation bias* (Blous 1993), anyway?

Now consider that the awful time described in the previous paragraph was not all that long ago in the grand scheme of things. Really stretching the limits, at the time of this printing, it was, at the most, 25 years ago (Lumb 2015). And even then, only super geeks were able to figure out how to ask questions and get real answers on the *information superhighway* or the *World Wide Web* as it was initially called. Those early DOS interfaces had nothing in common with today's modern browser with a graphical interface, filtered results, and (usually) reasonable information to help our search. Multiple source verification can confirm facts and figures almost instantaneously. Agreed best answers are uploaded and shared with ever increasing certainty and constant update. Seemingly, much of the world's known information is at our disposal if we just *Google* it—the word itself now a noun for looking things up as well as the name of a behemoth tech company.

Yet, with all these advances, somehow the world of work and information about our products and history of our workplace seems mired in antiquated and difficult to navigate interfaces and crowded channels. Why do organizations take shortcuts putting data into systems that end up costing them in the end? Tagging data as it goes in so we can find it later is a foreign concept in big organizations. *Speed going in does not equal speed in discovery.* Maybe speed is not always the answer.

There is an upfront cost to such things that most companies choose to avoid. Perception of burden when front loading any sort of technology endeavor is met with skepticism, difficult adoption, extensive training, and no shortage of financial implications. Though the actual text of documents can be discovered inside their systems, metadata and tagging documents and information on their way into systems certainly helps bring searches into reliable focus.

Filing systems in our mind before the Internet and search engines only went skin deep, so to speak. We could trace things down to a folder or a few layers of folders, but any deeper than that, and there was just no way to get all the detail we needed. General filing and organization techniques were limited by a risk benefit profile of workload. In some ways, that is still the case, but the technology enables much more to get done much faster—but front loading is required. This is the straw that eventually breaks the camel's back in an organization. The benefits far outweigh the risks, but the only way over this obstacle is through.

Now, back to The Library of Alexandria (Eagleman 2013). This is a chapter about knowledge management, not history. But in a way, *everything that has happened counts as history. In addition, history is meant to teach. Knowledge management is what should happen as history happens.* What happens when we learn from each other? We progress as a species. We progress as an entity. What if the secret to curing cancer was discovered in Egypt prior to the last big fire at the Library of Alexandria? No one is alive to tell us about it. If they wrote it down, it's gone. This is a concern, of course. How does one preserve all the information forever and make sure that future generations can benefit? How does a corporation that becomes a huge intertwined collection of individual contributors establish itself as a viable entity with a *memory*? *How do we save our most valuable asset on the planet—our time on the planet—by not repeating the mistakes of the past or spending that precious time figuring out something which had already been figured out? The answer, of course, is knowledge management practices, systems and the mindset of a learning organization.*

There is not a definitive account or inventory of the information that was once contained in the library—so we will never know what we have lost. Also, by most accounts, who was *actually* responsible for the destruction is not known. There are stories that link at least one of the fires to an accident in which, in an attempt to burn the port to keep out an enemy, inadvertently, the *defensive fire* spread from the port to the library (MacLeod 2004). So, over the course of history, while some of the destruction was purposeful, some was inadvertent and self-inflicted. How different, then, is the knowledge that organizations generate, lose, destroy, misplace, and sadly, *never capture in the first place.*

We Are Never Going Back

A few more questions to ponder …

- If given the choice between a manual typewriter and use of a word processing program would you go back to the typewriter?
- When was the last time you saw a functioning payphone? Did you use it? What if you could not use your mobile phone anymore and had to use payphones again?
- When you want to send information to someone quickly, would you like to send them a fax?
- When you want to connect to the Internet, would you prefer to have to *log on* through the telephone or just start working and assume that you are online?
- Has social media rendered the high school reunion obsolete?
- Do you actually remember anyone's phone number or do you just keep it in your phone and back it up to the cloud?

If you answered *no* to any or all of those questions, then you will one day be able to ask your colleagues that resisted (when working out loud and the use of KM integrated systems is ubiquitous as mobile phones, word processors, and e-mail) "Would you prefer we sit around and hope we can find this information by chance, remake it from scratch or just go search the KM system?" There will not be one person who wants to go back to life before KM. No one in their right mind would want to burn down the Library of Alexandria, right?

References

Blous, S. *The Psychology of Judgment and Decision Making.* New York: McGraw-Hill, 1993.

Eagleman, D. APQC Knowledge Management Conference. *David Eagleman Keynote.* Houston, April 2013.

El-Abbadi, M. *Library of Alexandria.* May 12, 2016. http://www.britannica.com/topic/Library-of-Alexandria (accessed June 21, 2016).

Elmer E. Rasmuson Library, UAF. *Boolean Searching.* November 9, 2014. http://library.uaf.edu/ls101-boolean (accessed April 12, 2016).

Kessler, S. *The "Sharing Economy" Is Dead, And We Killed It.* September 14, 2015. http://www.fastcompany.com/3050775/the-sharing-economy-is-dead-and-we-killed-it (accessed April 14, 2016) International Conference on Harmonization, "Pharmaceutical Quality System - Q10," 2008. www.ich.org .

Lumb, D. *A Brief History of AOL.* May 12, 2015. http://www.fastcompany.com/3046194/fast-feed/a-brief-history-of-aol (accessed April 14, 2016).

MacLeod, R. *The Library of Alexandria: Centre of Learning in the Ancient World.* London: I. B. Tauris, 2004.

Microsoft. *A history of Windows.* October 2015. http://windows.microsoft.com/en-US/windows/history#T1=era2 (accessed June 21, 2016).

Morgan, J. *The Top 10 Factors For On-The-Job Employee Happiness.* December 15, 2014. http://www.forbes.com/sites/jacobmorgan/2014/12/15/the-top-10-factors-for-on-the-job-employee-happiness/ (accessed April 14, 2016).

Ritvo, E. *Facebook and Your Brain.* May 24, 2012. https://www.psychologytoday.com/blog/vitality/201205/facebook-and-your-brain (accessed April 14, 2016).

Senge, P. M. *The Fifth Discipline: The Art and Practice of the Learning Organization.* Doubleday/Currency: New York, 1990.

The Economist. The Rise of the Sharing Economy. March 9, 2013. http://www.economist.com/news/leaders/21573104-internet-everything-hire-rise-sharing-economy (accessed April 21, 2016).

Ward, L. F. *Glimpses of the Cosmos, volume VI (1897–1912).* New York: G. P. Putnam's Sons, 1918.

Webster, M. *Social Media.* n.d. http://www.merriam-webster.com/dictionary/social%20media (accessed April 14, 2016).

Wikipedia. *List of Social Networking Websites.* n.d. https://en.wikipedia.org/wiki/List_of_social_networking_websites (accessed April 14, 2016).

19

Rapid and Robust Product Development Powered by Knowledge Management Capability

Anando Chowdhury

CONTENTS

Over the past several years, the biopharmaceutical industry has faced ever growing complexity from internal and external pressures. This has been coupled with an increasing complexity in the underlying science, now involved in the development of innovative new products. This Merck & Co., Inc. case study showcases how the development of a standardized KM process and collaboration framework has resulted in sustained high performance in new product.

Editorial Team

Introduction

Merck & Co., Inc. (Kenilworth, NJ, USA; henceforth Merck) is a company that is devoted to discovering, developing, and providing innovative products and services that save and improve lives around the world. As a part of this endeavor, the company has spent the past decade maturing knowledge management capabilities within its core divisions. One key area of focus has been the management of knowledge generated during the development and supply of biopharmaceuticals. In this area, Merck's strategy has been to develop and mature four core KM capabilities in together and in unison. There are two capabilities in the explicit knowledge area around product-specific and cross-product technologies. There are two capabilities in the tacit knowledge area around retaining and harvesting critical knowledge from deeply experienced individuals and providing a means of connecting globally in a virtual technical network (Lipa and Guenard, 2014). Although investment in these individual's capabilities is foundational to Merck's KM strategy, integrated frameworks and processes must be created to harness these capabilities toward a focused organizational end or business goal.

In this chapter, we explore how an integrated framework was created to harness these four Merck KM capabilities in the act of new product development. Figure 19.1 shows the way in which these four KM capabilities needed to enable new product development.

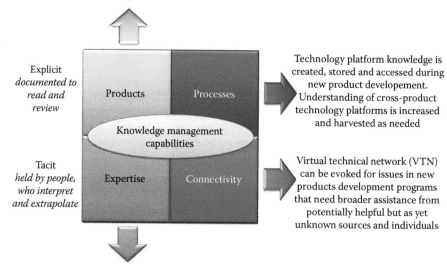

FIGURE 19.1
Merck's KM capabilities delivering to enhance new product development.

The Challenge of Managing Knowledge
in New Biopharmaceutical Product Development

In the late 2007, it was clear for Merck leaders that the previous practices used to manage knowledge in the new product development, and scale-up process were no longer ideally suited for the new challenges ahead. The key functions involved in bringing the physical offering and designing the supply chains for new pharmaceutical products saw an ever-growing mountain of complexities ahead of them. The pipeline size was growing, and often with acquired molecules or through lifecycle extensions of existing products. This change had the burden of putting development work on the critical path because in many cases the clinical programs to demonstrate efficacy were shorter in length. The work of chemical synthesis, formulation, packaging, and distribution of these products were as challenging as brand new molecules, but with much less time to do the work, and much less patience from the larger organization to create supply chains. The program workload saw an increase of more than 200% between 2004 and 2007 (Thien and Chowdhury, 2010).

The growing complexity of the actual chemistry and formulation was daunting as well. Close to 50% of the new molecules being commercialized were difficult in terms of solubility or stability making them harder to formulate into tablets. They required unique enabled formulations and technology began to trend toward specialized delivery mechanisms. A push for more continuous processing, inhalation, sterile devices, and drug-device combinations were the norm. Fast dissolving tablets and nonconventional dosage forms and devices demand were increasing as well (Eickhoff, 2007). All this was pushing demands in the packaging design arena as well, with the need to provide multiple formats (a growth of about 5% per quarter). The chemistry area was increasingly working with development houses externally with 80% of the initial synthetic routes being done externally and often outside the United States. In every arena, the push was to deliver as much value for a new product from the starting line as possible. This was in terms of improved cost of goods, heightened robustness, and low waste.

Finally the regulatory environment and expectations were evolving. With the advent of the risk-based and more holistic approaches to drug development, innovative companies like Merck were being asked more and more to show the multifactor science behind their manufacturing controls decisions. What risks were considered? What precautions were taken? And most importantly, what knowledge was brought to bear, either newly created or leveraged from prior efforts, to make these decisions?

All these sea changes called for a new way for the core functions in manufacturing development (chemistry, formulation, packaging, and distribution) to exchange functional knowledge much more seamlessly (Ansel et al. 2005).

It required leveraging the collective wisdom of the masses of experts and professionals on every program, whether they were assigned to that program or not. It called for nontraditional functions to get pulled in much more tightly with the core functions. The upshot being that hundreds of new connections where knowledge could flow needed to be created with thousands of new collaboration points, and potentially tens of thousands of pieces of information from every program needed to be viewed in aggregate, so that leaders and workers alike could manage a portfolio of ever increasing programs (Cross and Thomas, 2009a) (Cross, et al, 2009b).

It was clear that standard and conventional ways of connecting people would quickly devolve to a *hairball* of overconnectivity and inefficiency. The alternative of selective inclusion was also not an option, as harnessing the collective was vital to success. What was needed was an elegant and integrative solution that balanced getting as much expertise injected into the process with achieving the needed efficiency and momentum (Cross et al. 2003). A process was need that allowed the coupling of knowledge flow across the portfolio of new products, but did so productively and without wasting anybody's time. Merck needed a way that allowed its four KM capabilities to be brought to bear in the area of new product development. Merck had to come up with such a framework and it had to do it fast. But was such a thing even possible?

The New Product Development Process

The role of Merck's Global Pharmaceutical Commercialization group and the overall commercialization process is to develop, commercialize, file, and sustainably launch new products, taking them from achievement of *proof of concept* in Merck Research Laboratories to the hands of Merck Manufacturing Division's internal and external sites and supply chains. Each new product program is managed cross-functionally through an integrated development and supply teams (IDSTs). This structure is shown in Figure 19.2.

The main role of this diverse team is to harness the functional excellence that exists in each of its satellite functional *working groups* (Clark and Wheelwright, 1992). At the center of this process are the three core functional working groups of active pharmaceutical ingredient (API), formulation, and packaging. Each area brings to bear the deep functional expertise required. As the program goes through its development, each functional working group develops its body of knowledge of the new product, as it works through its technical challenges, creating the inputs from its domain for a larger virtual knowledge-containing *value object* (Katz, 1997). It is convenient to think about a value object as a collection of answers to questions about a new drug required to launch the product and sustainably supply the

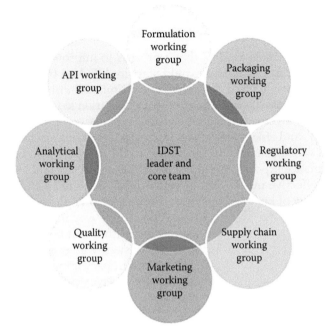

FIGURE 19.2
Merck's integrated development and supply team (IDST).

patient and customer. Knowledge built for any new product contains typically the following categories:

- Product knowledge (e.g., indication, dosage and side effects).
- Manufacturing process (safety, environmental profile, yields, impurity control, and analytical methods).
- Regulatory filling information (chemistry, manufacturing and controls, and product stability).
- Operational knowledge (from making and delivering the product).

For example, the API working group brings together all the deep expertise in chemical kinetics, thermodynamics, heat and mass transfer, organic synthesis, catalysis, automation and process control, reactor design as well as the API operational and quality specifics and a host of other knowledge elements to the table. Input from the three core technical areas often provides the bedrock around which the rest of the new product is based and built. This core functional knowledge is increasingly required to be managed across a wider array of teams involving a diverse and large set of individuals on increasingly more complex programs.

Building a Functional Knowledge Management Framework

The decision was made in 2007 that a process to improve knowledge flow for the core technical functions would start in API. Led by two of the API leadership team members, a design effort was initiated utilizing a process of transactional process design. The effort explicitly tried to focus on the target: a system that allowed API commercialization to *measure* progress, *value and prioritize* work, and efficiently *manage knowledge* over the project lifecycle. The process that was ultimately designed in the API space came to be known as high performance process development (HPPD) (Starbuck and Massonneau, 2007). Fully effective in 2009, the system quickly became an efficient way to manage the flow of knowledge in the API function. In reflective analysis, the success of the process was in large part due to the way the process handled Krackhardt's three organizational constraints (that will be covered later on in this chapter). Subsequently the HPPD system was expanded and fully effective in the formulation area in 2011 and in packaging and distribution in 2014. The second *P* in HPPD allowed for a clever customization to each area while keeping the acronym the same. Although the *P* stood for *process* in the API space, it effectively stands for *product* in the formulation space and *packaging* for the packaging area. Although each area customizes their approach slightly, the underlying system remained the same. At its core, each program in every functional area is broken simply into six consistent stages, representing the natural stages that happen throughout the course of commercialization, from initial target product profile (Stage 1) to support and final handover in commercial supply (Stage 6).

Six generic stages of HPPD are as follows:

- Stage 1: Entry and initial target product profile.
- Stage 2: Core product definition.
- Stage 3: Ancillary product and supply chain definition.
- Stage 4: Initial site readiness and filing.
- Stage 5: Final site readiness and launch.
- Stage 6: Monitoring of product in commercial supply and final handover.

Each program also has a set of important attributes that speak of the health of the program. Some are technical (e.g., in packaging we define a whole set of attributes around and activities around cold chain logistics), some are organizational (e.g., every area has an item called *partner interactions* to gage health of support needed from organizational units outside the group) and some are related directly to knowledge management (e.g., every area states explicitly knowledge repositories and knowledge deliverables at every stage). Each working group uses the HPPD checklists (of which there can

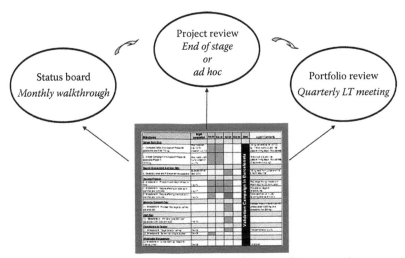

HPPD Stoplight chart – Development milestones = *f*(Project stage)

FIGURE 19.3
The HPPD program level checklist for a functional working group.

be six throughout a program's life, one per stage) to gage health of the programs, and through a stoplight mechanism light up risks and complexities for all to easily see. These individual program assessments are then easily viewable on an overall dashboard that allows the substrate around which true program-to-program knowledge can flow. A stylized representation of this system is shown in Figure 19.3. The process owner for the entire process in all three functional areas is a consistent individual (or pair of individuals) regardless of what stage of the process. The act of chairing and ultimate accountability for its outcomes resides with the functional head (also known as the department head) for that area. In all three implementations of HPPD at Merck, the process owners report directly to the department head.

Although the program checklists (that are all stored in a version controlled formats in a SharePoint system) are helpful in the sense that they provide *standard work* to people, it has the added benefit of allowing people to talk about problems in each other's programs. The aggregation of all programs in a dashboard format is where the true value comes into play. The dashboard is a way to have the key attributes of any program along the vertical axis and the programs in various stages of maturity along the top row. At the crosshairs of the matrix is a stoplight color. A stylized version of this dashboard is shown in Figure 19.4.

This dashboard then allows for a monthly review that has everyone in the function (and some partner functions) in attendance. HPPD provided a known time in the month when a whole department can connect in physical spaces and connected through virtual meeting media to the dashboard discussion. The monthly HPPD convergence also allows for the identification

	Stage 1				Stage 2							Stage 3					Stage 4				Stage 5
	MK-2653	MK-4763	MK-8521	MK-1122	MK-9922	MK-2289	MK-3457	MK-5566	MK-5566	MK-3221	MK-5566	MK-0912	MK-9122	MK-0921	MK-2232	MK-2511	MK-3333	MK-3142	MK-2201	MK-9227	MK-2201
Partner interactions	1	3	1	1	1	2	1	1	1	1	1	1	2	1	1	1	3	1	1	1	2
Primary packaging	1	1	1	1	1	1	1	3	1	1	1	1	1	1	1	1	3	2	1	1	1
Secondary packaging	2	2	2	1	1	2	1	3	2	1	1	2	1	1	1	1	3	1	1	2	1
Tertiary packaging	1	1	1	1	1	2	1	2	2	1	1	2	1	1	1	2	1	1	1	1	1
Distribution and logistics	2	1	1	1	1	1	1	2	1	1	1	1	1	1	1	1	2	1	1	1	1
Site operations	2	1	1	1	1	2	1	2	1	1	1	1	1	1	1	1	1	1	1	2	1
Materials management	2	3	2	1	1	1	1	2	1	1	2	1	1	1	1	1	1	1	1	2	1
Labeling	2	1	2	1	1	1	1	2	2	1	2	2	3	1	1	1	1	1	2	2	1
Regulatory	2	1	2	1	1	1	1	1	1	1	1	2	1	1	1	2	1	1	2	2	1
Quality by Design	2	1	1	1	1	1	1	2	1	1	2	2	1	1	1	2	2	1	1	2	2
Knowledge management	1	1	1	1	2	1	1	1	1	1	1	2	1	1	1	2	1	1	1	2	1

FIGURE 19.4
The HPPD dashboard for packaging and distribution.

of areas where the transfer of tacit knowledge would be helpful. The matrix affords the advantage of a working group leader being able to go *down* a column in regards to his or her program, talk about technical problems and issues, and seeking (in a speed dating kind of way) help or advice. In many cases, the functional members chime in proactively and ask Socratic or direct questions. This allows those connected to the program to gain knowledge from other programs on related issues. Easy and quick connections are made that assure knowledge flow to and from programs to problem areas to free up log jams. In the other direction, the matrix allows leaders to look *across* the different attributes to see if there are systemic patterns where potentially a lack of knowledge from a specific area is causing a systemic issue (Katzenbach and Smith, 1992). In Figure 19.4, an example of this might be the fact that there are 3 out of 4 yellows in the *secondary packaging* arena in Stage 1. Is this just a function of not knowing what we need at that point and that is normal, or do we need to build processes that create knowledge about secondary packaging in Stage 1 as an initiative for the area? Individuals are encouraged (even required) to ask knowledge-centered questions like *do we know that at that time?, is it knowable at that time?, who knows anything about this problem or has seen it before?* and *what know-how are we missing to solve this problem?* Taking this very knowledge-centered approach allows for tremendous focus to even generating the right future improvement projects for the area (Senge, 2006). The reason being, the problems are born out of real knowledge pinch-points on the actual portfolio across all the programs neatly laid out in time. The HPPD dashboard provides a remarkably efficient representation across a decade of time and quickly reveals patterns on where and how to generate improvements. Of course, none of it would work with a culture of transparency and a willingness to share in issues publicly in these forums. It is vital that the process owner and especially the department head make an environment where this willingness to speak up is easy. Reinforcing these expectations off-line after an HPPD review has proven quite effective and is an expectation that the organization has for its department leads (Skinner, 1974). We had the benefit of being able to observe different functional areas at different stages of implementation of HPPD. From our direct experience it was clear that knowledge flows less freely without such a process.

Since its implementation, HPPD has become a truly integrated and *in the flow* use of the full set of Merck's discrete KM capabilities shown in Figure 19.1. The KM capabilities provide the standards and foundation for knowledge flow in distinct ways, whereas HPPD provides the specific business context and the means of integrating the four capabilities into a seamless whole. In this way, HPPD acts as a *wrapper* that ensures, embeds, and guarantees knowledge flow by how it was originally designed and has evolved over nearly a decade.

Results and Why It Works at All

Since the serial deployment of HPPD from 2008 across the three main technical functions in MMD's Global Pharmaceutical Commercialization, the process has presided over close to 300 programs with incredible diversity and complexity. In this time frame Merck has launched 50 new products globally. HPPD coupled with other knowledge management systems in development (e.g., repositories of explicit knowledge of API and formulation known as TK-process and TK-product, coupled with a social community for problem solving and communication called VTN—virtual technical network) have greatly opened up knowledge flow and done so in a very efficient and consumable way (Bruno et al. 2012).

But why does this system work at all? It seems simple yet it solves a myriad of extremely complex knowledge management problems. On reflection, intuitively the teams that have created HPPD and grown it knew that designing the system was inherently fighting against three fundamental laws in organizations. These organizational constraints, described succinctly by Krackhardt are listed in Table 19.1 (Krackhardt, 1994).

TABLE 19.1

The Immutable Organizational Constraints when Design better Knowledge Flow and How a Viable Framework was Developed to Manage them

	Krackhardt's Organizational Constraints Paraphrased	How HPPD Works within Constraints
The Law of N-Squared	The number of possible links in a social system goes up approximately as the square of the number of elements in a system. With 80 people and 20 programs, the number of linkages is 9,900, whereas a system with just 8 people and 2 programs, the number of linkages is 90.	Create fixed forums in time when interactions and new (if only temporary) connections can be made to flow knowledge purposefully between people on different programs. This eliminates the need to have hyperconnectivity between all people and elements on the program.
The Law of Propinquity	The probability of two people communicating is inversely proportional to the distance between them (often known as the Allen Curve). In recent times, this law acknowledges that all forms of communication frequency reduce when there is a lack of face-to-face interaction.	Marshaling around a common dashboard regularly where every member of the collective function is together in the forum provides a means of sharing the common artifact as a substrate for proximity.
The Law of Oligarchy	Even the most egalitarian groups do rely on a small group of decision makers who have final authority and accountability. Often this authority is clear, but often it is the by-product of unspoken yet recognized expertise in an organization.	An inclusive culture in the fixed forums, assures a more level playing field for information flow, whereas decisions are clearly floated to the right individuals and accountabilities.

So how does HPPD solve for these constraints?

- In the case of the *Law of N-Squared*, which really refers to the problem of overconnectivity, HPPD allows often up to a hundred people to shop for the connections they need real time. Rather than needing to ping around constantly, the HPPD forum provides a single point in time when interactions and new (if only temporary) connections can be made to flow knowledge purposefully between people on different programs. There is no need to have hyperconnectivity between all people and elements on the program. The HPPD forum creates real-time links at regular intervals and allows them to disappear when the time is up for the value (Johnson and Johnson, 2000).

- For the *Law of Propinquity*, HPPD attempts to jar and knock loose the common affinities that people may develop due to the two largest centers of gravity, namely the program and the people in physical proximity. Since every member of the collective function is together in the forum, it provides a means of sharing the common artifact of the HPPD dashboard as a substrate for proximity (Allen, 1977).

- Finally, while formal or informal *oligarchy* is unavoidable and in some respects necessary in all organizations, the concept of HPPD is to make the act of asking questions completely agnostic to level, experience, and familiarity. Through building an inclusive culture in the HPPD forum the inputs and the leveling of the playing field is greatly enhanced, even though decisions and paths forward are decided by a select group of individuals on any given program. Conveniently, this culture of inclusion transformation was occurring in parallel in the Merck Manufacturing Division broadly and could be leveraged in HPPD quite naturally (Blenko et al. 2010).

Do–Learn–Plan—The Efficiency of High Performance Process Development Expansion

Any one of a myriad of design choices to HPPD could have propelled the ultimate product to fall into the limitations of any organizational knowledge business process provided by Krackhardt's three constraints. It is incredibly difficult to design a business process around these constraints that is capable of hitting a *Goldilocks' Zone* where process, tools, technology, and most importantly an inclusive culture overlap to be just right. With this in mind, the HPPD teams in all three technical functions took what can best be described as a Do–Learn–Plan approach (Kahneman, 2011). This approach was to implement the first design, run it, learn from it, and then change it purposefully for the next version. Within each function, it was also vital to *do the doing* for long enough, so that any deficiencies with the design were really deficiencies and not a function of initial learning and overcoming inevitable start friction. Having a means of collecting real-time improvements was vital (Makumbe, 2008).

Although learning and improving HPPD was vital, the concept of learning across functions during serial deployment was an added benefit. Having a consistent functional knowledge management business process across the three areas of API, formulation, and packaging creates a cohesion that then manifests itself in the largely IDST as well. In addition with learning from the prior function about how HPPD was deployed, simple mistakes could be avoided and learning could start from a higher point of knowledge (Nelson, 2003). Figure 19.5 shows the efficiency that was gained with each subsequent deployment of HPPD from API to packaging, showing the time needed from decision to implement to a fully working process. The figure shows a rough measure of times using a *change execution methodology* framework from Conner (Conner, 1993). The time intervals are from the approximate moment a management decision to implement HPPD is made to some evidence of what is known as *minimally acceptable realization*. Minimal acceptable realization is

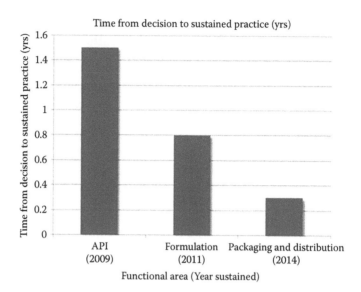

FIGURE 19.5
The tacit knowledge flow advantage of serial deployment of HPPD across functions.

a moment when the process has achieved certain predetermined endpoints enough to make the process itself *work*. It is acknowledged that at this point, the benefits of the change are outweighing the costs, while still some work is required to achieve *full realization*. Having members from the prior function on the new functions HPPD design team allowed for tacit knowledge to flow as well as having members of the new process sit in on an active HPPD dashboard reviews occurring in prior function. This tacit knowledge advantage is real and stark when made visible as it is in Figure 19.5.

HPPD was one of the key factors that enabled Merck to manage the tidal wave of complex programs and late stage product development. As an example, in 2015, the new products packaging function managed more than 70 programs simultaneously. Since the implementation of HPPD in packaging in 2014, the area has had 25 newly filed or launched pharmaceutical, vaccine, or therapeutic protein products. The process allowed for clear linkages to foundational KM capabilities and the knowledge, assuring that each program did its duty in recording and harnessing knowledge (Chowdhury and Thien, 2012).

Key Lessons and Looking ahead

With each successive implementation, enhancements, and new ways to harness the four KM capabilities have been introduced, both in the tacit and explicit categories shown in Figure 19.1. One recent improvement in the tacit knowledge category of connectivity is the integration of a way to track *assists*. Organizations are chronically bad at recording ways in which a colleague has helped at a crucial time but was not involved in the main train of the program. Merck's Global Pharmaceutical Commercialization (GPC) has instituted a peer-to-peer tracking system, which now is brought into HPPD reviews. By simply counting the number of *assists* through peer-to-peer cards, the impact of assists can be recorded and correlated to HPPD reviews themselves. It is clear that spikes of peer-to-peer assist recognitions go up right before and after a HPPD dashboard review (Guenard et al. 2013). This alludes to the potential of more purposefully timing HPPD reviews around a *decay curve* of assists. In addition, with the rise of more efficient ways and cleaner ways of data visualization allow for simpler and more streamlined ways to search, access, and contribute to the product knowledge and technical platform systems in our explicit KM capabilities during HPPD steps.

These enhancements are important and in the spirit of continuous improvement for a process that has been built. There are however some overarching takeaways from Merck's HPPD journey that can be used as universal learning across all knowledge management endeavors. These are listed as following:

1. Any organization serious about knowledge management should invest in a set of core KM capabilities (Kukura et al. 2007). At Merck, these discrete capabilities were built around the explicit knowledge categories of processes and products and the tacit knowledge categories of expertise and connectivity. These capabilities need to be built with great rigor and depth in partnerships with KM subject matter experts and those knowledgeable about the needs and practices of the core business.

2. Discrete KM capabilities alone cannot provide the full value to an organization without some process that helps to integrate those capabilities toward a focused and measurable business end. In this study, HPPD grew to be just such a *in the flow* system. It provided a canvas and a framework on which the capabilities could be arranged, connected, and perfected toward the goal of managing a growingly complex portfolio of new product development programs. These types of integrative business processes are an extremely effective way of harnessing discrete capabilities and can largely be owned by the line business.

3. When designing such an integrative business process, it is important to recognize that any *in the flow* business process is bound by well-studied organizational constraints. HPPD in using a very structured design process (in its case the process of transactional systems engineering or Design for Six Sigma) allowed for intelligent consideration of design choices in light of these constraints. A key learning therefore is while these constraints are ever present and organizationally ubiquitous; there are indeed good design choices that ease the

limitations to knowledge flow caused by these constraints. Equally so, there are bad design choices that actually amplify the resistance that can be felt by these organizational constraints to knowledge flow, but being thoughtful about them can help one to avoid them.

4. Serial implementation of an integrative business process like HPPD across disparate functions can be achieved with ever increasing efficiency. This is especially true if there is willingness and a desire from the area of the business new to the implementation to shed any local insularity and departmental ego, and truly embrace learning from their colleagues in another department who have already implemented and have relevant experience. That said, time scales for organizational change can be long, particularly the larger the organization is (Argyris, 1993). HPPD took a larger part of a decade to evolve to its mass influence, and it required an iterative coevolution between the business process and the underlying KM capabilities.

5. Finally, maintaining a common business process and nomenclature across different areas with similar remits is very beneficial. This is particularly true if those functions in turn participate in cross-functional teams together with other functions. For HPPD, a stage definition for instance stayed fairly consistent across all three technical functions in development (API, formulation, and packaging), allowing for a simple short hand of communication and simple alignment around things like risk and complexity.

It is hoped that this implementation and expansion of the HPPD framework demonstrates how integrated organizational processes can be effectively built to harness investments in core KM capabilities.

Acknowledgments

Special thanks to the inventors of HPPD: Cindy Starbuck and Viviane Massonneau. Thanks to the thought leaders and experts in KM: Samantha Bruno, Robert Cross, Adam Duckworth, Robert Guenard, Marty Lipa, Jodi Schuttig, Renee Vogt, David Vossen, and Jean Wyvratt. Thanks to the sponsors of HPPD process and KM: Liam Dunne, Gary Hoffman, Craig Kennedy, Joseph Kukura, Scott Reynolds, and Michael Thien. Thanks to my colleagues who helped embed HPPD beyond the first function: Marcus Adams and Frank Witulski. Thanks to those intelligent and passionate partners who led the embedding of HPPD into packaging: Frank Giuliani, Anita Shaw, and David Walker.

References

Allen, T.J., (1977). *Managing the Flow of Technology.* MIT Press, Cambridge, MA.

Ansel, H.C., Loyd, A.V., Popovich, N.G., (2005). *Ansel's Pharmaceutical Dosage Forms and Drug Delivery Systems* 8th Edition. Lippencott, Williams & Wilkins, New York.

Argyris, C. (1993). *Knowledge for Action: A Guide to Overcoming Barriers to Organizational Change.* Jossey-Bass, San Francisco, CA.

Blenko, M.W., Mankins, M.C., Rogers, P. (2010). *The Decision Driven Organization.* Harvard Business Review, Cambridge, MA.

Bruno, S., Guenard, R., Lipa, M., (2012). Combining Social Computing and Organizational Development Efforts into a Virtual Technical Network. *APQC's 2012 Knowledge Management Conference (Collection),* Houston, TX.

Chowdhury, A.A., Thien, M.P., (2012). Applications of technology roadmapping to planning and building systems for medicine and vaccine manufacturing. *Proceedings of the 11th International Symposium on Process Systems Engineering,* Singapore.

Clark, K., Wheelwright, S.C., (1992). Organizing and leading "heavyweight" development teams. *California Management Review,* 34: 201–215.

Conner, D., (1993). *Managing at the Speed of Change.* Random House, New York.

Cross, R., Parker, A., Sasson, L., (2003). *Networks in the Knowledge Economy.* Oxford University Press, New York.

Cross, R., Thomas, R.J., (2009a). *Driving Results through Social Networks: How Top Organizations Leverage Networks for Performance and Growth.* John Wiley & Sons, New York.

Cross, R., Singer, J., Zehner, D., Vossen, D., (2009b). *Global Pharmaceutical Commercialization Organizational Network Analysis Final Report.* Merck Internal Document, Merck & Company, Inc., Kenilworth, NJ.

Eickhoff, M. et al. (2007). *Decision Tree for Insoluble Compounds.* Merck Internal Document, Merck & Company, Inc., Kenilworth, NJ.

Guenard, R., Lipa, M., Bruno, S., Katz, J., (2013). Enabling a new way of working through inclusion and social media: A case study. *OD Practitioner,* 45(4): 9–16.

Johnson, D.W., Johnson, F.P., (2000). *Joining Together: Group Theory and Group Skills* 7th Edition. Allyn and Bacon, Boston, MA.

Kahneman, D., (2011). *Thinking, Fast and Slow.* Farrar, Straus and Giroux, New York.

Katz, R., (1997). *The Human Side of Managing Technological Innovation: A Collection of Readings.* Oxford University Press, New York.

Katzenbach, J.R., Smith, D.K., (1992). *Wisdom of Teams.* Harvard Business School Press, Boston, MA.

Krackhardt, D., (1994). Constraints on the Interactive Organization as an Ideal Type. In C. Heckscher and A. Donnellan (Eds), *The Post-Bureaucratic Organization.* pp. 209–230, Sage., Beverly Hills, CA.

Kukura, J., Starbuck, C., Chowdhury, A. et al. (2007). *Merck Commercialization Technical Forum White Paper on a Framework and Guidance for Distinguishing Core and Non-Core Technologies.* Merck Internal Document, Merck & Company, Kenilworth, NJ.

Lipa, M., Guenard, R., (2014). A practical approach to managing knowledge: Making knowledge flow in Merck's manufacturing division. *NASA Knowledge 2020*, Houston, TX.

Makumbe, P.O., (2008). Globally Distributed Product Development: Role of Complexity in the What, Where and How. Submitted to the Engineering Systems Division in Partial Fulfillment of the Requirements for the Degree of Doctor of Philosophy in Engineering Systems, Massachusetts Institute of Technology, Cambridge, MA.

Nelson, R.R., (2003). *Physical and Social Technologies and Their Evolution*. Columbia University working paper available from the author, Columbia University, New York.

Senge, P., (2006). *The Fifth Discipline: The Art & Practice of the Learning Organization*. Currency Doubleday, New York.

Skinner, B.F., (1974). *About Behaviorism*. New York Press, New York.

Starbuck, C., Massonneau, V., (2007). *High Performance Process Development (HPPD), Application of Transactional DFSS to Chemical Process Development and Commercialization*. Merck Internal Document, Merck & Company, Kenilworth, NJ.

Thien, M.P., Chowdhury, A.A., (2010). *Commercialization and Quality by Design: Towards an Improved Model for Pharmaceutical Development, Launch and Supply*. American Institute of Chemical Engineers, Salt Lake City, UT.

20

Knowledge Management Case Study: Using Near Real-Time Data Analytics and Performance Metrics to Ensure a Robust and Resilient Supply Chain

Eda Ross Montgomery, Mani Sundararajan,
David Lowndes, and Gabriele Ricci

CONTENTS

Shire's knowledge management strategy is inherently linked to its strategy for a robust and resilient supply chain. The key driver is utilizing their knowledge to enable the maintenance of the *state of control* for their products. Read about how Shire leverages a holistic knowledge framework, using both data analytics and smart performance metrics, to improve process capability while focusing on decreasing product defects.

Editorial Team

Introduction

Knowledge management is a key part of Shire's strategy for ensuring a robust and resilient supply chain. For Shire, with a complex network of more than 50 contract manufacturing organizations on four continents that supply product to more than 70 countries worldwide, understanding the factors that affect product performance and their impact on supply is essential. It is also a regulatory expectation: ICH Q10 (ICH, 2008) and FDA's 2011 process validation guidance both define maintaining a state of control over the life of the process as a measure of assurance of continued process performance and product quality. Applying the principles described in this case study to Shire's small molecule portfolio have resulted in an improvement in product performance (process capability), increased lot acceptance rate, and decreased deviation rate, while at the same time product and process understanding have increased, leading in turn to a more robust and resilient supply chain. The framework for Shire's knowledge management strategy that has led to this success will be discussed. It includes the following topics:

- People—the importance of teams and team dynamics.
- Process—the use of Lean Six Sigma and other tools to help identify and make decisions based on the data and the knowledge of how the manufacturing process is performing.
- Technology—the development of product dashboards and real-time data analytics.
- Governance—the importance of the working teams and the committees that make decisions and take actions on the data and knowledge of the manufacturing process.
- Results and lessons learned.

People

A critical component of any successful strategy is the people. In Shire's case, a technical function has been established within the commercial organization with deep, broad, and diverse expertise in materials science, manufacturing technology, pharmaceutics, and measurement disciplines and spanning product development, global registration, and commercial manufacturing. This technical function works closely with the pharmaceutical development organization starting at product acquisition or late Phase 2, whichever is later. The goal of this collaboration is to ensure both successful product registrations and to ensure that the factors impacting robustness of the intended commercial product are known. During preparation of the product registration dossier, product and quality technical teams are established within the commercial organization. The purpose of these teams is to ensure readiness for product launch and to support ongoing commercial manufacturing. The expertise on these teams is in four core functional areas (drug substance, drug product, analytical, and packaging), which serve as subject matter experts in their respective areas, along with commercial product technical knowledge management. The commercial product technical knowledge management team, which is part of the technical function within commercial manufacturing, facilitates, and champions knowledge management between and among the four core functional areas by serving as subject matter experts in statistics and data analytics along with the technology best used to deploy these tools. Members from the commercial product technical knowledge management team sit on each product and quality technical team. The product and quality technical teams also have access to the process excellence group, which serves as subject matter expert in advising on and facilitating the use of Lean Six Sigma and other operational excellence tools to each of the four core functional areas. Also providing subject matter expertise on the product and quality technical teams are representatives from quality and technical support representatives from the manufacturing site (e.g., contract manufacturing organization).

The Process

The processes behind Shire's successful deployment of knowledge management have evolved from 2010 to today. In 2010, when Shire's knowledge management journey began, product performance was not regularly measured or reported. As a result, 100% of products had process performance index (Ppk) < 1.33, and 75% of Shire's small molecule products had process performance index (Ppk) < 1.0. This poor product performance manifested itself in risks to the supply chain, which were mitigated by carrying larger inventories, some as large as several months. Shire started by establishing the cross-functional product and quality technical teams described above for each of Shire's small molecule commercial products to monitor product performance, gain an understanding of the factors leading to poor performance, and identifying and implementing actions to improve product performance and in turn supply chain robustness. Team leads were identified to provide direction for each team, facilitate collaboration among team members, and to identify barriers impacting their ability to improve product performance. To support these teams, a training program was established that included analysis of each team member's emotional intelligence followed by more in-depth training on topics such as leading teams, building collaboration, influencing, and strategies for improving team performance. A leadership team sponsor counsel individual team leads on specific issues and helps to address common issues. One tool used by the teams is the Drexler/ Sibbet Team Performance™ Model, which the teams use semiannually to assess their effectiveness on a scale of 1 (orientation) to 7 (renewal). Using this model gives the entire product and quality technical teams a common language to describe current and desired future state, as well as to quantify improvements in team performance. In 2015, all teams achieved a minimum score of 4.5, indicating that they have moved from the creating phase to the sustaining phase of team performance (Figure 20.1).

In addition to using the Drexler/Sibbet Team Performance™ Model to assess team performance, the product and quality technical teams have

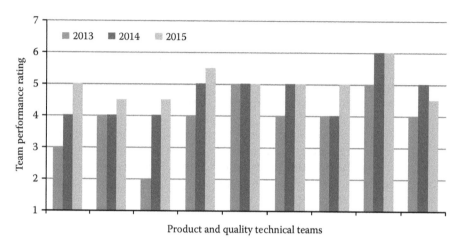

FIGURE 20.1
Team performance ratings.

product-specific team goals and goals common to all teams. The product-specific goals, which also encourage collaboration, depend on the lifecycle of the product and can range from obtaining regulatory approvals in new markets, to increasing capacity by bringing on a new manufacturing site, to improving efficiency in the manufacturing and/or testing process. The common goals are the following: provide visibility on product performance to the broader commercial manufacturing organization, resolve technical deviations in a timely and robust manner, maintain and improve both product-specific knowledge and general knowledge (new manufacturing and measurement technology, regulatory guidance, etc.), improve processes and reduce costs through improvement plans supported by data trending and analytics, and ensure newly commercialized products meet minimum standards of process and method performance. These are included in each team member's individual goals, and product and quality technical team performance accounts for 30%–40% of an individual's year-end performance rating.

Through data monitoring and data analytics, the product and quality technical teams monitor critical quality attribute data, determine product performance, and identify trends, variability, or shifts in the data that pose a potential risk to the supply chain. These risks are identified by using process capability data for critical quality attributes with poor product performance to calculate the predicted failure rate for each product. Predicted failure rates above a threshold limit represent a risk to the supply chain and trigger additional actions, including identification of this risk along with the actions to mitigate it. To reduce these risks, or to further product understanding and improved product knowledge, the teams perform hypothesis testing based on run charts of input material attributes, process parameters, critical quality attributes, or other variables in order to develop improvement plans. These activities are supported by the statistical analysis and data modeling provided by the commercial product technical knowledge management members on the product and quality technical teams. For example, one product and quality technical team performed a Kepner–Tregoe analysis in collaboration with a major contract manufacturing organization (CMO) to understand the cause for sporadic but recurring low yields for one strength of this oral solid dosage form. The commercial product technical knowledge management team member provided real-time data analytics to test the team's hypotheses, including possible correlations between input material attributes and process parameters and the low yield. As a result, the Shire–CMO team was able to identify a combination of material attributes and parameters that, taken together, were consistent with the low yield; the team is further investigating these factors.

To ensure that product understanding from all sources is captured and to drive sustainable improvements, the product and quality technical teams are required to document any knowledge gained from investigations and all other sources (e.g., design of experiments) in technical reports. In some cases these reports are stand-alone; for products with significant risk or value, the teams, supplemented with subject matter experts from Shire and the contract manufacturing organization, combine the conclusions of the stand-alone reports and other sources of information in a single document that reconciles all of the product knowledge gained to date. This criticality analysis document, which is maintained throughout the product lifecycle, includes input material attributes, process parameters, and in-process control ranges studied, their respective impact on the product's critical quality attributes, an assessment of the criticality of each variable studied, and references and/or scientific rationale to support the assignment. The criticality analysis helps to convert tacit knowledge to explicit knowledge and is used to evaluate proposed process improvements, assess deviations for potential impact to product quality, and to plan technical transfers.

Although product performance and predicted failure rate are important measures, they represent only part of the picture. Key to supply chain robustness is controlling the actual failure rate. The product and quality technical teams are also charged with understanding the differences between actual and predicted failure rate and implementing measures to reduce any gaps. For example, one product and quality technical team had a higher actual failure rate for uniformity of dosage units than expected based on the predicted failure rate for the product. Although the dosage form is a low drug load oral solid formulation with a potential for segregation, in a prior improvement project the team had proactively implemented measures to mitigate the risk of segregation during the mixing and transfer steps, and had observed an improvement in product performance. While investigating the higher than expected actual failure rate, they hypothesized that the assembly of the baffle lids, if done incorrectly, could cause segregation. The team arranged for permanent markings to be placed on the baffle lid indicating the correct lid orientation. Since these markings were installed, no failures for uniformity of dosage units have been observed, and the actual failure rate more closely matches the predicted failure rate.

Along with actual failure rate, the product and quality technical teams are responsible for timely resolution of deviations, eliminating repeat deviations, and reducing deviation rates relative to the previous year. Although root cause analysis is part of the process for all deviations, a more data-driven, holistic approach to reducing deviations was taken in partnership with a major CMO. When the deviations for all Shire products manufactured at the CMO were analyzed for root cause, the highest frequency root causes were manufacturing errors. When Shire's and the CMO's subject matter experts and operators were interviewed, they reported that the batch record instructions were not consistent from product-to-product, there were redundant batch record entries, and the sequence for data entries did not always match the manufacturing sequence. A joint Kaizen was performed to redesign the batch records. In the process, the team implemented a standard batch record format that clearly described the data entry format and where batch record entries should be made, further reducing opportunities for errors or missing data entries. This redesign resulted in batch records that followed the manufacturing process, up to 30% reduction in batch record entries, and a reduction in deviation rates attributable to manufacturing errors.

A separate Shire-CMO team performed a Kaizen to reduce laboratory testing errors and laboratory cycle time. This team implemented laboratory cells based on technology platforms that utilized trained specialists dedicated to Shire products and to each type of test. The cells were organized so that the documentation, equipment, reagents, and materials were adjacent to one another. Material and people flow was optimized to minimize errors and maximize efficiency in receiving, preparing, and testing samples and evaluating and reporting results. The result was an improvement in cycle time for laboratory testing from 27 days to 2 days, with a corresponding but less dramatic decrease in laboratory errors.

Technology

Although the people and processes implemented to establish Shire's knowledge management framework are critical to success, equally important is implementation of technology enabling the people to execute the new processes. To provide visibility to product performance, the actions being taken to improve product performance, and the status of those actions, a product technical dashboard was introduced in 2013. This dashboard includes the product performance for each critical quality attribute for each product at each manufacturing site, the activities that are in-process to improve product performance for each poorly performing CQA (Ppk < 1.33), and the current status of these activities. The technical product dashboard also shows historical product performance, the predicted failure rate of the product, and the statement of risk to the supply chain based on the predicted failure rate

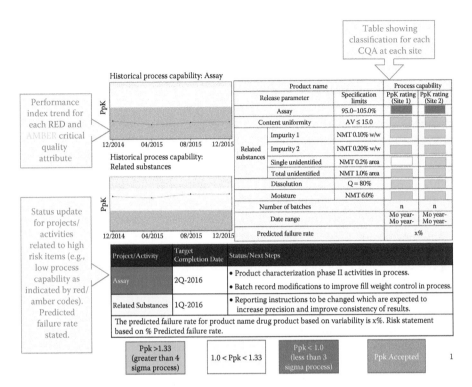

FIGURE 20.2
Technical product dashboard.

(Figure 20.2). Implementing the technical product dashboard has standardized the format for reporting on projects, and enables the organization to prioritize projects with the highest impact on product performance. It has also identified situations where product performance presents a potential risk to the supply chain, so that risk mitigations (e.g., interim controls, alternate source of supply, or additional inventory) can be put in place.

To further provide the product and quality technical teams with a workspace that links them to the products, given that Shire's small molecule portfolio is 100% outsourced, a technical team room was established at Shire. All product and quality technical teams, along with technical leadership and functional teams, meet in this technical team room. Joint Shire–CMO teams use the technical team room's video conferencing capabilities along with web-enabled meeting tools to increase effectiveness of their meetings and to share data and presentations. The product technical dashboards for each product are posted in the technical team room, so that the other product and quality technical teams, commercial manufacturing management, and other stakeholders have ongoing visibility to each product's performance and current activities to improve product performance and reduce deviations.

Another factor that is greatly influencing the success of Shire's knowledge management strategy is the availability of electronic data. The product and quality technical teams use a validated, shared data analytics tool to test for correlations, understand the impact of lot genealogy on product variability, perform routine statistical analysis, identify out of trend results, and routinely prepare automated reports describing product performance. Team members can use predefined queries or can design their own queries to test specific hypotheses using a current, common data set. The data source for this tool is an electronic data feed that extracts data from the LIMS and SAP systems from one of Shire's major CMOs into a database within the CMO. Within one day of batch release, the CMO database is queried and all data for all Shire products made at the CMO, as well as all data for all materials used in Shire products, is extracted and sent to Shire through a secure data transfer tool. Once the data files have arrived at Shire and are verified to be error free, they are written into a Shire database. The Shire database is then used as the data source for near real-time data analytics. For the small amount (<10%) of data that are not available electronically at the CMO, but must be monitored, personnel at the CMO enter the data into the database at the time of batch release, so that the data set is complete. The electronic data transfer tool was fully implemented in March 2014. Since then, data for more than 10,000 batches has been transmitted to Shire using this approach. Out of trend results are shared with Shire and the CMO simultaneously through a configurable, automated notification process. This automated, personalized, and flexible approach has greatly improved the teams' ability to improve product knowledge.

Governance

The governance of the knowledge management framework has several components. In addition to informing the organization of supply chain risks due to product performance, the technical product dashboards are also part of quality oversight. During each month's quality review of product-specific metrics, the technical product dashboards and the specific activities that are in progress to improve product performance become part of the quality record. More importantly, since the dashboards include a statement of risk to the supply chain based on predicted failure rate, products with high risk are highlighted for more in-depth discussions including, if necessary, the need for short-term mitigations, and escalation to senior management. Review of the technical product dashboards is also part of the annual product quality review process.

Product performance is also shared with the contract manufacturing organizations manufacturing each of Shire's small molecule drug products. This review typically takes place in conjunction with a periodic business review meeting. However, for some CMOs, Shire's strategy is to have a closer working relationship where the technical, quality, and business representatives from each company share responsibility for product stewardship and performance. This approach aligns the organizations at the sponsor and manufacturer on priorities that enable both companies to meet their quality and business objectives.

For one of Shire's major CMOs, the shared responsibility is evidenced by interactions at four different levels: the product and quality technical team, the joint execution team, and technical review committee, and the joint executive steering committee. For this CMO, the subject matter experts in the technical services organization are members of the product and quality technical teams for each Shire product made at the CMO. As such, they have equal visibility to the data, process improvement projects, team goals, and product understanding to the Shire personnel. The Shire–CMO product and quality technical teams meet regularly (typically weekly) to review current product performance, review open deviations, and discuss planned or ongoing projects.

To ensure successful execution of projects and initiatives and to review schedules for current and planned manufacturing across the Shire product portfolio at this major CMO, a joint execution team chaired by the business representatives from each company meets biweekly to review and address issues with timelines, deliverables, and costs versus budget. To align the two companies from a science and technology perspective, a technical review committee consisting of leaders from each company's technical and quality organizations meets (typically monthly) to review product performance, deviation rates, and process improvement projects across the Shire portfolio at the CMO. This team also sponsors and oversees joint projects to implement new technology or improve existing technology. A recent project overseen by the technical review committee was the implementation of hand-held Raman spectroscopy for incoming material identification testing, along with the electronic data feed described above. The strategy for the Shire–CMO interactions is the responsibility of the joint executive steering committee (JESC), which is made of senior management from the business, technical, supply chain, quality, and R&D organizations of each company. The JESC meets monthly to set strategy, review progress versus major business milestones, and approve projects and budgets.

Results and Lessons Learned

The results of the people, process, technology, and governance described in this case study has led to

- Improved response time to FDA and other queries.
- Decreased frequency of quality events.
- Faster closure of quality events.
- Ongoing, sustained improvements in product performance.

- Decreased inventory levels.
- Increased capacity.
- Decreased cycle time for manufacturing and product release.

Specifically, response times for FDA queries have decreased from an average of five days to less than one hour as a result of implementing the data analytics tool with the automated electronic data feed. Cycle time for product investigations has also decreased, along with deviation rate, due to the increased ability to make data-driven decisions to eliminate unlikely hypotheses and more quickly focus on the most probable root causes. For example, deviation rates at the CMO discussed above are more than 60% less in 2015 relative to 2013.

The availability of near real-time performance metrics has had significant, sustainable impact on supply chain performance. Today, four years after Shire's knowledge management journey began, 75% of Shire's small molecule products have process performance index (Ppk) > 1.0 and 30% have process performance index (Ppk) > 1.33, and the batch acceptance rate for the portfolio is currently at 99.6%. This has allowed Shire to carry lower inventory levels, and the lower deviation rate and high lot acceptance rate have increased manufacturing capacity at the major CMO. In addition, the availability of electronic data alone saves approximately two FTEs per year for both Shire and this CMO. Finally, the overall cycle time for Shire products manufactured at this CMO has also been reduced by more than 50%, resulting in reductions in both their inventory levels and working capital.

Beyond the direct business benefits described above, the approach described in this case study has led to other benefits. Shire has gained increased understanding of the CMO's systems, and the teams have been able to evaluate variables that they could not have otherwise. For example, one team was able to proactively evaluate the impact of combining raw material lots from different vendors versus single lots from single vendors of the same material for an oral solid drug product; they showed that there was no impact on product performance or variability. By providing the CMO's technical and quality organizations real-time access to out of trend notifications and automated reports across the Shire portfolio at the CMO, both organizations have increased visibility to product performance. This has further strengthened the working relationship between the companies.

Similarly, product and quality technical team behavior has changed from 2010 to today. The product and quality technical teams meet regularly, team members work collaboratively to solve problems, and routinely use Lean Six Sigma tools (e.g., after action reviews, peer reviews, and lessons learned) to identify ways to further improve their team performance. Problems are prevented because the teams proactively address changes in product performance, and they are starting to demand broader availability of electronic data and data analytics.

Although the FTE reductions due to the availability of electronic data may seem modest in comparison, consider the resources associated with expanding a data analytics program using a manual data source. As the data analytics tool is implemented for more products, the number of batches to be entered in the tool increases, the resources needed to input and verify the data increase, and the resources to *monitor the monitoring program* to ensure that the batch data are entered in a timely manner increase. With the electronic data source, resources are required up front to develop and validate the interface. However, once the interface is in place, the data analytics tool can be expanded to more products with no increase in resources, regardless of batch volume. Although concerns about confidentiality of client data, security of IT systems, and incompatibilities in technology presented challenges to establishing the electronic data stream at the outset, these challenges were successfully overcome. Shire does not directly access the CMO's source data (the data is pushed to Shire), the data transfer has been successfully validated through both companies' security and firewall technology, and a simple approach to data transfer (i.e., flat file) and data analytics (i.e., shared web-based platform) addresses differences in technology.

The electronic product-related data available from the CMO represents 75% of the volume of Shire's small molecule portfolio. Shire leverages several other technologies for exchanging key inventory, shipping, and financial data with many of our partners (including U.S. lot-based traceability requirements). Shire intends to optimize and expand our collaboration strategy as new technologies become available and the capability of our partners improve, so that electronic product-related data are available for 100% of the small molecule portfolio, and to further integrate with our CMOs on a strategic level. By doing so, changes in product performance will be addressed at the time of manufacture, quality events (e.g., deviations or change controls) will be correlated with changes in product performance, and Shire and CMO resources will be deployed where they have the greatest impact to the business.

In conclusion, a successful knowledge management program has been deployed at Shire. Critical to this deployment has been ensuring that the people with the right expertise are organized so that cross-functional collaboration and data-driven decision making are the norm. The processes that support the knowledge management program reinforce this cross-functional, data-driven approach. Technology, especially electronic data sources, is essential, as is linking these sources so that data analytics tools have near real-time access to the source data. Last but not least, governance processes at both sponsor and contract manufacturing organizations need to ensure that resources and priorities stay aligned throughout the implementation of the knowledge management program.

Reference

ICH Harmonized Tripartite Guideline: Pharmaceutical Quality System (Q10), Step 4 Version, 4 June 2008, International Conference on Harmonization of Technical Requirements for Registration of Pharmaceuticals for Human Use.

21

Knowledge Management Elements in Support of Generation of Chemistry, Manufacturing, and Controls Regulatory Documentation

Beth Junker

CONTENTS

Chemistry, manufacturing, and controls (CMC) submissions across multiple global markets are complex, time consuming, and knowledge intensive. However, with first-to-market pressures and new fast-track approval pathways, the need to be timely, effective, and accurate is driving new knowledge capture and knowledge transfer processes. This case study describes how a formal knowledge-focused framework can be used to manage the overall regulatory submissions and product licensing lifecycle from R&D to post-market surveillance.

Editorial Team

Introduction

A comprehensive knowledge management approach is indispensable for the reliable generation of chemistry, manufacturing, and controls (CMC) regulatory documentation, which in turn communicates the development and execution of a robust regulatory strategy leading to product approval. These latter activities are highly cross-functional (spanning from discovery to development to commercialization) as well as cross-disciplinary (inclusive of drug substance, drug product, analytical, quality, and regulatory subject matter experts [SMEs]).

The desire to be fast-to-market to serve the needs of patients worldwide has been further enabled by the rapid approval pathways (e.g., fast track, breakthrough, accelerated and priority [FDA, 2014], and "pathfinder" [sakigake*]) newly created by multiple health authorities. Consequently significant CMC timeline and resource pressures often rapidly erupt once encouraging clinical results are received. Moving forward, such pressures are likely to be the rule rather than the exception. They are further exacerbated by the current widely varying submission content expectations across the significant worldwide markets. Readily accessible explicit as well as tacit knowledge supports all stages of CMC regulatory documentation preparation, including post major market submission activities typically occurring after the SMEs associated with development are replaced by those from manufacturing science or technical operation functions.

This chapter describes elements selected as critical to knowledge management for generation of CMC regulatory documentation by addressing the following key areas:

- Content generation, from gathering of available information to delivery of final drafts.
- Systems, both IT-based and business processes.

---

* http://www.mhlw.go.jp/english/policy/health-medical/pharmaceuticals/140729-01.html.

- Timeline, from kickoff of preparation activities to final approval.
- Quality, for regulatory submission sections and their associated source documents.

Specific knowledge management practices are described that support activities (such as storyline development, authoring, technical reviews, draft revisions, status reports, quality checks, and document searching/archiving) for the generation of both the initial CMC regulatory submission and its associated source documents. Additional knowledge management practices are described that support post-submission activities such as responses to regulatory questions, regulatory submission preparation for subsequent markets, and post-approval updates during the product lifecycle (e.g., stability extensions and process or analytical method change implementations).

Mitigations of pain points and opportunities for improvements to address threats/risks, derived from informal and formal lessons-learned discussions, were used to improve these practices for subsequent CMC regulatory documentation preparation. Robust establishment of these knowledge management practices permitted introduction of further time- and resource-saving measures and best practice harmonization across multiple product modalities (i.e., small molecules, antibodies and therapeutic proteins, and vaccines). Controls as well as metrics to evaluate the impact of knowledge management practices on the generation of CMC regulatory documentation are proposed to sustain implementation.

Background

One major goal of CMC regulatory documentation is to share specific knowledge about product and process understanding with health authorities, both at the time of regulatory submission of the license application, as well as throughout the product lifecycle commonly via post-approval submissions. Specifically, product and process understanding is the basis for the control strategy, which is defined as "a planned set of controls, derived from current product and process understanding, that assures process performance and product quality" (ICH Q10, 2008). It evolves as knowledge is gained and risk management activities are undertaken, thus always reflecting the current state of product and process understanding (FDA, 2015).

A subset of the control strategy is included as part of the license application, and a subset of the license application is proposed to be designated as established conditions (FDA, 2015). The definition of established conditions has been proposed as the "description of the product, manufacturing process, facilities and certain equipment, specifications and elements of the associated control strategy, as defined in an application, that assure process performance and quality of an approved product." This regulatory submission information is proposed to be a binding commitment to the health authority, thus clarifying what is and is not considered a postapproval change requiring health authority notification in some format.

In addition to the draft FDA guidance about what CMC information is considered to be an "established condition" additional ICH guidance is being assembled to develop a globally harmonized approach to "regulatory commitments" including delineation of the appropriate level of detail and information necessary for regulatory assessment and inspection (ICH Q12, 2014). As with the FDA guidance, ICH Q12 also plans to emphasize the use of the control strategy as a key component of the applicant's regulatory commitment, aiding as well as globally aligning identification of future changes (ICH Q12, 2014).

A better understanding of CMC regulatory commitments in turn improves communication between applicants and health authorities throughout the product lifecycle. It also leads to a better understanding of what knowledge supports these regulatory commitments, both simplifying application of risk management principles (ICH Q9, 2005) to assess postapproval changes and highlighting what information should minimally be included in knowledge management efforts (ICH Q10, 2008).

Knowledge management is defined as the "systematic approach to acquiring, analyzing, storing and disseminating information related to products, manufacturing processes and components" (ICH Q10, 2008) and is a key enabler of a robust Product Quality System (PQS). Recognized knowledge management implementation challenges indeed may have led to additional proposed global efforts to "clarify expectations and reinforce the need to maintain a knowledge management system that ensures continuity of product and process information over the product lifecycle" (ICH Q12, 2014). Better knowledge management directly translates into clearer communication between applicants and health authorities, particularly via CMC regulatory submissions, which in turn optimizes utilization of both industry and health authority resources. Ultimately, these significant improvements in effectiveness positively impact accessibility of patients globally to the medicines that they need.

Knowledge Management Elements for CMC Regulatory Documentation Generation

Four knowledge management elements—content generation, systems, timeline, and quality—have been identified as critical for CMC regulatory documentation generation. Each one is defined, including a description of current practices, and then evaluated in terms of pain points, mitigations, threats/risks, and opportunities.

Content Generation

Definition and Current Practices

Content generation begins with gathering of available information and ends with the delivery of the final drafts of all source documents and regulatory submission sections, prior to quality checks. Information typically has been generated over several years of process development and clinical manufacturing activities. These activities potentially occurred at different sites and in functional areas, and in some cases at different companies. Source documents for this explicit knowledge vary in formality based on the stage of process development, ranging from laboratory notebooks, technical reports, non-Good Manufacturing Practice (GMP) process documentation to GMP batch records/laboratory information management system (LIMS) data, final process/method descriptions, and validation protocols/reports. Once gathered into a comprehensive "listing", information gaps are assessed relative to the desired submission content, which has been translated from tacit to explicit knowledge via a storyline document. Storyline development identifies threads and interconnected messages and plans their clear and consistent descriptions in one or more sections of the regulatory submission. The storyline is especially critical for key cross-disciplinary information that supports comprehensive product and process understanding and often builds on Quality by Design development and execution efforts (ICH, Q8).

Authoring begins according to the approved initial version of the storyline and often uses past submission sections or section templates as a model where available to ensure alignment in level of detail, location of information, and incorporation of previous internal and health authority feedback. Generators of source information are identified and made accessible to authors (if they are not already the authors) for their tacit knowledge, which often is a critical component to assuring scientific accuracy. Ideally, authoring proceeds based on completed and approved source information. However, draft source documents often are used to maintain authoring timelines raising the importance of systems to track source document versions for subsequent quality checks. Terminology harmonization decisions are documented to assure the use of a single term in the regulatory submission, yet bridge the selected term to similar terms used in development or manufacturing sites to support quality checks as well as pre-approval inspection (PAI).

Reviews occur at different points during the authoring process, recognizing that more than one review and revision cycle is likely needed for most source document or submission sections especially if source information is changing. Several cycles are likely needed for complex sections that are cross-functional or cross-disciplinary, which often contain the most sophisticated messages of the storyline and must simply but effectively synthesize the various threads of source information.

Technical reviews are conducted by the authoring team, and other functional SMEs, at various points during authoring of the source document or regulatory submission section. These reviews focus on logical, correct, and complete presentation, especially checking for consistency in data, information, level of detail, and terminology within and across functional sections. Cross-checks are also conducted with the approved storyline and differences are highlighted for resolution.

When a significant number of submission sections are drafted and through authoring team review and comment resolution, the regulatory submission is assembled and reviewed by a cross-functional and cross-disciplinary team consisting of authors, product development team members, "next level" functional management, and other technical and regulatory SMEs. Through their comments and subsequent discussions to resolve them, these reviewers ensure incorporation of tacit knowledge not able to be made fully explicit in the form of approved source documents or source document/regulatory submission templates. Selected functional area leaders are added as reviewers. Major comments impacting the approved storyline are reviewed with all functional leaders to secure their approvals to proceed. Naïve reviewers, not intimately familiar with the product development history or the storyline, also review the submission (both before and after reading the storyline) with a focus on clarity, consistency, and readability.

Review comments are designated by the reviewer as minor (e.g., optional editorial improvements) or major (e.g., mandatory to address or discuss further) to aid in prioritization of comment resolution activities. Comment resolution meetings consisting of authors and reviewers establish the path forward for addressing the major, more complex comments as well as reconcile conflicting comments. Strategies for comment resolution are built on prior knowledge of regulatory experiences as well as technical knowledge. The storyline and terminology harmonization documents are updated to reflect and communicate the decisions made. The authors revise both supporting source documents and regulatory submission sections.

After authoring for the initial submission, significant modifications are often necessary for submission to subsequent markets to meet expectations of country-specific health authorities. Primary authoring and review responsibilities shift away from functional authors to CMC regulatory, except possibly in markets such as Japan where new information or significant changes in presentation are needed.

Responses to health authority questions, received as soon as two to four months post-submission rely heavily on efficiently locating previously used as well as additional source information. This knowledge is used to develop supplementary explanations, or to provide further technical justifications. A focused strategy (i.e., storyline) for each response is quickly developed, aligned cross-functionally, checked against available source information, and documented for communication. Authoring, review, and comment resolution activities are also compressed compared with authoring of the regulatory submission itself to meet the timelines requested by the health authority. These timelines can be reduced to as little as a few days or weeks for the newly implemented rapid approval regulatory pathways, thus requiring quick accessibility to the relevant explicit and tacit knowledge.

Once market approval is received, significant CMC changes are submitted to health authorities through a separate submission of the impacted regulatory sections. Additional or revised source documents are developed and approved. Authoring, review, and comment resolution activities are focused on those regulatory submission sections requiring update or replacement.

Evaluation of Pain Points, Mitigations, Threats/Risks, and Opportunities

A summary of pain points, mitigations, threats/risks, and opportunities for content generation is shown in Table 21.1.

Systems

Definition and Current Practices

Solid, reliable systems, covering both IT-based and business process implementations, are often the critical knowledge management enablers for CMC regulatory document generation. Regulatory documentation generation centers on the selected platform(s) for collaborative authoring and review for both the regulatory submission sections and supporting source documents. Collaborative platforms permit multiple authors and reviewers to work simultaneously, specifically viewing, editing, or commenting (and responding to comments) in the same document. Thus, all edits, comments, and resolutions are visible in real time to those team members with access, as well as archived with the various document versions for future reference. In addition, final drafts of submission sections with source document maps showing links to the location of content in source information are archived.

Strong collaborations among multiple authors and reviewers, typically associated with initial CMC regulatory submissions, benefit from additional systems to describe and communicate the state of content generation. Tracking sheets (especially for authoring responsibilities, open items and assignments from decision logs, and identified risks and associated mitigation activities) can readily highlight not only where content generation activities are lagging but also where functional SMEs with tacit knowledge might be overburdened, thus permitting early intervention. Project management tracking systems showing assignments, progress and timing for outstanding source documents, and regulatory submission sections become critical to enable complete review and revision cycles.

A major IT system addresses centralized source document storage, permitting searches for a specific source document or all source documents associated with a specific submission section, and archiving of source document versions previously used in authoring. Prior source document versions become important when content generation occurs for additional market or post-approval submissions. Standard procedures are used for loading

TABLE 21.1

Knowledge Management Elements for CMC Regulatory Documentation: Content Generation

Pain Points	Mitigations
Gathering:	**Gathering:**
• Incomplete list of required source documents	• Build complete list of information supporting regulatory submission
• Initiation of knowledge gathering activities well after information generated (e.g., studies performed)	• Identify and gather in "real-time" content associated with tasks that support regulatory submission as it is generated
• Non-standard format and incomplete content of source documents	• Create single location (with e-links to official repositories), which associates source documents with specific content requirements and submission sections, preferably using regulatory submission-based taxonomy
• Difficulty in assembling documentation from multiple storage locations (e.g., team sites, document management repositories, and local repositories)	
Authoring:	**Authoring:**
• Multiple number of source documents for non-GMP (e.g., development) information	• Create and update source document and regulatory submission templates illustrating desired format and content
• Multiple revisions of source documents at various stages of process development history, potentially creating misaligned source content	• Use more content directly from source documents with minimal modification
• Inconsistent terminology between clinical/commercial manufacturing sites and development, or within functional development areas	• Identify model content from prior source documents and regulatory submissions
• Incomplete or unavailable prior examples or templates depicting desired format and content for source documents and regulatory submission sections	• Minimize repeated content in source documents and regulatory submission
	• Create selected additional source documents to combine and describe multiple separate content sources
• Unawareness of content that requires consistency among the various regulatory submission sections	• Perform team and (where applicable) cross-functional reviews of source document prior to its approval
• Substantial content revisions due to review comments including significant changes in authoring storyline	• Designate gatekeeper(s) for changes to the storyline
	• Pre-author where possible to permit "drop-in" of late arriving content for rapid authoring and review
• Finalization of source documents delayed until after regulatory submissions are nearly final	• Create subsections for larger regulatory submission sections to permit advancement in advancement to the next content generation step when authoring is complete

(Continued)

TABLE 21.1 (*Continued*)

Knowledge Management Elements for CMC Regulatory Documentation: Content Generation

Pain Points	Mitigations
Reviewing:	**Reviewing:**
• Reviews by various stakeholders generating large number of comments, some of which are conflicting	• Build in time for authoring team reviews prior to formal review cycles
• Incomplete or out-of-date storyline available to aid in review	• Comprehensive, "end-to-end" review by designated reviewers
• Multiple rounds of review especially for complex sections	• Generate clear guidance for designation of mandatory or discretionary comments
• Repeated review comments after comments had been resolved	• Generate decision log detailing resolution of major comments and update storyline as appropriate
• Subsequent review comments contradicting revisions leading to reversions to prior content	• Track risks due to content and review comments unable to be resolved for the submission as potential health authority questions for advance preparation of draft responses
General:	**General:**
• Updates or shifts in regulatory strategy	• Develop and update storyline especially focusing on any interconnecting cross-functional themes
• Unexpected content generation outcomes resulting in significant changes to storyline	• Develop and update cross-functional terminology harmonization document
• Late arriving (or delayed) content generation resulting in compressed authoring and review, and incomplete assessment of impact on related content	• Create a map showing related regulatory submission sections
• Lack of clear expectations for authors at each stage of content generation	• Cross-functional and cross-disciplinary reviews of selected source documents, especially those supporting related submission sections
	• Assemble all expectations for deliverables at each stage of content generation into a checklist and distribute to authors at the start of authoring

(*Continued*)

TABLE 21.1 (*Continued*)

Knowledge Management Elements for CMC Regulatory Documentation: Content Generation

Threats/Risks	Opportunities
General:	**General:**
• Speed-to-market pressures including rapid approval regulatory pathways causing key content to arrive late in timeline (e.g., stability, comparability, process, and other validation studies)	• Comprehensive global initial submission and post-approval regulatory strategy
• In-licensing of products, or acquiring of products via acquisitions or mergers, with informal or incomplete content transfers (especially with respect to documentation sufficiency)	• Regulatory guidance delineating required content (e.g., FDA guidance on established conditions, ICH Q12)
• Ongoing regulatory feedback (e.g., same or other product submissions and regulator presentations at conferences)	• "Right-size" target level of detail to match health authority as well as internal expectations
• Varied and evolving market-specific regulatory submission requirements	• Database of market-specific regulatory submission requirements
• Further functional development of content concurrently with source document or regulatory submission content generation	• Identification and aligned understanding of storyline risks among stakeholders
• Different/changing stakeholder views about content inclusion requirements	• Impact assessment and communication of significant storyline changes
	• Identification, exchange, and capture of tacit knowledge within and across programs, as well as functions and disciplines

the document itself, or more commonly the electronic link (e-link) to the document's official storage location to avoid introducing additional storage locations for the same document. Metadata is associated with each document from a taxonomy that includes items such as the regulatory submission section(s), specific regulatory submission, and document version. Thus, the latest version of each document can be searched for and readily located when subsequently needed for authoring or quality checks for a specific regulatory submission section as well as for post-submission activities. Highly trained "super users" assure consistency and improve efficiency, especially with loading large numbers of source documents.

A final IT-based system addresses centralized regulatory documentation storage, including initial submissions, partially or fully revised submissions, responses to health authority questions/assessments, prior-approval supplements, and other health authority communications (such as meeting requests, backgrounders, and meeting outcomes). Ability to locate as well as search this information becomes increasingly complex as the number of health authority interactions increases for a given product. When regulatory documentation is filed for multiple products as well as multiple treatment modalities within a single company, these challenges are augmented further which in turn limits timely performance of explicit knowledge gathering and increases reliance on tacit knowledge recall by SMEs.

Evaluation of Pain Points, Mitigations, Threats/Risks, and Opportunities

A summary of pain points, mitigations, threats/risks, and opportunities for systems is shown in Table 21.2.

Timeline

Definition and Current Practices

Timelines for the preparation of regulatory documentation span several years from the kickoff of regulatory submission preparation activities to final approval of all images and supply chains in all the target commercial markets. Assembly of the timeline is based largely on tacit knowledge compiled from past or similar experiences. Roles and responsibilities are distributed among the functional SMEs, and cross-functional collaborative tasks are identified. Project management develops and maintains a single execution plan incorporating resource and time estimates for each function. A separate but integrated execution plan is developed for late arriving information relative to the expected review cycle timing for the regulatory submission drafts. A third execution plan is developed for information arriving post-submission, based on the applicant's regulatory strategy for mitigation activities for expected health authority questions and post-marketing commitments. Interim milestones, in the form of draft

TABLE 21.2

Knowledge Management Elements for CMC Regulatory Documentation: Systems

Pain Points	Mitigations
• Access reliability challenges leading to authoring outside the collaborative system • New team members using the system without training or understanding of desired best practices • IT system failures or suboptimal performance due to equipment limitations (e.g., older computer models and limited bandwidth internet) • Lost or delayed work if too many simultaneous users working on the same document • Multiple versions of the same source document • Similar content in multiple source documents • Multiple source document repositories including local systems • Inability to search for source documents associated with specific regulatory submission sections for specific health authorities • Lack of awareness of assignment, or its scope and timing • Too many assignments for the same individual in a too short timeframe • Data loss risks with multiple users of Excel-based tracking logs • Maintenance interruptions or system downtime prior to deliverable milestones limiting last minute activities	• Full time (24 hours, 7 days) rapid response IT support • Live demonstration and onboarding of new system users • Training and designation of super users • Clear best practice documentation and communication of system use expectations • Designated Project Managers responsible to ensure update of tracking sheets • Comprehensive identification of potential issues/risks, including identification of early intervention mitigation activities • Single, searchable, source document repository for all supporting documentation (e.g., ability to query source documents supporting each common technical document [CTD] section of each regulatory submission) • Single, searchable, regulatory documentation repository (e.g., ability to query submissions for technical information for all products and all markets) • Aligned and communicated taxonomy for loading source documents into repository
Threats/Risks	**Opportunities**
• Partner organization (e.g., contract manufacturing/research organizations, CMOs/CROs) system access limitations • Aging document management software platforms	• Development of robust, "new generation" collaborative authoring platforms • Standardized systems across all functions, sites, and organizations involved in regulatory documentation preparation (including CMOs and CROs) • Integrated risk identification and evaluation tools that deliver risk roll-ups to support prioritization of mitigation activities • Improved search tools to find relevant knowledge documentation across existing storage locations • Easier management of regulatory submissions to various health authorities, along with supporting source documents, throughout their lifecycle

regulatory document reviews and revisions, are established to assure the desired delivery timing of the final submission.

The status of content generation is assessed in "near real-time" via informal or formal update meetings. Accuracy and timeliness of status information are enhanced when these systems are implemented and leveraged fully.

Evaluation of Pain Points, Mitigations, Threats/Risks, and Opportunities

A summary of pain points, mitigations, threats/risks, and opportunities for timeline is shown in Table 21.3.

TABLE 21.3

Knowledge Management Elements for CMC Regulatory Documentation: Timelines

Pain Points	Mitigations
General: • Tight and aggressive timelines due to speed-to-market pressures • Critical activities concentrated into short periods close to the key milestones • Delays and rushes near deadlines (i.e., drama and heroics) • Insufficient assigned trained and experienced resources • Reduced bandwidth or unavailability of resources (e.g., vacations, travel, and outside of normal work hours) • Burnout of individuals in high demand as resources • Scope creep in deliverables (especially post-approval supplements)	**General:** • Harmonized, structured, business process for project management for all impacted functions • Initial planning meeting to select required tasks from a comprehensive menu of potential tasks relevant to the specific regulatory submission • Accurate and timely tracking of progress, issues, and risks to interim milestones • Check-in meetings to review risks to timeline and escalate issues, matching frequency to the intensity of work during a given period • Advance notice of estimated timing for intense working periods (or likely downtimes) • Utilize vacation tracking sheet to plan resources several months prior to intense working periods • Leverage of time zone differences to augment business day compared with a single time zone • Ensure timeline permits reasonable timeframe for authoring and reviews (e.g., calculation of number of pages per day of review)
Threats/Risks	**Opportunities**
General: • Compression of timelines due to rapid approval of regulatory pathways • Competing activities and assignments for functional SMEs as resource loading becomes increasingly lean • Turnover of functional SMEs (both authors and reviewers) to unrelated jobs within the company, retirement, or jobs outside the company (including CMOs or CROs)	**General:** • Ease of assembly and communication of a "single version of the truth" in "near real-time" • Ability to rapidly identify and create visibility to potential and actual issues in complex and interconnected activity schedules with sufficient runway for useful mitigation

Quality

Definition and Current Practices

Quality for regulatory submission sections and their associated source documents is focused on technical data accuracy and information consistency (an evaluation of explicit knowledge), as well as the robustness of scientific arguments and conclusions (an evaluation of tacit knowledge). The quality of source documents directly impacts the accuracy of the regulatory submission sections that they support. High quality thus is essential starting from the earliest stages when activities are planned that eventually result in source information.

Quality checks are conducted using second scientist/person reviews by functional SMEs and often followed by documentation audits by the quality function. Accuracy of information can be traced from the regulatory submission section back to supporting source documents using (1) primary sources that are directly linked to "on the floor" documentation (e.g., electronic notebooks, certificates of analyses, LIMS, batch sheets, and SOPs) or (2) secondary sources that are compilations of primary sources with additional commentary optimally using formats more directly suited for regulatory submissions (e.g., technical reports and process descriptions). A source document map, indicating exactly where content is located in source documents, is created for each common technical document (CTD) section ideally at the time of authoring and used for these quality checks. Resolution of quality check findings results in correction of the regulatory submission, update of the source document map, or revision to the source document.

Quality checks for statements comprising general scientific knowledge, backed up solely by technical expertise (specifically statements of tacit knowledge), pose a special knowledge management challenge. These statements need to be evaluated carefully during quality checks to assure their scientific accuracy. Although no literature references likely exist for such statements, it is usually possible to obtain a few literature examples that are consistent with and thus suitably test the statement's accuracy.

Typically, quality checks are conducted after all reviews, comment resolution, and revisions are completed for an individual regulatory submission section. Thus, each section proceeds separately through the quality check process. Consequently, it can be more difficult to perform quality checks for consistency among sections with related content. Changes in primary or secondary information sources, as well as updates during the preparation of the regulatory submission, may result incomplete identification of the change's impact across related regulatory submission sections. Careful assessment of consistency during reviews confirms the technical and scientific consistency of the knowledge being communicated to health authorities.

Efforts to assure quality of regulatory submissions continue after the initial submission of sections. Errors and inconsistencies are identified during preparation of subsequent regulatory documentation or by health authorities in their questions and assessments. Tracking of these items assures that their impact is evaluated by CMC regulatory SMEs, and that appropriate actions are taken to correct or clarify regulatory documents. Thus, ongoing communication of accurate knowledge to health authorities is assured.

Evaluation of Pain Points, Mitigations, Threats/Risks, and Opportunities

A summary of pain points, mitigations, threats/risks, and opportunities for quality is shown in Table 21.4.

TABLE 21.4

Knowledge Management Elements for CMC Regulatory Documentation: Quality

Pain Points	Mitigations
General:	**General:**
• Compressed timeframe for quality checks due to content generation delays • Last minute revisions to source documents or regulatory submission sections causing source document map inconsistencies • Out-of-date or conflicting source information • Errors and inconsistencies potentially missed by quality checks • Churn over how to perform quality checks for statements based on scientific expertise (i.e., tacit knowledge)	• Authoring based on approved and latest version of source information whenever possible • Up-to-date terminology harmonization document to bridge terms used in source documents to submission • Quality check expectations communicated and documented, along with training and "spot-checks" where applicable • Special focus on quality checks for non-GMP source information and training for its providers • Source document map creation training and best practice communication to minimize questions arising from quality checks • Aligned business process for quality checks of statements of scientific expertise (e.g., using literature references consistent with statement) • Source document storage repository for version control and easy retrieval of source documents during quality checks • Additional reviewer (or potentially quality) checks based on map of related sections with high probability of similar information
Threats/Risks	**Opportunities**
General:	**General:**
• Compression of timelines due to rapid approval regulatory pathways • Competing activities and assignments for trained resources performing quality checks • Manual transfer of data between IT databases and source document	• Expectation for right-first-time outcome supported by timeline and resources • Flexible number of trained resources for quality checks to support unexpected peak requirements (both within a program and for the portfolio) • Risk-based approach for quality checks (e.g., focus on established conditions and regulatory commitments) • IT systems for validated data exports (e.g., from LIMS directly into source document)

Knowledge Management Improvement Opportunities for Preparation of CMC Regulatory Documentation

Evaluations of mitigations and improvement opportunities for knowledge management practices for the preparation of CMC regulatory documentation are focused in the areas of content generation, systems, timelines, and quality (Tables 21.1 through 21.4). Several of the mitigations listed were developed, applied to subsequent efforts, further improved, and then reapplied to the same or other programs. The success of implementation of a specific mitigation was directly related to how well the mitigation was placed in the flow of planned regulatory document preparation activities (Lipa et al., 2013) and expectations communicated.

In contrast, the ability to implement the opportunities listed appears limited by the desired level of resource investment, especially when many of the same resources (both SMEs and systems) are actively engaged with the initial, ongoing, and lifecycle management of several regulatory submissions. Timely completion of draft guidance (i.e., draft FDA guidance on established conditions, ICH Q12) by health authorities has the ability to catalyze and prioritize implementation of improvement opportunities for knowledge management practices.

Controls and Metrics

Optimally, controls to sustain implementation of the knowledge management elements for preparation of CMC regulatory document generation are placed within the project management framework established for the program. Initial establishment of a comprehensive project plan, inclusive of interim milestones for all tasks and for all functions, forms the basis for customization of suggested tracking tools. Ultimately, it is the collective outputs of these tracking tools that provide the necessary controls, both for a single program as well as across the portfolio.

Metrics assess the impact of the knowledge management elements, and from these metrics the goals of further improvement opportunities are obtained. Since metrics incentivize behaviors, selection of the metric as well as its target must be done carefully to avoid inadvertently incentivizing undesired along with desired behaviors (Table 21.5).

TABLE 21.5

Example Lagging and Leading Metrics for Impact of Knowledge Management Elements

Lagging Metrics to Evaluate Outcomes	Leading Metrics to Evaluate Milestones
On-time regulatory document submissions, application validation, and approvals in all markets filed	Adherence to interim milestones for timing and quality of deliverables (both source documents and regulatory submission sections)
No significant health authority questions or post-marketing commitments leading to new studies beyond those anticipated in regulatory strategy	Actual resources required within estimated values; no unplanned significant resource spikes and sufficient resources assigned
No major errors or inconsistencies identified post-submission	"Right-first-time" for quality checks pre-submission
Minimal number of minor errors and inconsistencies identified post-submission by either health authorities or by functional SMEs	

Conclusions

The four elements of knowledge management selected—content generation, systems, timelines, and quality—fully support the preparation of regulatory documentation for multiple submission activities. Collectively, these four elements, as illustrated in Figure 21.1, assure the suitability of the knowledge communicated to health authorities. The mitigations and opportunities assembled provide a comprehensive menu for prioritization of improvement efforts, which comprise significant but beneficial resource investments. Although transferred in the explicit format of a regulatory submission, this knowledge depends on the continued availability of major amounts of tacit knowledge during both regulatory submission preparation as well as the post-submission period of product lifecycle management. Knowledge management must last for about a decade longer than 10–15 years typically associated with loss of patent exclusivity for a product. This long time period substantially exceeds the employment timeframe of most functional SMEs associated with the product, and this disparity is likely to worsen as job changes increase in frequency.

Consequently, additional opportunities for long-term management of tacit knowledge are urgently needed to supplement the improvement opportunities already identified that primarily focus on explicit and short-term tacit

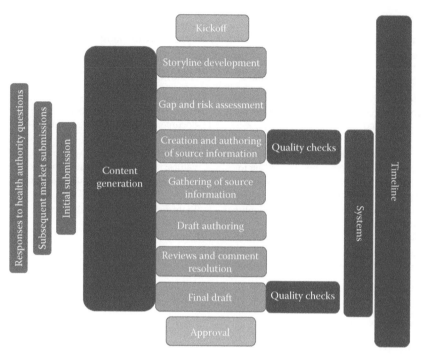

FIGURE 21.1
Illustration of the four elements of knowledge management (dark blue) supporting generation of CMC regulatory documentation (purple). Details of content generation (green) span from kickoff to approval (light blue).

knowledge management. Ideally, these additional opportunities should focus both on converting tacit to explicit knowledge and on significantly improving retention of tacit knowledge in its tacit form. Investments in improvements to both explicit and tacit knowledge management collectively speed up global patient access to medicines by sharpening communication of knowledge supporting regulatory commitments between applicants and health authorities.

Acknowledgments

This work summarized the ideas, experiences, and efforts of many colleagues over several years of regulatory submission document generation. Particular acknowledgment (in alphabetical order) is extended to: Elise Avram, Marc Bastiaansen, Tamas Blandl, Michel Chartrain, Ryan Cooper, Marissa Fleishman, Wout van Grunsven, Cathy Hoath, Rachel Howe, Weining Hu, John Lepore, Gargi Maheshwari, Anne-Marie O'Connell, Colette Ranucci, and Tracie Spangler.

References

FDA, Established conditions: Reportable CMC changes for approved drug and biologic products: Guidance for Industry, Draft Guidance, US FDA, May 2015, http://www.fda.gov/downloads/Drugs/GuidanceComplianceRegulatory Information/Guidances/UCM448638.pdf

FDA, Expedited programs for serious conditions—Drugs and Biologics: Guidance for Industry, US FDA, May 2014, http://www.fda.gov/downloads/Drugs/ GuidanceComplianceRegulatoryInformation/Guidances/UCM358301.pdf

ICH Q9, Quality Risk Assessment, November 2005, http://www.ich.org/fileadmin/ Public_Web_Site/ICH_Products/Guidelines/Quality/Q9/Step4/Q9_Guideline. pdf

ICH Q10, Pharmaceutical Quality System, June 2008, http://www.ich.org/fileadmin/ Public_Web_Site/ICH_Products/Guidelines/Quality/Q10/Step4/Q10_Guideline. pdf

ICH Q12, Final Concept Paper: Technical and Regulatory Considerations for Pharmaceutical Product Lifecycle Management, dated 28 July 2014, endorsed by the ICH Steering Committee on 9 September 2014, http://www.ich.org/ fileadmin/Public_Web_Site/ICH_Products/Guidelines/Quality/Q12/Q12_ Final_Concept_Paper_July_2014.pdf

Lipa M., Bruno S., Thien M., Guenard R., 2013. A practical approach to managing knowledge—A case study in the evolution of knowledge management (KM) at Merck. Pharmaceutical Engineering, November/December, pp. 1–10.

22

A People Approach to Managing Knowledge: Why Do You Come to Work and Who Are You Working For?

Siân Slade and Catherine Shen

CONTENTS

> Bristol–Myers Squibb faced a challenge in providing what their customers wanted—at the right time—every time. This case study shares the BMS journey in establishing a marketplace for knowledge, which has increased *knowledge velocity* and continues to enhance the culture of collaboration.
>
> **Editorial Team**

This case study provides an overview of *Vanguard*, a project developed at Bristol-Myers Squibb (BMS) based on building a knowledge marketplace, where people would gather to exchange knowledge. Our objective was to illustrate knowledge sharing as an organizational culture augmented by specific tools to provide systematized knowledge transfer globally, supporting the mission of Bristol-Myers Squibb to discover, develop, and deliver innovative medicines to help patients prevail over serious diseases (BMS 2016b).

Knowledge is a key asset in any organization and transfer of knowledge between individuals is of critical importance to the organization both internally and externally. Every employee of a pharmaceutical organization is, de facto, a knowledge worker. All employees need to recognize this as part of their role. Our purpose (Sinek 2014) with *Vanguard* was to ensure that we could equip colleagues across 159 locations worldwide with the ability to access the latest knowledge in order to enable excellence of service to our external customers. In ensuring swift knowledge transfer internally, the broader intent was to enable provision of current, relevant information to external customers, quicker than Google.

As with any large global organizations, the internal company challenge at a country level is ensuring access to the latest information to ensure decisions related to strategic product planning, resourcing and patient care can be made. Examples include development planning for phase III programs or commercialization planning related to reimbursement submissions.

Of key importance and directly related to working in a medicine-related, health care organization, the external challenge is to ensure continued provision of up-to-date, timely, accurate, and relevant information to health care professionals responsible for making decisions about patient care with a key focus of enabling safe and appropriate use of medicines. In this age of 24/7 digital access, the imperative is to provide exactly what the customer is looking for when they need it—right information, right customer, right time, every time (Figure 22.1).

The ability to achieve this directly fits with the Bristol-Myers Squibb Mission and as shown in the "Who Are You Working for Campaign" (BMS 2016a), why employees come to work: making a difference for the patient.

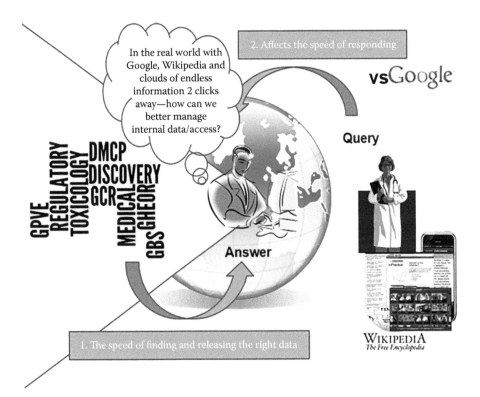

FIGURE 22.1
The speed to find, release, and respond.

Vanguard was developed in direct response to a business need first identified in markets. The pivotal question: "what knowledge do you need to be successful in your role?" was the guiding mantra. Lack of access to knowledge means delays in getting work done expediently, leading to frustration and time wasted in trying to find answers; in addition, the potential for rework due to lack of visibility of existing materials is unproductive and a resource drain, and importantly there is a concern of creating the potential for risk in using out-of-date information.

Vanguard is an example of a project that was developed and led from within the organization, showcasing the importance of listening to employees and empowering people, irrespective of their organizational role to come together, discuss, and attempt to resolve business challenges and develop scalable solutions.

The following sections describe the project genesis, the change journey, the key learnings, and the practical tips to others who may be considering embarking on similar challenges. Following identification of the business challenge, the genesis, there were three distinct phases: building the foundation, enhancing capability, and embedding capability.

Key learnings: Anyone in an organization can identify a business need

Key enablers: A company culture that supports innovation

The Genesis of *Vanguard*

In early 2011, within the Global Medical Information group, a newly created organization function was set up entitled, Core Content and Knowledge Management. The mandate was the creation and dissemination of content to enable Bristol-Myers Squibb Medical Information colleagues worldwide to answer customer enquiries with one global voice.

Content is an example of *explicit* knowledge, that is, it is written down. It can be stored, shared, and updated. It can be translated into other languages. It can be version controlled. It can be assigned with key words or metadata. It can be searched. Content is a *tangible* deliverable.

But what about knowledge? Knowledge (Oxford University Press 2016) is defined as facts, information, and skills acquired through experience or education; it is not data or information though these words are often used interchangeably. The key aspect is the word experience and thus a human component in how to interpret and apply data and information. Knowledge is thus more than content. In any organization, imagine the potential value of being able to tap into all that experience or tacit knowledge within your teams!

Content is a deliverable without the overlay of knowledge. If two individuals were asked to present a slide deck on a drug mechanism of action, they both would have access to the same content, but one may have more experience of the topic area than the other and thus be able to speak with more credibility and authority. Experience is thus important.

Exploratory discussions with senior leaders across functions yielded animated discussions on managing knowledge with a focus on actively accelerating the body of knowledge, that is, ensuring *knowledge velocity*. This concept of *velocity* adds emphasis to the need for accelerating the speed of knowledge transfer, an action not explicitly represented by the term *knowledge management*. There was a clear recognition of the organizational challenges of both building explicit knowledge in the form of content and embracing the critical importance of tacit knowledge. Imagine the power of downloading tacit knowledge, for example, a human thumb drive, to create an artificial intelligence system with an *Ask Me Anything* user interface.

Coincidentally, accelerating organizational interest in knowledge platforms was emerging in different parts of the company and helped create a sense of change. A major knowledge management project, developed a decade earlier, provided key insights into both learning from past experience and also understanding that it was not a *one and done* event that can be ticked off a *to-do* list.

At the time of thinking through the above, two other factors are added to the impetus and need to act: the first was technological and the second was societal.

Technological

Bristol-Myers Squibb introduced Microsoft SharePoint as the key collaboration platform worldwide. Although introduced as a sharing platform with clear governance regarding site layout and purpose, invariably, they moved to not just being sharing platforms but also repositories with no clear governance regarding review/archiving. The challenge for any individual, either new to the organization or new to the therapeutic area, was the ability to navigate easily to, and indeed have security access, the information they needed at their fingertips. Without a map or navigation, it was unclear which SharePoint site held which information and as a side-effect consequence, you had no means of knowing if you were accessing the latest information.

Societal

In March 2011, following the tragic events of the Japanese tsunami, people in Japan were advised to take potassium iodate to prevent the uptake of radioactive iodine (associated with thyroid cancer). An urgent enquiry came in from Japan Medical Information asking what, if any Bristol-Myers Squibb drugs, reacted with potassium iodate. Who would be the right person to provide this type of information and how would you find them? In fact, the answer was quickly obtained through immediately escalating to the head of R&D to find the correct individual—but the solution was person dependent versus system or process dependent.

Leaders put forward different viewpoints as to how to solve for building knowledge capability ranging from the importance of people, collaboration, and knowledge networks to the use of technology including availability of sophisticated search tools with reach across multiple databases and systems.

Thanks to the support from the Chief Medical Officer that commitment and resource were given to truly understand the challenges presented to achieving effective knowledge velocity across the organization. It was agreed that this would be a pilot project, with worldwide scope, in a fast-moving highly scientific therapeutic area with assets entering phase III development with first launch worldwide planned for Japan.

The scene was set.

> Key learnings: Clearly articulate the business need, proposed next steps, and resources required
>
> Key enablers: Accessible senior management; a growing organizational interest in the capability

---

Building the Foundation: "The Journey of a Thousand Miles Begins with a Single Step"—Lao Tzu

Embarking on organizational change leadership is extremely challenging and requires belief, passion, and tenacity. As with any journey, the right equipment is mandatory. An orientation to leading strategic change, an internal training program, provided an understanding of the complexity of leading change, the key phases of a change journey, and the critical importance of stakeholders, advocates, and skeptics. Ongoing coaching support was also provided. Kotter's eight-step process for transformational change (Kotter International 2016) provides an excellent and practical orientation to change leadership. This chapter will illustrate the importance of a change framework—what to plan, what to expect, and how to overcome inevitable challenges.

It is widely acknowledged that *70% of change efforts fail* (Kotter International 2016). In order to be part of the successful 30%, the need to engage individuals across the assigned therapeutic area worldwide was imperative. Although senior management had endorsed *Vanguard*, it was these individuals, working within the functions and geographies, who would enable this project to succeed or otherwise.

A dynamic Project Manager, assigned from the Development and Medical Business Operations Group, developed a briefing deck. This was presented to the development leadership team to gain their agreement regarding the business need as well as support. This was a key enabler to engage others who could see the interest in the project and wanted to be part of the *guiding coalition* (Kotter International 2016).

Interviews were conducted with 40 colleagues in both global and market functions from across the matrix and focused on the three key elements: people, process, and technology. The current state was described as technology heavy with lack of process clarity regarding knowledge sharing.

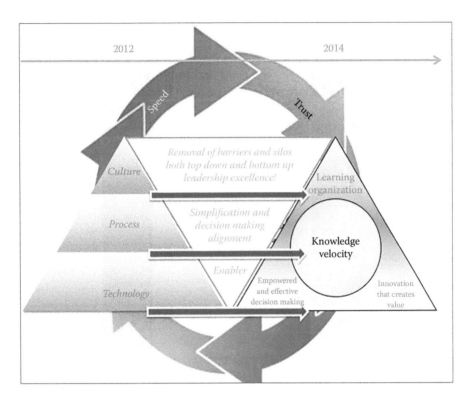

FIGURE 22.2
Achieving *knowledge velocity*.

The default position was to err on the side of caution. The goal of achieving *knowledge velocity*, that is, rapid knowledge transfer was considered to be people focused and technology enabled (Figure 22.2).

A strong customer-focused culture was outlined as critical driver to achieve organizational success.

The primary objectives were determined:

1. Increasing the ease and speed of finding and releasing information for internal/external use.
2. Addressing the knowledge model to be used across the project scope.
3. Developing a sustainable, scalable approach to be used across other therapeutic areas.

Achieving the aforementioned objectives would require streamlining and optimization of the existing knowledge flow processes, understanding how technology could be used, and would require the need for continued connection and collaboration.

Throughout the project the importance of people was strongly recognized. *The Dancing Man* (Sivers 2010) as well as the *Diffusion of Innovations* (Rogers 2003) were useful reminders about the importance of engaging people in a common purpose, to *join the Vanguard* and to be the part of a solution to lead change. Of course, there were skeptics as well as advocates but also full agreement, across all stakeholders, that there was a burning platform to be addressed.

> Key learnings: Clearly articulate the burning platform, proposed
> next steps, and resources required
>
> Key enablers: Accessible senior management who actively listen
> and seek to understand

Vanguard and Vision: A Call to Action

At project launch, many questions were received about the name *Vanguard*. The name was a deliberate move, based on the dictionary definition of first movers leading the way, a call-to-action for the organization to think critically about knowledge as a key asset and mobilizing the power that knowledge has internally and ultimately externally. In our minds, it was not *just* a pilot.

The pharmaceutical industry is a knowledge industry. It undertakes innovative research and rigorous development in a highly regulated setting. New drugs are developed to meet unmet therapeutic needs, and time to market is imperative to bring innovation to customers expediently. Consulting firms and news corporations were identified as analogous knowledge industries. Both rely on codifying knowledge and enabling quick access of both explicit and tacit knowledge.

It was the worldwide news service that was our guiding light. Our vision was to set up an internal approach analogous to the British Broadcasting Corporation (BBC). *The BBC mission is to enrich people's lives with programs and services that inform, educate, and entertain* (BBC 2016). The BBC has a news service, available worldwide 24/7, through multiple channels (TV, smartphone applications, and website), updated regularly in English and other languages. Our vision: "What If... You Could Instantly Access BMS Knowledge... anytime, anywhere, anyplace..."

Key learnings: Identify analogs in other industries and sectors and learn from their experience

Key enablers: An impactful project name created curiosity and engagement

Before You Move Further, Clearly Understand the Current State

The first important deliverable was to provide a visual map of the key knowledge documents by function.

In 1933, the original version of the famous and distinctive London Underground map appeared. Originally, the different rail lines of the underground were run by different companies, each with a map, of different styles with no integration or common format. The fully integrated London Underground map was developed to encourage subway travel and to address concerns about poor financial results. This map is a key tool enabling travelers to traverse London. The publication *No Need to Ask!* (Leboff and Demuth 1999) was cited regularly in *Vanguard*. A map meant individuals could be self-sufficient and navigate their way round the content (and associated authors and subject matter experts) versus ask a *policeman* to tap into their tacit knowledge of the organization. As part of related research work, an organizational network analysis identified the *policemen* or knowledge go-to people with organizational knowledge who could easily direct the enquiry in terms of an appropriate subject matter expert.

A visual illustrative knowledge map provided a preliminary navigation and inventory of where all the asset knowledge resided and provided a guide to the totality of the asset content. The initial map presented has undergone subsequent iterations, and Figure 22.3 is provided as illustration only. The map was developed by working with individuals from across the product lifecycle stages, within discovery, development, and delivery. The delivery of the map provided fresh impetus and interest and provided points of intersection with other groups across the organization.

Focus was on key knowledge deliverables, for example, investigator brochure, the corresponding databases, areas of functional responsibility, and the stages in the development cycle that they would be produced.

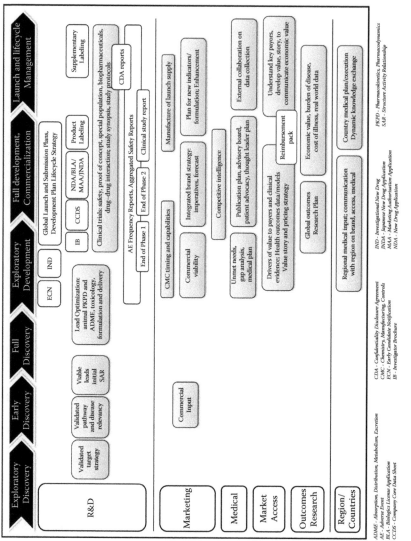

FIGURE 22.3
Knowledge map (illustrative).

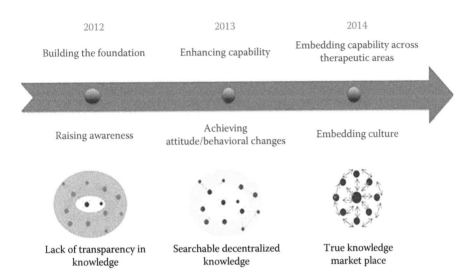

FIGURE 22.4
Vanguard three-year knowledge road map.

Collaboration platforms and user access eligibility were captured as well as functional content owners were identified. Respondents were asked about knowledge sharing processes and the governance rules and standard operating procedures that applied.

The knowledge, or *i-map* (short for information map, as it was known), which was developed was a pivotal document that provided a clear scope of the totality of the knowledge to multiple stakeholders across functions. For many people, particularly those who had only ever worked in one function, it provided a clear picture of the entirety of the body of knowledge across the product lifecycle. For those new to the company, it provided an immediate orientation as to what information was available. Leveraging information about the current state and knowledge map, a three-year knowledge road map was developed for *Vanguard* (Figure 22.4).

Key learnings: The importance of understanding the totality to determine the scope of the challenge

Key enablers: Collaboration by individuals, the tenacity, and discipline of the Project Manager

Concurrent Work and Exploration

Concurrent to the map development, short-term support was accessed from a knowledge consultancy who had worked with partners in the early development organization. Work was undertaken to understand knowledge flows related to four specific areas: protocol amendments, congress activity, reimbursement materials, and strategic planning.

A series of interviews with functional members located worldwide provided greater clarity on the challenges of timely knowledge sharing. Again, the same question was asked "what knowledge do you need to be successful in your role?" A pivotal part of this analysis focused on scientific congresses, covering the pre-, peri-, and post-congress periods. This work became a spin-off as a separate but related project not part of the broader *Vanguard* effort. The other three areas of assessment were used to inform the corresponding organizational capabilities, and the insights gained enabled clear focus on the markets and receiving and recording specific questions.

> Key learnings: Testing different avenues to decide on what direction to take, or not take a project, forward
>
> Key enablers: Working with external consultants who establish an immediate rapport with the markets

Enhancing the Capability: Achieving Attitude/Behavioral Changes

Bristol-Myers Squibb is a place to thrive (BMS 2017). Bristol-Myers Squibb recognizes the importance of people making a difference, thinking diversely, and developing the workforce. The success of the *Who Are You Working for Campaign* (BMS 2016a) highlights the high level of engagement toward the mission and serving patients. All employees are expected to demonstrate (and also are measured on) the biopharma behaviors (decide and act, innovate and improve, connect and collaborate and grow and engage) in their everyday work, and these behaviors are recognized as the way to drive and build a high-performing culture and deliver the company mission. Individuals often work in matrix structures, collaborating across the global footprint and one of the key achievements that *Vanguard* achieved was helping to accelerate the building of the global community.

People: Community

Recognizing the clear importance of people and an engaged community across a global footprint, a collaboration expert with skills in knowledge management as well as an experienced contractor was added to the team to fulfill the combined role as the Knowledge Lead. A critical part of this role was building a formal (and filterable) directory of all the therapeutic area stakeholders and continuing to add and include new employees joining the therapeutic area worldwide.

In addition to building the directory, the focus was on understanding real-time knowledge challenges, making them visible companywide and addressing and posting up the solutions. Using the themes presented by Malcolm Gladwell in the Tipping Point (Gladwell 2002), in essence the two individuals fulfilled the role as the *connectors*, ensuring that individuals worldwide could be connected with *the mavens* (information specialists) (Merrill et al. 2007). An organizational network analysis was undertaken to understand who the key go-to people within the therapeutic area were.

> Key learnings: Conduct organizational network analyses early in the project to determine who the mavens are
>
> Key enablers: Dynamic, engaging knowledge leads facilitate visibility of market individuals, promoting voice

Process: Knowledge

We have already outlined the importance of both explicit and tacit knowledge in an organization.

Work to further develop the knowledge or i-map was undertaken. This involved identifying and outlining the top 20 key knowledge documents for each of the assets in development. This was then subdivided to an individual functional level to identify a top 10 list of documents per function, for example, the top 10 documents for Medical Information. This visibility of all the content deliverables is applicable to every therapeutic area across the organization. It remains a significant step in enabling full line-of-sight of functional content infrastructure to new colleagues joining the organization. The above was solely for *explicit* knowledge.

In an attempt to share tacit knowledge in real time in a fast-moving scientific therapeutic area, an internal social media platform was utilized. The approach involved actively engaging individuals attending a sequence of highly regarded international scientific congresses in 2013. Senior leadership actively supported the first congress, promoting activity and scientific exchange on the platform to role model and encouraged others to participate. Key data were transmitted worldwide in real time. Although some element of traction with certain parts of the community was seen, this was certainly an opportunity to further develop dynamic knowledge sharing.

In the second congress, practical tips included helping colleagues download the smartphone application and also to encourage posting as well as managing concerns relating to network access, potential costs, and the challenges of data roaming charges.

Key learnings: Subject matter experts have high impact but require encouragement to collaborate on social media

Key enablers: A map of key company content is an invaluable and time-efficient resource

Technology: Scientific Search

As previously highlighted, similar to many other companies in 2011, SharePoint technology had been widely introduced as a company collaboration tool. SharePoint was widely being used as a de facto content repository and often the first comment made to individuals was *"is it on the SharePoint site?"* For the assets in question, 25 SharePoint sites were identified as being used by colleagues within the project scope. An ongoing challenge revealed that individuals working within the same therapeutic area had different levels of security enabling different depths of ability to access knowledge. The security model to assign access was by role or level but not by knowledge need. The problem was resolved in the form of a highly experienced and motivated scientific search scientist joining the project and developing a search facility which would enable search visibility across the totality of the content on the SharePoint sites regardless of security access. To preserve security, results would only be returned as a top-line excerpt for those colleagues without full security access. Interestingly, GSK presented data (Bio-IT World 2013) on their own search tool, Socrates, and it was interesting as well as supportive to know that we were not alone in our endeavors and to learn from companies handling similar organizational challenges.

Key learnings: Ensure that development tools are sufficiently robust before testing widely to keep audience engaged

Key enablers: Search tool that is able to ensure visibility of key resources while preserving security/governance

Putting It All Together

As highlighted in the opening paragraph, knowledge sharing is an organizational culture aided by tools. The specific deliverable for *Vanguard* was the creation of a virtual marketplace providing *systematized* knowledge transfer.
The deliverable consisted of three key elements (Figure 22.5):

1. A fully searchable directory, with filters of all the people worldwide as well as an internal social media tool.
2. A knowledge map, outlining all the key knowledge deliverables, purpose, location, and document owner.
3. A scientific search tool.

Orientation training was provided virtually, in time zone, to individuals. Early adopters in markets championed the virtual tool and became part of the growing user community across the geographies. Weekly project management and metrics tracking and question resolution took place in addition to stakeholder feedback.

Specifically for search, the focus was on repeat usage across multiple functions within the development and medical organization. Usage was initially high, but repeat usage was quite fragmented partly due to a number of challenges in search function development.

A marketplace providing "systemized" knowledge transfer

Community
Look for people

Community connects people which forms a market place of knowledge supply and demand

Knowledge
Discover key documents

Knowledge creates transparency of key documents, their purpose and ownership

Search
Try it now

Search provides visibility to the right documents at the right time

Ensuring discipline and making it stick

Behavioral change from supply side to share knowledge that could be useful to others

"Knowledge manager" ensures discipline around updating content and appropriate use of tools

FIGURE 22.5
Vanguard.

Rewarding the Early Adopters

Reward and recognition awards were put in place to recognize the efforts of the developing super user community, and their efforts were recognized both in terms of small monetary awards and recognition by senior leaders on the internal social media platform. These super users themselves acted as advocates within their own geographies to lead the *Vanguard* locally. What had been in part essentially a push model of information from global headquarters, started to develop as a pull model with people asking the Knowledge Lead for assistance rather than asking a local colleague. This served to ensure clear visibility at a global level of areas where knowledge was needed and the ability to address these gaps to strengthen the support model in a global-market model.

Notably, the countries that most accessed the service were the most distant to the global group, the top three user communities being Japan, Brazil, and Korea with a high degree of access, repeat usage, and active engagement.

Given the specific importance of Japan as first launch market, an internal secondment was enabled through an internal Talent Promotion Program. The successful candidate from Japan was based in the global headquarters for six months. The ability to make this happen was enabled by senior management. Clear objectives were set for the secondment, which was in addition to the individual's existing clinical role, to aid knowledge transfer and connectivity between teams in Japan and the USA.

> Key learnings: Knowledge sharing is multidimensional, and the de facto position is to see tacit knowledge
>
> Key enablers: Strong and active support of individuals within the markets to provide them a voice and platform

Embedding Capability across Therapeutic Areas

In 2014, the project learnings were reviewed and as per the three-year road map, a decision was made that there was enough experience and evidence as well as organizational need to warrant scaling up of the approach.

The project was resourced to include a second portfolio with a dedicated knowledge lead to further refine and build the capability. The project was renamed and provided with dedicated functional support to continue to evolve the capability.

Key learnings: The importance of leadership/resourcing and being embedded within a team to ensure success

Key enablers: The project name was changed to allow a generic application across assets/therapeutic area

Conclusion

More than 70% of change efforts fail. Why? Because organizations do not take consistent, holistic approach to changing themselves, nor do they engage their workforces effectively (Kotter International 2016). This project enabled us to identify a clear business need and to experience firsthand both the opportunity and challenge of leading global change across an organization.

With a clear overarching purpose to always focus on the customer, there were two common learning themes:

1. Enabling colleagues to quickly build community with other colleagues worldwide, both in market and global roles, building trust and a culture of tacit knowledge sharing, including the systems and associated business rules to facilitate this.

2. Orientating individuals to explicit knowledge in terms of content inventories together with access to tacit knowledge in terms of a people directory.

Our focus over the three years was to achieve a true knowledge marketplace. The challenge for an organization is that managing knowledge is not just a tool, it is a culture. *Vanguard* was about creating awareness of the importance of knowledge velocity and creating a culture of knowledge sharing, as well as the importance of business rules and governance. As per the quote from Kotter, our experience taught us that culture change requires significant effort, passion, and resilience but is at least in part achievable even over a relatively short project. The challenge at an organizational level is to achieve a sustainable approach to continue to evolve and develop.

Cultures are created by people, attitude, and behavior. Key to leading change is engaging people in championing the cause, creating the solution, and rewarding the short-, mid-, and long-term objectives. People are critical factors to ensure that projects receive attention, support, and maintenance of organizational pressure to continue to challenge the status quo. There is no quick fix to knowledge sharing and at the root of all solutions lies planning, discipline, and rigor. Process and technology are standardized approaches which apply to all therapeutic areas. Culture cannot easily be standardized. It is strongly directed by a sense of purpose and related to this, the building of a collective community and responsibility to drive knowledge velocity— to better support the end user in fast-moving therapeutic areas.

Building a worldwide knowledge capability is a strategic organizational imperative that enables individuals, irrespective of location, to access information, either tacit or explicit, when they need it for decision making. The overarching intent of developing *Vanguard* was directly focused on enabling internal knowledge sharing, leadership, and excellence to create an internal approach that would indirectly benefit external customers in line with the Bristol-Myers Squibb mission.

In recent years with the rise of digital technologies, big data, and ever increasing need for speed, agility, and focus on reducing time to market, we have seen a renewed interest in this core capability as companies look to streamline approaches and aid the ability to find and integrate data and information expediently, and thereafter to access and use with appropriate understanding and to understand the power of technology as a key enabler.

The opportunity to develop and build this project was a pleasure, a privilege, and a unique learning experience. We would like to acknowledge all the colleagues across the Bristol-Myers Squibb worldwide organization that contributed and made it possible.

References

BBC. Inside The BBC—Mission and values. 2016. http://www.bbc.co.uk/aboutthe-bbc/insidethebbc/whoweare/mission_and_values (accessed March 16, 2016).

Bio-IT World. Searching for Gold: GSK's New Search Program that Saved Them Millions. June 5, 2013. http://www.bio-itworld.com/2013/6/4/searching-gold-gsk-new-search-program-saved-millions.html (accessed March 16, 2016).

BMS. 2015 Annual Report. *BMS.com*. 2016a. http://s2.q4cdn.com/139948097/files/doc_financials/annual%202015/2015-BMS-AR.pdf (accessed March 16, 2016).

BMS. BMS—Our Mission & Commitment. 2016b. http://bms.com/ourcompany/mission/pages/default.aspx (accessed March 16, 2016).

BMS. More than just a place to work—a place to thrive. 2017. http://bms.com/careers/Our_Culture/Pages/default.aspx (accessed March 14, 2017).

Gladwell, M. *The Tipping Point: How Little Things Can Make a Big Difference*. New York: Back Bay, 2002.

Kotter International. The 8-Step Process for Leading Change. 2016. http://www.kotterinternational.com/the-8-step-process-for-leading-change/ (accessed March 16, 2016).

Leboff, D. and T. Demuth. *No Need to Ask! Early Maps of London's Underground Railways*. Harrow Weald, Middlesex, U.K.: Capital Transport Publishing, 1999.

Merrill, J, S. Bakken, M. Rockoff, K. Gebbie, and K. M. Carley. Description of a method to support public health information management: Organizational network analysis. *Journal of Biomedical Informatics*, 40(4): 422–428, 2007.

Oxford University Press. Knowledge. 2016. http://www.oxforddictionaries.com/definition/english/knowledge (accessed March 16, 2016).

Rogers, E. M. *Diffusion of Innovations*, 5th Edition. New York: Free Press, 2003.

Sinek, S. *Start with Why*. TED Talk. March 3, 2014.

Sivers, D. First Follower: Leadership Lessons from Dancing Guy. February 11, 2010. https://www.youtube.com/watch?v=fW8amMCVAJQ (accessed March 16, 2016).

23

Developing a Lessons Learned Process—Where Lessons Are Learned: A Case Study of Pfizer Pharmaceutical Sciences

Phil Levett

CONTENTS

Capturing lessons learned is a key to enabling knowledge flow and reuse of the learnings, but typically how effective are such activities? In this chapter, Pfizer shares how they developed a standardized knowledge led process for not only capturing these lessons but actually enabling learning from these lessons within their product commercialization teams.

Editorial Team

Introduction

Since 2012, the Pharmaceutical Sciences division within Pfizer has created and embedded a lessons learned process that has evolved over this period into a key component of divisional operational excellence. This case history tells the story of this evolution, from early approaches to today's third generation Learning Management System (LMS) (Lessons Learned 3.0). The key objective is to enable the reader to implement an effective LMS within their own organization while benefitting from Pfizer's learnings.

A key component of the drug development cycle at Pfizer is the transition from development into commercial manufacturing. This technical transfer, named Co-Development at Pfizer, is from the Pharmaceutical Sciences division to the Pfizer Global Supply division. Co-Development occurs over several years with gradual transition of knowledge, experience, and project accountability, starting during the commercial process design phase and completing postlaunch. The workflows accompanying Co-Development are well documented at Pfizer and have historically undergone periodic refresh cycles, the last of which was in early 2012. That refresh established 17 core activities that required contemporization, one of which was effective lessons learned.

Prior to the 2012 refresh, in what is now referred to as Lessons Learned 1.0, each Co-Development team was expected to undertake a single *end of Co-Development* lessons learned session. The aim was to capture in slides their learnings, present them to the business as continuous improvement opportunities, and then file those PowerPoint slides to the Co-Development lessons learned SharePoint site for other teams to access and learn from. This expectation was not a managed process and so compliance was variable. Many teams did spend 1–2 days face-to-face trying to recollect experiences from their previous 2–4 years together, and did capture an extensive list of learnings *summarized* over 50+ slides. The presentation of the lessons to the business often generated good discussion and some verbal commitment for business process improvement from leadership, however many of the suggestions were either not actioned or had no follow up mechanism. What was definitely not happening was any reviewing of the lesson slide decks once they entered the SharePoint site; *SharePoint was the place where lessons went to die!*

Lessons Learned 2.0

To initiate the lessons learned refresh, the team went back and reviewed those slide decks. In total they found more than 700 lessons spread across 20+ projects. They systematically reviewed the quality of those 700+ lessons by considering how actionable each lesson would be for either a continuous improvement team or a Co-Development team and concluded that approximately 120 (<20%) were high value lesson substrate. Much of the remainder was well intended but poorly constructed, often consisting of either bland statements of the obvious [*good communications/planning are important for achieving program success*] or unrealistic expectations of the unaffordable [*invest more time, dollars, and resources earlier to increase chance of program success*]. This convinced them of the need to include both a quality gate and *what makes a good lesson* guidance within our revised lessons learned process.

It was at this time that Pfizer was also fortunate to glean expert advice from the APQC, a world leading knowledge management research organization based in Houston, Texas. Dr. Carla O'Dell (CEO, APQC) advised that

- "Knowledge is sticky, without systematic processes and enablers, it won't move"

and reiterated the importance of

- "teaching in the teachable moment."

Cindy Hubert (Executive Director, Client Solutions, APQC) championed the criticality of

- "embedding knowledge management solutions within the flow of work"

and the APQC also reinforced the value of

- "completing the Knowledge Flow Process Framework learning cycle in order to realize business value" (Figure 23.1) [1]

These four principles resonated strongly with the refresh team and with senior leaders within Pfizer's Pharmaceutical Sciences division that led to their use as the foundational principles on which to build a lessons learned process.

The team re-examined Pfizer's Co-Development workflow and decided to break the process down into 5 distinct drug development phases that included some unavoidable chronological overlap (Figure 23.2)

- Process design
- Process development and clinical manufacture

FIGURE 23.1
Illustrates the APQC's knowledge flow process framework.

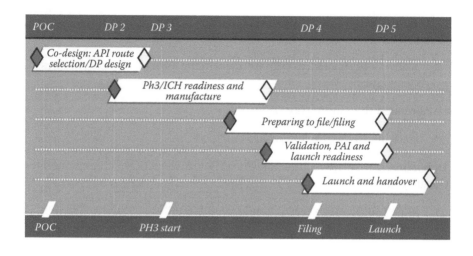

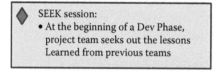 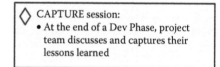

FIGURE 23.2
Illustrates the 5 phases within Pfizer's Co-Development workflow.

- Regulatory filing
- Commercial process validation
- Commercial launch

The next decision proved to be particularly valuable. As a Co-Development team completes each of the 5 phases of development, they were *required to complete a knowledge CAPTURE lessons learned session.* This was made a required component of the Co-Development workflow. This has resulted in teams capturing lessons when they are fresh in the memory with the added benefit of important lessons being made more rapidly available to other project teams. These CAPTURE sessions used the well-established U.S. Army's after action review approach of asking *What actually happened?*, *What should have happened?* and *What is the learning?* To keep this activity within the flow of existing work, these CAPTURE sessions have been run as one hour agenda items built into the regular weekly team meetings as required (approximately twice per annum per project team). Theses CAPTURE sessions usually occur at the Co-Development sublevel, split separately into active pharmaceutical ingredient (API) lessons and drug product (DP) lessons.

Similarly, when a Co-Development team commenced each of the 5 phases of development they were required to review *just those lessons learned associated with the development phase they are about to commence* in what we call a SEEK session. Within a SEEK session the project team would do an initial rapid pass of each of the relevant lessons within the lessons learned log repository (this will be discussed in more detail later in this case history) usually filtered on API or DP and also the phase of development they are about to commence. As the teams are only SEEKing lessons relevant to the work they are about to do (within approximately the next six months), they are *learning within their own teachable moment and thus are engaged and interested to learn.* Co-Development teams are passionate about successfully delivering their projects, so raising their awareness of relevant lessons, including contextual information and key contact names for follow-up, has proven that a strong catalyst for ensuring those lessons are appropriately considered within their own project plans. In addition, SEEK sessions also included attendance from members of a project team that had recently completed that phase of development, in order that they can share their experiences and address questions about the latest technical, workflow and regulatory challenges, and associated good practice solutions.

Of course these CAPTURE and SEEK session did not happen without a systematic process being in place. Each of the Co-Development project teams was led by a Pharmaceutical Sciences Team Leader (PSTL–pronounced *pistol*) and it was their responsibility to ensure that the CAPTURE and SEEK sessions were occurring. To facilitate this, one of the PSTLs kindly volunteered to develop and own the CAPTURE and SEEK schedule, which they updated on a quarterly basis by checking in with each PSTL individually on project progression. This CAPTURE and SEEK schedule proved to be a great trigger for prompting timely project team engagement with the process, particularly as it is the PSTLs themselves that are defining the optimum timing for each of their CAPTURE and SEEK sessions.

The other major feature of the 2012 Lessons Learned Process 2.0 was the introduction of a *Lessons Learned Network (LLN)*. This consisted of approximately 20 highly experienced colleagues, one representative from each of the Pharmaceutical Sciences lines (Analytical, Process Chemistry, Formulation, Quality, Regulatory, Operational Excellence, Clinical Manufacture, Clinical Supply, Project Management, and Learning and Development) and their respective partners in Pfizer Global Supply. The lessons learned network had the following *three key responsibilities*:

- First it managed and communicated the effectiveness and continuous improvement of the lessons learned process.
- Second it was accountable for promoting the desired learning culture throughout the division.
- And finally the LLN was the quality gate, responsible for ensuring that only high quality lessons entered the lessons learned log (LLLog) for the benefit of the project teams.

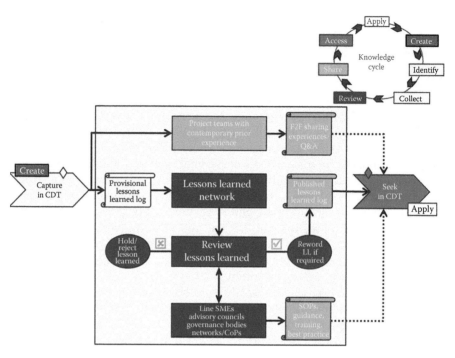

FIGURE 23.3
Illustrates the lessons learned network review process.

To achieve this, a lessons review process was developed (Figure 2.3).

All lessons entering the LLLog, from the project team CAPTURE sessions, were manually screened by the LLN administrator and distributed to the most relevant LLN line representative for review. Depending on the lesson details, the LLN line representative would review the lesson (often with some consultation with the original CAPTURing project team) and do one of the following: (a) reject the lesson, (b) put the lesson on hold (particularly for very specific lessons that may not have transferable relevance), (c) approve the lesson for publication in the LLLog (usually with some rewording to improve lesson clarity), or (d) engage with other subject matter experts to consider the lesson in greater detail. Sometimes business process improvements were required (i.e., standard operating procedure updates) before the lesson could be approved and published in the LLLog.

As part of introducing the refreshed Lessons Learned Process 2.0, guidance was also shared with project teams on what constituted a valuable lesson. In particular it was found that targeting the PSTLs for this training was most effective as these were the colleagues actually entering the lessons into the LLLog during their CAPTURE sessions. The guidance was simple but highly effective when followed:

- Be specific rather than vague when recommending lessons.
- Ensure the lesson is something that can be practically implemented.
- Before submitting, ask oneself if you would have found this lesson of value if you had received it?

The team also encouraged a philosophy of

- Less is more, whereby a few high quality lessons were preferred to multiple mediocre lessons.

Teams appreciated this guidance; it meant the lessons learned CAPTURE session should be a short, highly focussed discussion where it was clear that high value work was being undertaken for the benefit of others. Of course the LLN also benefitted from a more manageable volume of lessons to review with less wasted effort on poor quality lessons.

To embed these changes with colleagues early in the programme, the AIM (Accelerated Implementation Methodology) change management principles were employed. For instance, Co-Development teams were strongly encouraged to engage in the Lesson Learned 2.0 process from senior leaders of Co-Development (as reinforcing sponsors); indeed, this was a clearly expected deliverable of each team.

So from 2012 through to 2014, the Lessons Learned 2.0 process, as described above, was the mode of operation. During that period approximately 600 lessons were processed with just over 300 being approved and published in the LLLog. Biannually interesting anecdotes of valuable lessons were shared across the division. Procedures and workflows were being updated and often completely redesigned before being re-embedded back into the line's workflows. Lines were also adopting similar lessons learned approaches to manage their own technically detailed lessons; it was clear that culturally lessons learned was becoming more entrenched in the way pharmaceutical sciences colleagues worked. The process was clearly adding value. However, there were also learnings about the lessons learned process, providing several opportunities for significant improvements. Issues included the following:

- The process required considerable manual tracking by the LLN, resulting in some lessons not being reviewed and progressed as desired.
- The review system and outcomes were not transparent to the project teams, they wanted to understand what had become of all the lessons they had proposed.
- The process required teams to use SharePoint software which had been set up with a poor user interface.
- It was becoming apparent that the SEEK sessions were not ideal for sharing learnings back with the teams.

So, late in 2014 Pfizer Pharmaceutical Sciences Division embarked on designing and implementing a new and improved Lessons Learned 3.0 process.

Lessons Learned 3.0

For Lessons Learned 3.0 the following four central pillars of enabling any strong knowledge management initiative were considered: people, process, technology, and governance [2]

- During the three years of operating Lessons Learned 2.0 cultural acceptance of the value of undertaking lessons learned activities was achieved, such that its use was spreading organically throughout the business (People ✓).
- The process had been successfully embedded within the drug development workflow rather than operating as a stand-alone noncore activity (Process ✓).
- The well-respected Lessons Learned Network had successfully overseen and managed full process implementation (Governance ✓).
- The one area the team felt could be further leveraged to make substantial enhancements to the process was the lessons learned log (Technology ✗).

From the review of Lessons Learned 2.0 it was clear that the technology solution needed to have the following capabilities:

- Ability to track each lesson through its own workflow.
- Automatic forwarding of differing notifications (for awareness or action) to various recipients dependent on outcomes.
- Simplified user interface.
- Functionality to enhance the SEEK sessions.

There was no budget to introduce a bespoke informatics solution to enhance the lessons learned log, so the team were reliant on utilizing an enterprise solution. Due to its extensive configurability, without the need for coding expertise, they decided on utilizing SharePoint 2010 (SP), with the freely downloadable SharePoint Designer 2010 (SPD) to introduce the more advanced functionality not available directly *out of the box*.

TECHNICAL DETAILS FOR DEVELOPING
LESSONS LEARNED 3.0

1. *Defining the workflow of a lesson.* In order to track a lesson through its own workflow the first task was to define what that workflow should be. After much consideration the following workflow was settled on (Table 23.1).

2. *Building the SharePoint list of lessons.* First a new SharePoint team site was created to host the SharePoint list of lessons, followed by the generation of a new custom list that was titled *lesson learned log.* Probably the key activity when building a SharePoint list is deciding the column purpose and type, which stores information about each item (lesson) in that list. The topic and configuration of each of these columns was determined in order to (a) provide the desired lesson attributes for those reviewing and SEEKing the lessons and (b) enable the automated tracking and notification functionality required of Lessons Learned 3.0 (Figure 23.4).

3. *Designing and building the InfoPath entry form.* Once the columns within the custom list had been chosen and configured, selecting

TABLE 23.1

Describes the 6 Stages of the Lesson Workflow

Activity	Responsible Person	Automated Outcome
A. Enter Lesson in LLLog	PSTL	Automatic Alert to LLN Administrator to Assign LLN Reviewer
B. Assign LLN Reviewer	LLN Administrator	Automatic Alert to LLN Reviewer to undertake Initial Lesson Review
C. Undertake Initial Lesson Review	LLN Reviewer	Automatic Notification to PSTL of the Initial Review outcome
D. Assign Subject Matter Expert for Expert Review (if required)	LLN Reviewer	Automatic Alert to Subject Matter Expert to complete Expert Review
E. Complete Expert Review (if required)	Subject Matter Expert	Automatic Alert to LLN Reviewer for Final Lesson Approval
F. Final Lesson Approval	LLN Reviewer	Automatic Notification to PSTL of the Final Lesson Approval outcome

Guidance for Entering Lessons: Please complete the GREEN sections ONLY		
SECTION A - ENTERING THE LESSON (PSP1/CDT)		
Entry Date	Date field	
Product	Free text field	
API or DP LL?	Choice field	API | DP | API and DP
Formulation Type	Choice field	not applicable to LL | Solid IR Tablet | IR Capsule | Solid MR | Spray Dried Dispersion | Liquid | Parenteral | Topical | Sublingual
Development Phase	Choice field	Pre-POC (R) | Route Selection / DP Design (D) | Ph3 / ICH Readiness & Manufacture (D) | Validation Readiness & Manufacture (D) | Preparing to File (D) | Launch Readiness (D) | Business Development (R and D)
What Should Have Happened?	Free text field	
What Actually Happened?	Free text field	
Learning / CI Recommendation?	Free text field	
Project Lead	Person field	
Key Contacts	Person field	
Guidance for Reviewing Lessons: ONLY Lessons Learned Network members complete the ORANGE sections		
SECTION B - NOMINATING LLN REVIEWER (LLN ADMIN) - please complete SECTION B within 3 days of Entry Date		
Date LLN Reviewer Assigned	Date field	
Which Line Owns LL?	Choice field	CRD | ARD | DPD | DPS | GCMC | GCS | QA | OpEx
LLN Reviewer	Person field	
SECTION C - INITIAL LL REVIEW (LLN REVIEWER) - please complete SECTION C within 14 days of Date LLN Reviewer Assigned		
Initial Review Date	Date field	
Initial Review	Choice field	
Initial Review Rationale	Free text field	
SECTION D - ENABLING EXPERT REVIEW (LLN REVIEWER) - please complete SECTION D within 28d of Date LLN Reviewer Assigned		
Date Line SME Assigned	Date field	
Line SME	Person field	
Line SME: Agreed Actions	Free text field	
CI Expertise Required?	Choice field	CI Group Involvement NOT required | CI Group Advice requested
Expert Review Due Date	Date field	
Guidance for Expert Review: ONLY Line SMEs or LLN members complete the BLUE section		
SECTION E - EXPERT REVIEW (LINE SME) - please complete SECTION E by "Expert Review Due Date"		
Date Expert Review Complete	Date field	
Status of Agreed Actions	Free text field	
Approval Recommended?	Choice field	Approval Recommended | Approval NOT Recommended
Guidance for Approval: ONLY Lessons Learned Network members complete the ORANGE sections		
SECTION F - FINAL APPROVAL (LLN REVIEWER with LINE SME endorsement) - please complete SECTION F by "Expert Review Due Date"		
Final Approval Date	Date field	
Final Approval	Choice field	Approved | Not Approved | Lesson Embedded
Final Approval Rationale	Free text field	
APPENDIX (ADMIN &/or LLN REVIEWER)		
Comments (for LLN)	Free text field	

FIGURE 23.4
Illustrates each of the columns within the InfoPath entry form including the drop down options.

the customize form option from the SharePoint list ribbon automatically opened InfoPath and generated a *new item* entry form. This entry form displayed the selected columns and was easily formatted within InfoPath to have the look and feel of choice. The entry form was designed to reflect the 6 stages of the lesson workflow A–F (Figure 23. 4 illustrates each of the columns within the InfoPath Entry Form including the drop down options).

4. *Building the lesson workflows into SharePoint using SharePoint Designer.* To build the 6 lesson workflows A–F the additional functionality built into SharePoint Designer was required, which can be opened using the *Edit in SharePoint Designer*

option found within the standard SharePoint application. Once in SharePoint Designer, navigate to Workflows > List Workflow > Lesson Learned Log to start building the 6 workflows. When building these 6 workflows, what was effectively happening was instructing SharePoint to send an e-mail to a designated recipient(s) when a field in the lesson entry form is changed. Although building the workflows in SharePoint Designer did not require coding skills, it did help to have familiarity with Boolean logic and how this is utilized within the SharePoint Designer workflow module. Newcomers to SharePoint Designer are well advised to study a few *how to build a SharePoint workflow* online tutorials or seek support from an IT savvy colleague as such guidance is beyond the scope of this case study.

5. A screenshot for workflow B *Assign LLN Reviewer* has been provided (Figures 23.5 and 23.6). Please also note the following suggestions that may be of value: (a) when completing the first workflow A, only select the *Start workflow automatically when an item is created* option, (b) when completing the subsequent workflows B–F, only select the *Start workflow automatically when an item is changed* option, (c) set up each workflow to *only* send out an e-mail when a *specific* field (column) is changed, not when *any* field is

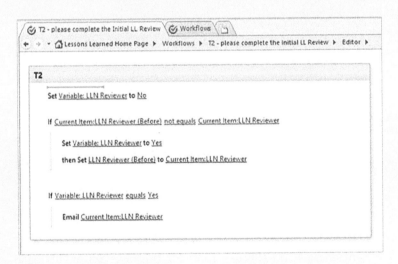

FIGURE 23.5
Illustrates the completed workflow.

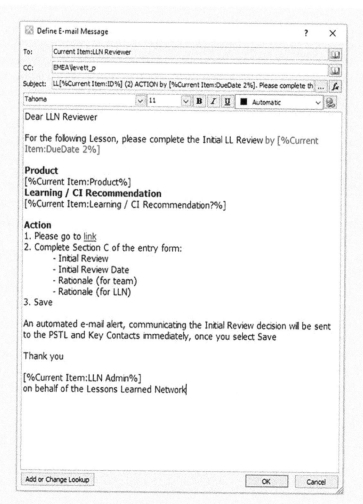

FIGURE 23.6
Illustrates the preprepared e-mail for workflow B.

changed (to achieve this it is recommended to set up a dummy field in the list example *LLN Reviewer (Before)* to achieve a before and after comparison (see Figure 23.5 for example), (d) use the *Add or Change Lookup* functionality to include lesson specific details within e-mails, (e) include within the e-mail responses a direct link to the lesson new item entry form (Figure 23.4) for the next actionee's ease of use, and (f) make the required actions very clear and precise for the recipient(s) and include a due date (based on any date field from within the list.

6. *Building the tracking and performance reports into SharePoint.* To assist the LLN and the Co-Development teams, several different SharePoint views were created (Table 23.2).

TABLE 23.2

Illustrates the Purpose of each SharePoint View

View/Report	Purpose
Review All Lessons	For the LLN to immediately see the status of all lessons that have not yet completed the lesson workflows. Categorized by reviewing line to immediately see where any hold ups in the review process are occurring. This list is reviewed at the monthly one hour LLN meetings.
Key Performance Indicators	Given all the 6 workflow stages have been designed to capture both a due date and an actual date, it has been possible to measure the on-time performance of the Co-Development teams, PSTLs, LLN reviewers and subject matter expert reviewers. This enables one to identify hold up points in the process and provide support and guidance as required.
SEEK Lessons Learned	This view enables Co-Development teams to access all approved lessons and filter them based on their needs (i.e., phase of development and API vs DP) to support their SEEK sessions.

The new Lessons Learned 3.0 process was reviewed one year after being implemented. During that period there have been several marked improvements including the following:

- The initial lesson review time has more than halved and the percentage of incomplete lesson workflows has been maintained below 5%.
- Project team members have expressed delight at the transparency of the new automated notification system that is having the added benefit of sharing back with the teams the rationale for any lessons not progressed; this is proving to be an effective learning loop for what makes a valuable lesson.
- The instant access to the lesson entry form, from the SharePoint homepage and the alerting e-mails has received considerable positive approval from the Co-Development teams, LLN reviewers, and subject matter expert reviewers; it has streamlined the process considerably for all users.

Future Plans

There is still an outstanding objective from the Lesson Learned 2.0 review; to enhance the value of the SEEK sessions for the project teams. Initially the SEEK sessions were put in place to share all relevant lessons from a particular phase of Co-Development. In practice there are effectively two types of lessons emanating from the CAPTURE sessions. There are the *specific recommendation* lessons that can be immediately implemented by teams. In addition, there are the *please improve this broken part of the business workflow* lessons that require subject matter experts to examine the issues experienced by the team, establish route cause, and if appropriate define and embed a solution, often using continuous improvement tools. These latter type of lessons, by the nature of their continuous improvement focus, usually successfully embed effectively into the business workflows without the need for the SEEK sessions. Effectively the team had stumbled on a mechanism to harmonize the Lessons Learned 3.0 process with their divisional continuous improvement activities. They have extended this synergy by including a question on the entry forms asking subject matter experts if they would value additional support from divisional continuous improvement experts to help resolve complex lessons. To further improve the SEEK sessions it is intended to link the SEEK scheduling dates to the lessons learned log, so that *relevant* lessons can be automatically pushed to project team members at the *teachable moment* without colleagues even having to open and search through the LLLog. Every step that moves lessons learned deeper into the existing flow of work, teaches at the teachable moment, and enables that the completion of the learning cycle is a step toward delivering a more effective Lessons Learned 3.0.

References

Carla O'Dell, C. H. (2011). *The New Edge in Knowledge: How Knowledge Management Is Changing the Way We Do Business.* Wiley: Hoboken, NJ.

Stephanie Barnes, N. M. (2014). *Designing a Successful KM Strategy: A Guide for the Knowledge Management Professional.* Information Today: Medford, NJ.

24

Capturing Critical Process and Product Knowledge: The Development of a Product History File

David Reifsnyder, Kate Waters, and Kayhan Guceli

CONTENTS

Mergers and acquisitions within the pharmaceutical sector have become the norm in recent years and bring with them complex product portfolios involving long or even extended product lifecycles. Organizing and preserving prior knowledge can be a daunting task in the post-acquisition environment. Learn how Roche has approached organizing prior knowledge in product history files, which has enabled efficient access to legacy content for all involved in the product lifecycle.

Editorial Team

Background

The past 10 years have been very interesting, not only for our company, but for the pharmaceutical industry as a whole. In March 2009, Roche acquired Genentech that resulted in an immediate increase in biologic and small molecule commercial products as well as a growing pipeline of new products in clinical development. In parallel, heath authorities were aligning with ICH Q10 (2008) and expanding opportunities for utilizing product knowledge and decisions made in development throughout the product lifecycle. Quality by Design concepts allowed for the capturing and leveraging product and process knowledge, the framework to demonstrate product and process control through internal quality management systems. This case study will describe Roche's approach for identifying, indexing, and sharing critical process and product knowledge throughout the product lifecycle, the product history file (PHF).

Introduction

The Roche knowledge management (KM) approach is built upon the principles described in ICH Q10. Our vision is characterized in three words: connect, share, and learn. It starts by *connecting* people with content and people to people. Systematic processes for *sharing* critical knowledge and core competencies through communities of practice and expert forums are key mechanisms to minimize knowledge silos. Finally, our *learn* component focuses on the recognition and capture of our experience and expertise and its transformation into shared organizational learning.

Initial surveys on information sharing showed that greater than 50% of the respondents wanted improved search capabilities to find information across the global network. When asked how much time people spent searching for product and process information, survey results suggest each employee spends, on average, 2.5 hours each week looking for information (Figure 24.1).

Improved document sharing would yield significant productivity value. A strategy to improve identification of product lifecycle documents was developed with the initial focus on the source documents that support regulatory submissions. This strategy evolved into the PHF, which is a cornerstone in fulfilling how Roche connects people to information, a key tenant to our knowledge management strategy, and promotes a culture of learning and sharing.

Our history of developing lifesaving drugs started before the personal computer age. Consequently, much of the documentation related to legacy products are paper based and often hard to locate (legacy products refer to products approved prior to the existence of the Pharmaceutical Quality Management System). These challenges did not diminish after personal computers became integrated into our business practices. In many ways,

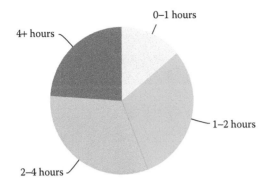

FIGURE 24.1

How many hours per week do you spend searching for files and information inside of Roche/Genentech?

it became easier to generate isolated collections of information throughout our company. The PHF not only streamlines the document flow for new products, but also provides a tool for systematically discovering and indexing critical documentation supporting legacy products.

Knowledge management is a broad topic and difficult to handle unless brought down to tangible, real life examples. The PHF is a good example that addresses a fundamental need for the business: a systematic approach to gathering and sharing product information. PHF helps retrieve product information quickly and accurately and facilitates a more streamlined approached for submissions, inspections, and investigations.

Roche Pharmaceutical Quality System and Foundation for the Product History File

The PHF evolved out of a need to provide a more systematic approach for collecting and categorizing documentation that supports commercial products and processes. As a result of the merger, the quality systems from both companies were combined into one system (Roche). Redundant documents were consolidated, and practices across companies were harmonized into a new Pharmaceutical Quality System (PQS). The document hierarchy for the Roche PQS is depicted in Figure 24.2.

Business processes and common requirements are described in approximately 100 Quality Requirements (QRs) and Global Standards and Practices (GSPs).

The QRs and GSPs are organized relative to common business practices. For example, there are document families for analytics, stability, and risk

In alignment with ICH Q10:

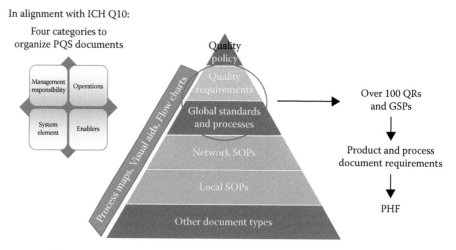

FIGURE 24.2
Roche pharmaceutical quality system document hierarchy.

management to help people execute their daily jobs. An example for the product and process lifecycle family of documents is shown in Figure 24.3. The GSPs for the different processes (depicted in yellow) provide specific guidance in functional areas that are needed to support a project as it moves through various phases of clinical development through commercial manufacturing. Embedded within each one of the functional areas were specific

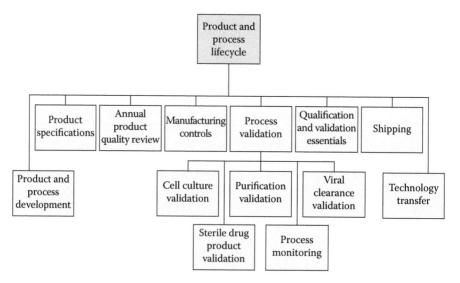

FIGURE 24.3
Product and process family of documents.

documentation requirements that were needed in order to eventually support licensure and routine commercial manufacturing.

One of the challenges facing the teams that were supporting the clinical programs was that there were so many PQS documents that needed to be referenced in order to have a complete list of the documentation requirements. By collecting the PQS documentation requirements that were identified into a single source, the PHF serves as a road map for critical process and product information.

The product history file (PHF) provides an index of the PQS documentation requirements to support commercial products. Product knowledge comes from various sources that collectively provide a detailed and comprehensive product and process history. The PHF indexes the accumulated explicit knowledge from pharmaceutical development through commercialization to product discontinuation. Historically, Roche has been successful at collecting information during development and launch in preparation for submission packages. However, once a product was approved and commercialized, sharing product and process information that accumulated during routine manufacturing became a challenge: documents were created and shared within a site or functional area but not effectively shared across sites. Consequently, the process development scientists did not fully appreciate the impact the process might have once it was transferred to a commercial site. Similarly, site technology teams lost visibility to the development history and occasionally ended up repeating small scale experiments that had already been performed in order to close out investigations. There was a need to be more efficient in prospectively identifying and collecting documents in preparation for submissions and inspections. The documentation challenge was escalated as Roche-Genentech became a global organization following the merger with multiple products being manufactured at multiple sites (internal and external). As a global company with operations shared across the network, there was a shift from a site-based to a product-based organization. An improvement in communications across sites and departments to better support our products was needed. During this time, more emphasis was being placed on product knowledge management and process understanding by worldwide health authorities. It became crucial to develop a more systematic approach to collect and share product information across our network.

The basic structure of the PHF is shown in Figure 24.4. The left side of the figure depicts activities that occur during the process development phase of a product and includes validation and transfer activities and the documentation that supports a submission to the health authorities. This information is consolidated and condensed in product registration (e.g., BLA or MAA). The right side of the panel depicts systems that exist to support commercial manufacturing. Quality risk management and knowledge management are enabling activities supporting product development and commercialization.

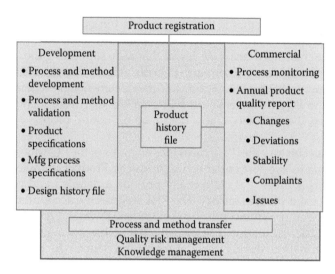

FIGURE 24.4
Product history file layout.

Initial Implementation

Once the PHF was approved into our quality system, an implementation strategy was needed. The requirements had been defined without an underlying business process, resulting in considerable anxiety about what needed to be done when, and who would do the work. In order to more quickly gain support across the global network, a cross-functional implementation team was formed to develop and execute an implementation strategy.

A pilot program was proposed to create PHFs for five products spanning the Roche product portfolio to develop the business process and identify resource requirements for the program. However, it was soon realized that five products was unrealistic and the approach was simplified, focusing instead on three products with readily accessible documents. The pilot program focused on two marketed products (large and small molecule) and one new product that entering its first submission for approval. The pilot was intended to identify *quick wins*, realize the value to be gained from the creation of the PHF, and provide more visibility for the PHF program.

Once the pilot was endorsed, the next objective was to define the PHF document strategy and understand our current document landscape. By locating relevant product documents and moving the information into the Roche electronic global document management system (EDMS), the PHF could provide structure, global access, and search capabilities. In addition, since this was an existing global documentation system, the PHF should be easy to create and maintain.

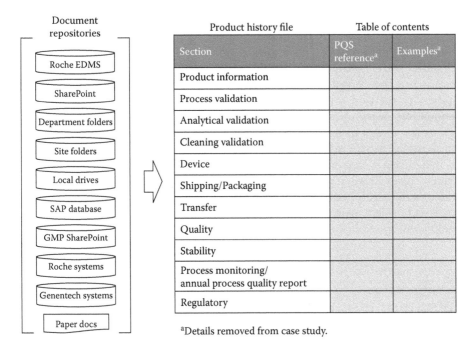

Document repositories	Product history file Section	PQS reference[a]	Examples[a]
Roche EDMS	Product information		
SharePoint	Process validation		
Department folders	Analytical validation		
Site folders	Cleaning validation		
Local drives	Device		
SAP database	Shipping/Packaging		
GMP SharePoint	Transfer		
Roche systems	Quality		
Genentech systems	Stability		
Paper docs	Process monitoring/ annual process quality report		
	Regulatory		

[a]Details removed from case study.

FIGURE 24.5
Document repositories and the product history file.

The initial expectation was to find most of the documents in a few storage locations. In reality, documents were stored in multiple systems and document repositories (Figure 24.5).

In addition, the documents were in multiple formats and languages. Many documents were available only in hard copy versions or could not be located. In cases where hard copy documents had been scanned, they were not scanned into a format that supported text searches or that had been verified against the official copy. For many of the electronic documents, they were stored in proprietary systems with limited access, and often there was no metadata, or the metadata was not useful.

Undaunted, the team started the process of identifying documents and copying the documents into our new repository. During the initial execution of the pilot program, it was noted that copying documents was a mistake because it enabled the use of an unofficial copy rather than the original source document. In addition, since there were so many document storage locations, it was not feasible to migrate documents into a common repository. Therefore, the team modified the plan and focused on document identification. An approach was developed in which the first step for PHF consisted of creating a product document index, which included important aspects of the document (e.g., title, author, date, storage location, and so on.). The goal was to build a system that would provide direct access to documents, where the first step was to identify the documents and generate the metadata. During document identification, it became apparent that most of our projects had detailed project plans that include extensive document lists (a big win for capturing the complete inventory). Even though the project plans were generally focused on a single aspect such as process validation, method validation, or technical transfer, they served as a valuable input to the PHF. These document lists accelerated our ability to build the PHF, and we started to see patterns where documents could be found. Another interesting aspect is that once a systematic approach was applied to indexing documents, it was much easier to see gaps in missing information. We did not rely solely on document lists. It became very important to interface with the various products and project team members to tease out their tacit knowledge in order to help identify explicit information.

Product History File Tools and Basic Concepts

In order to ensure consistency across all the product teams, a formal business process was developed, and frequent communications on the implementation status were utilized. A PHF template for indexing documents was created to drive consistency. The template was constructed using a common software tool (Microsoft Excel) and was made available throughout our company and posted in a SharePoint collaboration forum. As shown in Figure 24.6, the PHF is organized into distinct tabs including product information, process validation, analytics, and etc.).

Deliverables	Product stage	Sources document ID	Version	Site of execution	Document title	Sources storage location	Record owner	Author
Product specification file	Drug substance and drug product	Cure PFS 1	1.0	South San Francisco	Product specifications for curemab used during IMP development	SharePoint	QA	Helen Vondrea
Development report	Drug substance	Cure DR 197	2.0	South San Francisco	Optimization of the downstream purification for curemab	Genentech system	Process Sciences	Phil Weber
Technical transfer master plan	Drug product	Cure TTP 01	1.0	Basel	Technical transfer of the curemab DP process from SSF to basel	Roche ED MS	MFG Sciences	Heinrick Ulmer
Validation master plan	Drug substance and drug product	Cure VMP 02	1.0	Vacaville	Curemab process validation master plan	Roche ED MS	QA	Mary Worth

FIGURE 24.6
Product history file template.

Wherever possible, pull-down menus are used to provide consistency. In its simplistic form, the first phase of building a PHF is nothing more than creating an organized collection of document lists. The second phase is in connecting the document list to the actual documents. Since the documents (either in hard copy or electronic format) that are identified in the PHF are located in multiple document repositories throughout the company, it was not practical to move the documents into a single repository. However, it was possible to provide links to the documents in PHF, expand the metadata, and thereby connecting people to information.

Once the team completed the pilot program and started to expand the PHF to more products, we faced an organizational challenge. Many of our products had post approval changes such as multiple sites, process version, and new formulations. How should we capture each of these product changes in the PHF? It was ultimately decided to organize these changes for a product within a PHF product family. The PHF product family includes all of the information for the product, and the metadata allows the flexibility to select for specific product attributes during a search. New project teams can add documents to the PHF as the project moves through the various phases of clinical development, technology transfer, or commercial manufacturing. This generates a *working version* of the PHF, which will exist during the lifecycle of the product. At a certain point in time (e.g., prior to the initial inspection that will support approval for new biologic products), the working version is published into our validated document management system to create a permanent record.

Another challenge has been building PHFs for multiple-marketed products, which already exist at Roche. A risk-based approach was utilized to help define our strategy for legacy products. Since these products were approved years ago before our current PQS was established, a smaller core set of requirements was identified for existing marketed products. It was not practical to create PHF for all the existing products at the same time, so a multiyear plan for creating PHFs that was based upon priorities such as medical need or new transfer activities was utilized.

Maintenance of the Product History File

The ownership of the PHF evolves over lifecycle of the product. For new molecules, the ownership resides with the development team who support the product throughout clinical development through the initial product approval. After approval, ownership transitions to the commercial organization. Regardless of who is designated the owner of the PHF, there is a common thread throughout the processes in that all the teams need to have representation from technical, quality, and regulatory organizations.

Although the PHF was created out of the quality systems, it is not owned by quality. Rather, the PHF is a good example of a quality-engrained philosophy that spans across our sites and functions. In addition to these teams, there needs to be a core group that can provide tools and business systems to facilitate PHF creation. Since the core group has visibility across the entire PHF portfolio, they provide valuable oversight to the teams and assist senior management in PHF governance.

The PHF is a living system that is updated continuously throughout the product lifecycle. Two versions of the PHF are maintained. One version is considered the *working version*, which is the living PHF and will contain the most up-to-date information and reflects continuous improvement activities supporting the product. The second version is published and stored in the Roche electronic documentation management systems. The timing for publishing the PHF is based on key product events (e.g., annual product quality reports, technology transfers, and version changes). The published version is approved by the product team members and serves as a *snapshot in time* of the living PHF. Publishing the PHF provides more formal review and documented evidence of the PHF.

Conclusion and Future Plans

The PHF provides a realistic approach and reinforces the ICH Q10 concepts for knowledge management. Although the PHF was started as an internal quality requirement, it is driven by cross-functional teams and benefited from a grass roots approach. Use of a smaller pilot program provided time to implement lessons learned as the program evolved and expanded across the network. The PHF benefited from changes in our culture, which included a product-based focus on our marketed products and development of technical product teams to support marketed products. The PHF aligns with the marketed products technical team's mission to ensure product health. With an increasingly complex supply chain and multisourcing of products across the network, improved information sharing was needed to enable a product-focused approach and support global operations. Sharing critical process and product information is vital in order to improve efficiencies during transfer between sites, to support process improvements, and to aid in investigations and inspections.

The future plans include identification of fewer repositories to consolidate documents and facilitate use of improved search tools. It should be possible to preidentify document types so that they can be automatically incorporated into a PHF. It is also critical to advance oversight and maintenance of the PHFs to improve consistency and sustainment throughout the product lifecycle.

The PHF is a step in connecting people to content. It promotes a global and product-focused system for sharing information and provides a valuable tool to document product health. Through identification of information throughout the lifecycle of a product, the PHF increases product understanding and thereby builds a foundation for knowledge.

Acknowledgments

The authors would like to recognize John McShane and Ann Gillan for their contributions in developing the Product History File and thank Becky Nagel for her assistance in preparing and reviewing the case study.

Reference

International Conference on Harmonization (2008). Pharmaceutical Quality System - Q10, www.ich.org.

25

Communities of Practice: A Story about the VTN and the Value of Community

Renee Vogt, Joseph Schaller, and Ronan Murphy

Creating healthy communities in the workplace can be elusive for a variety of reasons—once established they do not just *go viral*. See how the manufacturing division of Merck & Co., Inc. (Kenilworth, NJ, USA), known as MMD, developed a holistic approach to establishing and sustaining virtual communities that harness the energy of the workforce and also deliver tangible business results.

Editorial Team

Problems Big and Small: In 2009, knowledge did not flow seamlessly throughout the global science technology & commercialization (GSTC) organization—a group of approximately 1,000 highly skilled scientists, engineers, technicians, and support persons who are dispersed across more than 50 locations in over 20 countries. GSTC is the technical organization of the manufacturing division of Merck & Co., Inc. (MMD) whose primary function is to perform late stage product development, launch and ongoing technical support of the manufacture of all our pharmaceutical products (Lipa et al. 2013).

The lack of knowledge flow made it difficult to effectively resolve both big problems (like discarded batches and major investigations) and day-to-day problems (like equipment downtime and material shortages). This was magnified when Merck & Co., Inc. acquired Schering–Plough (itself an inhomogeneous product of other recent M&A activity), nearly doubling the size of the company, and further slowing the wheels of our innovative problem-solving machine. The result was a seemingly unsustainable environment where the personal networks that once enabled knowledge flow became fractured; issues became less visible, knowledge became less accessible, and similar problems flared in different parts of the network without a common way to connect them—GSTC needed a new way of working that would improve the flow of knowledge and enable us to solve the same problems only once.

Mike Thien, the organization's leader, was keenly aware of these challenges that prevented our organization from reaching our full potential. A deeper dive into the *why* surfaced numerous cultural, organizational, geographical, and technological barriers, including the following:

- The absence of fit-for-purpose communication channels through which information could be shared rapidly.
- A reticence to widely sharing knowledge and information across functions.
- A resistance to asking for help.
- A pervasive hierarchical mindset that made problem solving, finding information, and collaborating very difficult.

Finally, there were increasing economic and business pressures that threatened to make the situation even more difficult—headcount reductions, budgetary and travel restrictions—forcing everyone to *do more with less*.

The solution is not about technology: The goal was to enable the flow of knowledge and enhance collaboration in order to improve GSTC's ability to solve technical problems and accelerate the development of more robust products in a globally diversified company. The solution was a multifaceted knowledge management approach built on the foundation of sound change management. A targeted knowledge management program, based on extensive benchmarking and research with APQC (American Productivity and Quality Center) and companies leading in KM, was conceived, designed and installed in GSTC. The program started with four initial capability *pilots*, all targeted at increasing systematic access to both explicit and tacit knowledge. One of these pilot capabilities was the virtual technical network (VTN)—a collection of virtual communities of practice centered around key technical topics that were core to the work in GSTC, supported by a fit-for-purpose communication platform based on existing enterprise technology. These virtual communities were designed to harness the existing collective knowledge and experience (tacit knowledge) which was previously difficult to surface and access. Externally, these might also be referred to as *helping communities*—solving problems and serving as a place for best practice sharing and innovation (APQC 2001 and 2002). The approach to designing the VTN capability was anchored by the same four knowledge management pillars—people, process, content, and technology.

PEOPLE: People are critically important to the success of a capability like the VTN. Within our ecosystem, there are five critical roles:

- Originators
- Responders
- Viewers/consumers

- Stewards
- Sponsors

The *originator* is what most likely comes to mind first - the person asking the question or sharing the best practice. To engage these persons, we needed to encourage an environment that made it *safe* to ask a question or share a problem. Safety was important because Merck & Co., Inc. was a *knowing culture* that created reluctance for people to say they did not know. A set of community *ground rules* were established that normalized the code of conduct within these communities consistent with our company culture. We also solicited the help of well-respected early adopters to seed some posts to get the ball rolling, show others by example how to effectively use the tool, and demonstrate that there were no negative consequences for making an issue visible. To the contrary, people raising visibility to issues so they could be addressed by the community were celebrated.

Responders are a diverse blend of individuals ranging from those who share a similar issue, those with experience in having addressed the same issue before, and SMEs (subject matter experts) with definitive credibility in the relative field of inquiry. For these persons, the key was mostly around awareness that an issue was raised and needed attention. This was primarily accomplished by establishing alert functionality, where individuals could receive e-mails when discussions were initiated in a community/topic of their choice. Some functions with subject matter expertise responsibilities (Centers of Excellence) have gone so far as to setting specific expectations relative to this role, for example, responding to all posts in their field of expertise in <24 hours. Many of these enablers are baked into our business processes as described later in this section.

The *viewers/consumers* are the dark horse of this story—they are getting value from reading discussion threads, viewing content, downloading documents, and so on, but not asking or answering questions. These consumption type of data are inherently more difficult to track but we do know that the content in our communities is widely leveraged. A sampling of several discussion threads across a variety of our communities tells us that the ratio of unique views to posts can range from 10:1 up to 50:1, with views on a single post, at times, as high as 500.

Finally, each community is managed by a minimum of two *stewards* and one *sponsor*, both *part-time* commitments with clearly defined roles and responsibilities. The stewards are individuals within a function that are also likely SMEs for the respective topics and/or highly connected individuals in the social network of practitioners related to the topic, a knowledge broker. The community sponsor is likely someone in a leadership position within the organization and is able to set the direction and priorities for a community. In addition, the sponsor will also develop strategies to expand utilization, influencing others to help remove barriers to adoption across the organization, review community progress, and provide feedback to community stewards, applying both positive and negative consequences to improve performance.

We would be remiss not to mention the role of the community steward's manager as well; although, not directly involved in the day-to-day operations of a community, the manager can be an important liaison working with the sponsor to identify coaching and development opportunities for the steward, align on resource expectations, help the steward balance workload priorities or recognize a job well done. Affectionately referred to as our *community triumvirate* (Figure 25.1), each one with an important role to play in the community and how the community delivers value to the company, as well as with the personal and professional development of the community steward.

Community management is an emerging field and the path is still ill-defined; however, the community steward role has the potential to be very exciting and gratifying. We use the analogy of running a small start-up company when you are managing a community—you are in a position of leadership and responsibility:

- You must identify your mission
- Define your strategy
- Establish a unique value proposition
- Understand your customer segment
- Define metrics and more...

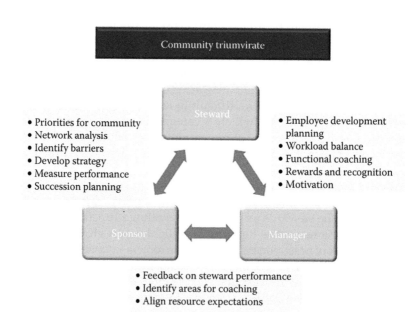

FIGURE 25.1
The community triumvirate.

In addition, it helps to have an entrepreneurial mindset, one who is:

- Comfortable with taking risks
- Independent
- Persuasive
- Able to influence and negotiate

This is one of those rare opportunities—in an otherwise highly structured and regulated atmosphere—to be empowered to make decisions, understand *the sell*, engage sponsorship, and truly own something on behalf of the broader organization.

PROCESS: Change cannot succeed without solid business processes built around it. Although our practices continue to evolve, the following processes have made the VTN what it is today:

- Strong governance and support: Since its inception there has been a core team composed of colleagues from the MMD knowledge management CoE, key leaders, and stakeholders from the business and IT partners who are responsible for setting the direction, establishing policies, and routinely monitoring, all in the name of enabling our communities to deliver maximum value back to the organization.
- In the flow: To enhance adoption and usage, the aim is to get the VTN *into how work gets done.* We have found that when the community is embedded in the flow of people's daily work, it is minimal effort for the user but with maximum impact—by solving those pesky, everyday problems in a faster and more efficient way. An example of this is during a common internal escalation procedure for day-to-day issues that inhibit progress of a planned activity (e.g., finishing a manufacturing run). The person conducting the escalation procedure will often ask the following question: *Have we consulted the VTN on this issue or posted a question about this issue on the VTN?* prior to escalating the issue to the next level.
- Another is alert functionality that allows messages to be sent to an individual's e-mail when updates are made to certain self-identified topics, penetrating what is arguably the predominant work space for many of today's knowledge workers.
- A subset of foundational behaviors and mindsets that have enabled us to work collaboratively include the following:
  - Change management—it was necessary to have a framework for addressing behavioral barriers to adopting this new way of working, these barriers ranged from *I do not want my manager to know that I do not know something* to *I am uncomfortable posting*

a question because English is not my native language to *I do not do Facebook, why would I do this?*

- To address these barriers it was necessary to engage senior leaders, having them set the expectation to leverage the VTN as often as possible but also *walk the talk*, showing by example that the communities are a safe place to seek help and to vet things. This was a mindset shift and was going to take some time. The other framework we leveraged was the DCOM® model, deployed in MMD in partnership with CLC (Hillgren and Jacobs 2010):
  - Direction: Make it clear what we want people to do, for example, when a problem arises, navigate to the community and post a question to the network.
  - Competency: Make sure people know how to use the underlying technology, how to set up alerts, how to post a question, reply to a question, and so on.
  - Opportunity: Leaders and managers need to allow the time for people to be on the community.
  - Motivation: Tap into the intrinsic motivation people have to help others and recognize/reward where appropriate.
- Inclusion as the How® (Katz and Miller 1995)—prior to the VTN, MMD had adopted a set of principles guiding *HOW* we would work together. We leveraged those same principles to create VTN's own *Rules of Engagement* (Table 25.1). These rules were designed to

TABLE 25.1

Rules of Engagement

VTN Rules of Engagement
Build trust in the community (honoring and respecting each other)
Be willing to share information
Be willing to put yourself out there and ask questions
Listen as an ally
• We want to be a learning community
• All questions are good (respond constructively)
• Do not talk about people's postings in a negative way
Create a sense of safety for yourself and others
• No negative consequences for speaking up
• Give positive feedback for exhibiting good behavior
Create a 360 degree vision and accept people's frame of reference as true for them
• Problems are complex and we want many perspectives on them to get the power of collaboration
Ask who else needs to be in the room
• Who else needs to be in the conversation to build out the solution

set a common expectation and create an environment that would motivate people to seek connections with others and share with them (Guernard et al. 2013).

- Lean Six Sigma (LSS)/Continuous Improvement—the company strives to be a high-performing organization and leverages many well-established methodologies including LSS and Lean manufacturing. A primary goal is to solve problems once and stop reinventing the wheel each time. The VTN enables us to do this as the problem and solution is posted visibly and distributed widely across our network of communities. The rich discussion as well as resolution is then saved as searchable content.

CONTENT: The content that resides on our network of communities is focused on key technical topics that are core to the work of GSTC. In this context, we have a few key priorities.

One is that the communities are effective at enabling information to flow, having the right people engaged across the board to ensure that issues are identified, resolved, and solutions available in a timely and effective manner.

Another is that the information that flows within the community is reliable and credible. A highly active and visible community will usually self-correct any misinformation on its own, but this may be a focus when fostering a less mature community.

Finally, understanding that the maintenance of these communities requires inputs from a limited pool of resources, we seek to ensure that the communities that we decide to create will add a sufficient level of business value.

Although the demand for these communities has continued to mount, and our number of communities has increased over five-fold since the original pilots, we continue to stay true to the original intent: if the topic is core to the work that we do, there will almost always be an eager pool of potential participants waiting to get engaged. In addition, if the topic is tied to business/strategic objectives then there will be tremendous value for the organization when the community goes live. To ensure that our communities are optimized in this way, each potential topic is run through a formal vetting and review process through which a final recommendation is made. We assess the prospective topics on the following main principles:

1. Adequate, broad interest in topic, relevant to current business priorities.
2. The internal knowledge and experience to advance the topic.
3. The topic is *sustainable* (e.g., enable ongoing dialog for >1 year) versus short term (e.g., exception-based issue resolution).

Once selected, existing communities are periodically reassessed against these criteria to determine continued viability. These assessments can result in many outcomes ranging from

- Reinvigoration: The topic is important/relevant but needs increased focus to be successful
- Sun-setting: The topic has lost relevance
- Merging with another community: A topic is important but lacks critical mass, or
- Spin-offs into multiple communities: A community has gotten too large/diverse to be effective

In this way, we ensure that our portfolio of communities is optimized to both organizational demand as well as ongoing business value.

TECHNOLOGY: Although the technology that the VTN *sits on* is important, we view it as secondary; a means to an end. We leverage a combination of 2 elements: A collaboration space in which the data is more structured and searchable, and a social application that has less structure to the data but makes connecting to people very simple. Both environments have their place. The technology needs to be simple enough that it is not a barrier to using but useful enough that people can find what they are looking for.

The Virtual Technical Network: Today the VTN is a vibrant, active network of 27 technical topic communities with more than 5,000 community members and 90+ community stewards who guide and facilitate the on-line discussions, resulting in direct savings of more than $15 MM captured through quantified success stories, and an indirect impact many times greater than this through improved connectivity of the organization, perhaps a value of 6–10X. In 2011, the VTN started out with over 1,600 questions and answers and that combined number has increased approximately 27% annually and resulted in over 4,300 questions and answers in 2015.

The reach of the VTN communities has expanded beyond that of GSTC, crossing into other divisions such as Merck Research Laboratories (MRL) and information technology (IT), and spanning the globe: >40% of our members reside outside of the United States. This high degree of activity has translated into real business impact, such as rapid and efficient problem resolution (due to the global nature of participation, problems often *follow the sun* and can be worked on for >16 hours/day when necessary), broad visibility to issues and best practices so that others can learn and avoid repeating mistakes and a more engaged and autonomous workforce. In 2015, Willie Deese, Executive Vice President of MMD, reflected on the business benefit of the VTN and the overall knowledge management (KM) program, stating ".........the VTN had a significant role in transforming the division......the return on that investment is immeasurable."

Some notable success stories include the following:

- Clinical supply of an emerging oncology product was at risk, the VTN was leveraged to talk through a cosmetic issue seen with the vials. Within 1 week, root cause was identified and replacement vials located all through discussions and connections made on the VTN. The poster states: "The rapid resolution was possible due to the VTN; the ability to quickly communicate and collaborate across the MMD network is one of the most powerful tools we have in our arsenal."
- A Top 10 product's supply was at risk of being disrupted when our global network engaged through the VTN rapidly to locate a replacement part (internally) for a piece of manufacturing equipment that had broken, eliminating the need to work through the equipment vendor and saving MMD 20 weeks and $3 MM. The poster states: "At first I was reticent to post but The VTN makes you feel like you're part of something larger."
- A laboratory instrument was needed urgently to meet a regulatory deadline with the FDA. By leveraging the VTN to see if the instrument was available across the network, there was a 99% reduction in time from what it typically takes to locate a piece of laboratory equipment. The poster states: "The ability to reach such a broad audience so quickly has eliminated the non-value added work for me and shown me the power of a Boundaryless organization."
- After the Schering–Plough acquisition, a small site in Ireland that itself had only recently been integrated from a previous acquisition, did not know any of the key SMEs within the new network. Key thought leaders at the site became early adopters of the VTN, quickly recognizing the value of the portal from geographically remote sites into the knowledge community of MMD. Another colleague from a remote site in Mexico cited "the VTN as their only way to connect to colleagues in a similar function elsewhere in the company."

The power of being able to solve problems autonomously and more rapidly than through one's own personal network cannot be understated—we find that ~50% of people who respond to a post are unknown to the poster. This type of connectivity is unparalleled in the more traditional problem-solving paradigms. It is also good for the soul—that sense of feeling better connected and working toward a common goal play into employee engagement and satisfaction. Data from a VTN survey tells us that *80% feel more included in the organization as a result of using the VTN*. The speed and power of our network connectivity has been recognized externally as well. In 2014, MMD was awarded the prestigious Manufacturing Leadership Award (Moad 2014) for the VTN in workplace leadership. The VTN was recognized as a *breakthrough project for helping to shape the future of connectivity and global collaboration across the manufacturing network.*

Reflections and lessons learned: Just over five years into our journey with the VTN, we have a recipe for success—a set of operating principles that have brought us to where we are today. We have learned what processes (i.e., strong governance and support from leaders and a core team) and key mindsets (i.e. inclusion and knowledge seeking/sharing) are necessary to truly drive value back into the organization. We have also learned a number of other lessons along the way.

Specifically, we have found when the following conditions are met, a virtual community of practice like the VTN, is a great solution:

- A distributed group or groups (not co-located)
- A critical mass of potential community members (>50 people)—greater opportunity if a broad, cross-functional network
- A desire and a need to collaborate and share among peers (not just one-way information *push*) and energy behind this
- Need for open and broad access to collaboration spaces and content
- Ability to link to relevant functional/divisional key performance indicators (KPIs)
- Clear and motivating sponsorship

Important lessons we have learned along the way:

- *Not everyone in the organization, regardless of how good they are at their job, will be a good community steward.* It requires soft skills, energy, enthusiasm, and a little bit of creativity. We find that the best community stewards are those who are already active in the community—posting and replying, sharing best practices, and so on, they are natural knowledge brokers and seekers and are well suited for this type of role. We have had less success with the involuntary assignment of subject matter experts who are assumed to be interested and capable of running a community, only to find out that they lack the soft skills, personal network, or interest and energy required to this unique job well.
- *Make an (ongoing) investment in your community stewards.* By investing in them you show: (1) you care, (2) this is an important job, and (3) an inherent reciprocity has been established. Investment means things like: creating opportunities for them to hone their community management skills (bimonthly phone calls and an annual face-to-face meeting where they have a chance to connect and learn from each other), representing the VTN communities at highly visible forums

both internally and externally, participating on the core team or leading mini projects, and rewarding them in their jobs, be it written, verbal, or monetary. For those who have taken full advantage of the role, it has been a tremendous career development opportunity. In fact, the role has evolved to the point where having a community steward role in the organization can become a key differentiator in the individual's performance evaluation.

- *Do not lead with the technology.* Of the four knowledge management pillars—people, process, content, and technology—we view the technology pillar as important but it is not the ultimate goal. We quickly learned that a focus on the *new, shiny* technology was misdirected and led to much wasted time and effort trying to keep up with the latest and the greatest—we were not in the business of creating IT solutions … we were rolling out a holistic KM solution. As we discussed, the technology is a means to an end, a fit-for purpose solution to enable people to connect to others and the expertise they need, in an easy and efficient manner, to solve their everyday problems.

- *Evaluate your communities regularly, but do not evaluate the success of a community on activity and membership alone.* It has been impossible to systematically and objectively evaluate 27 communities using the same measuring stick. Over time, we have learned that, despite an overarching goal of standardization, our technical communities are not created equally…and that is okay! We have had better success identifying meaningful metrics based on a community's individual goal or purpose instead of relying on the obvious general measures of activity and membership. For example, we have a community whose primary goal is *to teach the organization about topic X* and, as such, they will measure their value based on consumption metrics like number of viewers, page views, and downloads. However, we also have a few big, broad topic communities whose goals are to solve problems and make them more visible. In these cases, we are more focused on direct activity and broad reach (membership). Some of our metrics—activity, membership, and consumption are readily obtained, but others like quality of responses and value generation are more of a manual and subjective process. We still have work to do here but have found tremendous value in the mere exercise of routinely measuring the relative success and value of a community.

- *Track and record your successes.* Small successes are additive and build the momentum of the community.

- *Share your successes whenever you can.* Use success stories to enroll new members and secure sponsorship from senior leaders. Success stories are powerful as they give the value proposition context—for example—I can see how my colleague engaged on VTN (exhibited desired behavior) and achieved an important outcome (value for the company).

- *Refresh the Steward team.* Rotate people in for short assignments to bring new energy and also develop past stewards into future sponsors.

- *Look for new ways to use your community.* Information sharing, focal point for communities of practice, and so on.

- *Do not be afraid of change.* Merge smaller communities or retire others that have passed their *sell by* date. Add a topic to an existing community instead of creating a new community.

The need for connectivity and virtual collaboration will continue to increase, and the VTN will no doubt need to evolve to meet this demand. At the doorstep are features like mobile and the cloud and we will need to continue to improve accessibility and usefulness. The path ahead is a challenging one, but with a solid knowledge management foundation and a connectivity platform like the VTN, Merck & Co., Inc. is in a position to enable its employees to do their best work and help the world *Be Well.*

Acknowledgments

Over the years the VTN has been led by a remarkable group of leaders and sponsors, particularly Mike Thien whose courage, leadership, and vision got us to where we are today. In addition, Mike's leadership team has always been there for invaluable support and guidance. Marty Lipa, who runs the KM CoE and has been our trusty guide and navigator, a special thanks to Dave Vossen and Anando Chowdhury who have been great thinking partners and visionaries, Yunnie Jenkins and Douglas Arnold, our IT Partners, and the VTN core team who govern and guide as if the VTN were its own company. The stewards who have been with us from the beginning: Jim Velez, Jerry Wang, Louis Obando, and John Higgins, and those who have left their mark permanently on the VTN: Samantha Bruno, Rob Guenard, Brad Holstine, Ronan Murphy, Ray Rudek, Joe Schaller, and Chris Tung. We especially want to show our gratitude to all former and current community stewards whose effort and dedication to the VTN is a constant source of pride and inspiration and an impressive display of out of the box thinking and entrepreneurialism. And, finally, the 5,003 community members who are brave enough to make their problems visible, smart enough to understand the whole is greater than the sum of its parts and bold enough to challenge themselves (and all of us) to be better than we are.

References

APQC. *Building and Sustaining Communities of Practice.* Houston, TX: APQC, 2001.

APQC. *Communities of Practice: A Guide for Your Journey to Knowledge Management Best Practices.* Houston, TX: APQC, 2002.

Guernard, R., J. Katz, S. Bruno, and M. Lipa. Enabling a new way of working through inclusion and social media—A case study. *OD Practitioner,* 45(4):9–16, 2013.

Hillgren, J. S. and S. Jacobs. DCOM: A Proven Framework for Outstanding Execution. *clg.com.* 2010. http://www.clg.com/Science-Of-Success/CLG-Methodology/Organizational-Change-Tools/DCOM-Model.aspx (accessed May 27, 2016).

Katz, J. and F. Miller. *12 Inclusive Behaviors* (unpublished manuscript). Troy, NY: The Kaleel Jamison Consulting Group, 1995.

Lipa, M., S. Bruno, M. Thien, and R. Guenard. A case study of the evolution of KM at Merck. *Pharmaceutical Engineering,* 45(4), 2013.

Moad, J. *Winners of the 2014 Manufacturing Leadership Awards (ML Awards).* March 17, 2014. http://www.gilcommunity.com/blog/2014-manufacturing-leadership-award-winners-announced/ (accessed May 26, 2016).

26

Identification of Critical Knowledge: Demystifying Knowledge Mapping

Paige E. Kane
with a special contribution from Christopher Smalley

CONTENTS

> Identifying critical knowledge and where it can be found is crucial to enabling the effective flow of that knowledge to enhance timely decision making. This chapter provides examples of how to map your organization's knowledge and assess critical knowledge gaps in order to improve access, flow, and reuse of critical knowledge by the people who need it, when they need it.
>
> **Editorial Team**

What Is Critical Knowledge?

The harmonized guidance document ICH Q10 *Pharmaceutical Quality System (PQS)* (ICH Harmonized Tripartite Guideline, 2008) describes the key components of an effective Pharmaceutical Quality System as well as defining the product lifecycle stages as product development, technology transfer, manufacturing, and product discontinuation. Throughout these various product lifecycle stages a variety of data, information, and knowledge are created. Significantly, ICH Q10 was the first pharmaceutical regulatory guidance to highlight the need for knowledge management, which is listed as one of the two enablers for an effective PQS. The other PQS enabler is defined as Quality Risk Management (QRM).

Due to the globally regulated nature of the pharmaceutical industry, there are also clear expectations within the various regional regulations[*] regarding the management of data and information related to each product; whether in regards to licensing a new product, details relating to the facility, or about the manufacturing process. More recently, as the industry has matured, companies have increased their organizational capabilities for capturing and processing data and information. However, the question still remains as to how well the industry actually performs in regards to learning from what it captures and reusing data, information, and knowledge to create new insights. Are learnings feeding the product and process improvement? Are they improving the operations of facilities that manufacture and supply said products to the market? Are they ultimately passing on the benefits of the learnings and improvements to the patient?

For many years the pharmaceutical industry has been building capabilities to collect and synthesize data and information for use with new product development, regulatory submissions, event resolution, enhancing process capabilities, and to meet other regulatory commitments. Arguably, a key challenge inherent in these data and information capture is the identification and retention of *critical knowledge*.

Sally retired from the company; she was the *go to person* for anything to do with product XYZ as she worked on it for 20 years: "Sally seemed to know every scientist that ever touched product XYZ. When we lost her, we lost the link to that product knowledge, I'm sure we did experiments that I can't find now.... I'm frustrated I'll have to do them over....."

---

[*] For example good manufacturing practices (GMPs), good clinical practices (GCP), and good laboratory practice (GLP)—collectively referred to henceforth as *GxP*.

Does this example sound plausible? The scientists, development team, and process chemists are all *knowledge workers* (Drucker, 1999). The members of the quality team and all the experts instrumental in technology transfer activities (whether a new product introduction or moving an existing product to a new site) are also knowledge workers. Can they easily find the knowledge they need? As *knowledge workers*, the tools and process of the *day job* are quite different than 20 or 30 years ago. Has the industry kept pace with tools and processes suitable for the needs of knowledge workers?

This chapter seeks to understand what is critical knowledge? Is it the same as, or limited to, regulated *GxP* information? If not, then what it is, and why might it be different?

It is the opinion of this author, that critical knowledge includes content, information, and personal knowledge that can add value to the business or the patient. It can therefore be identified from a number of additional and varied sources beyond the traditional *GxP* lens, including the following:

- Lessons learned
- Product or process expertise
- Expertise or *know-how* of how things work, whether it be a technical or a business process

Typically business processes are not regulated by health authorities; however, the knowledge of *how things work* and *how things get done* is critical to an efficient and effective workflow and could have an impact on the ability to consistently deliver high quality medicines to the patient. Often times this knowledge is not recognized as *critical* until someone leaves their role or even more challenging, exits the company. By which point it is often difficult or impossible to capture or recover the respective knowledge. This dilemma of *knowledge loss* is not specific to the pharmaceutical industry; however, there may be a false sense of security regarding the ability to recreate such knowledge within the pharmaceutical industry due to the focus on retention of regulated data, records, and information.

Focusing back to ICH Q10, knowledge management (or managing what we know) is listed as one of the two enablers to an effective Pharmaceutical Quality System. ICH Q10 has set down clear expectations from the regulatory authorities that industry and companies should leverage and utilize the knowledge that they have in order to improve the products, the process, and the delivery to the patient. In order to achieve this improvement we need to understand not only what the critically regulated information and knowledge might be, but we also need to layer in the critical business and technology knowledge that may not be considered *regulated*.

Before we start let us review some of the following key definitions:

- *Data*: Symbols that represent the properties of objects and events (Ackoff, 1989).

- *Information*: Information consists of processed data, the processing directed at increasing its usefulness, for example, data with context (Ackoff, 1989).
- *Content*: The topics or matter treated in a written work.[*]
- *Tacit knowledge*: Knowledge that you do not get from being taught, or from books, and so on but get from personal experience, for example, when working in a particular organizational knowledge that you do not get from being taught, or from books, and so on but get from personal experience, for example, when working in a particular organization.[†]
- *Explicit knowledge*: Knowledge that can be expressed in words, numbers, and symbols and stored in books, computers, and so on: knowledge that can be expressed in words, numbers, and symbols and stored in books, computers, and so on.[‡]
- *Functional knowledge*: Knowledge created within a specific function within an organization (e.g., specifications created by the engineering organization, batch record review processes created by the quality organization).
- *Community or network based knowledge*: Knowledge that is leveraged or curated by a community or a network (e.g., listing of subject matter experts, documents within a particular area of practice, and online discussion boards for a community or network).
- *Process knowledge (business or technical)*: The knowledge gained from business or technical processes. An example could be reports, e-mail, or lessons learned from a technical transfer activity between two sites (business process), or process diagrams and reports providing process efficiencies for a viral removal step from a manufacturing process.
- *Knowledge*: In the purest sense, as defined by the Cambridge Dictionary[†‡]: awareness, understanding, or information that has been obtained by experience or study, and that is either in a person's mind or possessed by people. However, in the context of an organization, knowledge can be a combination of content, information, as well as explicit and tacit knowledge. For simplicity in this text, references to information, content, tacit, and explicit knowledge may be generally referred to as knowledge.

---

[*] Merriam Webster online dictionary.
[†] Cambridge online dictionary.
[‡] Cambridge online dictionary.

Introduction to Knowledge Mapping

Knowledge mapping is one useful knowledge management (KM) practice that is been successfully used in many industries to identify and catalog critical knowledge.

Grey describes a knowledge map as "a navigation aid to explicit (codified) information and tacit knowledge, showing the importance and the relationships between knowledge stores and dynamics." (Grey, 1999)

There are several approaches to knowledge mapping, which are discussed in an article published by Jafari (Jafari and Akhavan, 2009) examining multiple knowledge mapping techniques from an academic perspective:

- Yellow paging
- Information flow analysis
- Social networking analysis
- Process knowledge mapping
- Functional knowledge mapping

Although each of the noted methodologies has merit, this chapter will focus on *functional knowledge mapping* and *process knowledge mapping*. These methodologies require little or no capital investment and have proven effective in many sectors, including pharmaceuticals.

> Pinpointing these gaps and barriers [of knowledge] helps the KM core team develop a targeted plan to tackle them in the right way, in the right order, with the right resources.
>
> **APQC**[*]

APQC has conducted pointed research describing knowledge management maturity. Figure 26.1 shows the relationship of knowledge mapping to an overall KM program maturity and the impact of leveraging knowledge mapping as an enabler to progressing KM maturity.

Use Cases for Knowledge Mapping

A knowledge map can help generate the *lay of the land* for a functional group, a community of practice/network, or process (business or technical). The knowledge map provides a visual representation of the content and

---

[*] APQC (American Productivity and Quality Center) Not for Profit research organization based in Houston, Texas, United States.

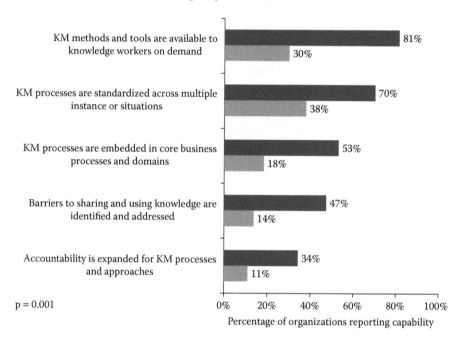

■ Organizations that use knowledge maps to identify knowledge needs/gaps (N = 70)
▨ Organizations without knowledge maps (N = 148)

FIGURE 26.1
How Knowledge Mapping relates to KM Maturity (APQC 2016).

knowledge generated, which can be beneficial to describe actual or expected contributions of a group or the current content and knowledge components of a process. These knowledge maps can be visually represented in a spreadsheet, a text document, or other format. Some examples of use cases for knowledge mapping are listed below:

1. *On-boarding new employees (new to group or an organization)*: When orienting new employees the knowledge map can be leveraged to describe outputs and knowledge created by the group.
2. *Improvement of internal knowledge capture and reuse processes*: Evaluation of existing content and knowledge management approaches can further develop group understanding and, in turn, lead to improved knowledge capture and reuse processes.
3. *Internal reorganization*: When groups are reorganized, technical and business processes that generate critical knowledge may move to new groups. A knowledge map provides a quick reference to ensure the knowledge is accounted for and is managed properly.

If a knowledge map does not exist, it could be a good opportunity to describe the knowledge created by a group and a map as to where it resides.

4. *Development of a new communities of practice (CoP) or network*: Knowledge mapping can identify the knowledge that already exists and the knowledge that is needed in the future.

5. *Merger and acquisitions*: Leveraging existing knowledge maps (or developing) provide efficiencies when describing knowledge created and curated by respective groups/organizations/communities and networks during merger and acquisition evaluation and following integration activities. Often, only documents are reviewed during these activities not taking in account of the *know how* and *know what*.

Understanding What Is Important

Identification of knowledge via knowledge mapping can also help identify its relative importance—as not all knowledge is equally important. Indeed the manner in which knowledge and content is stored and curated* should be commensurate with the relative importance of the knowledge. This concept is aligned with the recommendation in the ISPE GAMP© Electronic Records and Signature guidance (ISPE Guide, 2005) noting "application of appropriate controls commensurate with the impact of records and the risks to those records."

Knowledge mapping can also assist with *reverse engineering* the methodology and rationale of how information, content, or knowledge is currently captured (see use case 2 in the previous section). Current methods for capturing knowledge and content could be fit for the purpose for those immediately involved in the process; however, it may be difficult for others outside of the said process to find and leverage knowledge if customers of the knowledge have not been identified (refer back to the example of Sally having the knowledge of *know what* and *know who*).

When thinking about mapping knowledge it can be helpful to first categorize how knowledge in the organization is generated. The notion of categorization is intended to help later in determining how to initiate a knowledge mapping exercise and who should be involved.

---

* *Curate something* (especially on the Internet) to collect, select, and present information or items such as pictures, video, and music, for people to use or enjoy, using your professional or expert knowledge. Oxford Learners Dictionary online.

1. Knowledge is generated *within organizational structures or functional areas*, for example, development, manufacturing, quality assurance, engineering, operational excellence, learning and development organization.
   a. Knowledge generated by *short-term teams to address a project*—this is typically curated via an existing organizational structure.
2. Knowledge is generated by *longer standing cross-functional constructs such as communities of practice or technical networks* (the key here is these are not functional groups as noted in example 1).
3. Knowledge that is generated during a business or a technical process. Business and technical processes may span multiple functional groups or organizations (e.g., deviation management, technology transfer, and regulatory submissions).

The following sections describe considerations for knowledge mapping that is useful whether the knowledge in question has been:

1. Generated within an organization
2. Generated by a community of practice/network
3. Generated during a business or technical process

Mapping Functional Knowledge

Mapping functional knowledge is typically a *reactive* activity as the function in question already exists and has already generated and stored knowledge in some manner. Nevertheless, it is an effective tool to improve the capture, storage, and reuse of functional knowledge and to identify potential gaps and efficiencies. In addition, when mapping existing knowledge, it provides an opportunity to apply a lens from the customer standpoint; is the knowledge produced in a consistent and accessible format by all that need it?

Questions to consider for mapping functional knowledge (as well as limited duration team activity):

- What is the type of knowledge or content created? (Reports, evaluations, and decisions in the form of e-mails, memo's, procedures, and so on.)
- If the knowledge or content comes from somewhere else, who or where does it come from?
- What format is it in?
- Who are the primary customers for this knowledge?
- Are there any secondary customers for this knowledge? (e.g., those people that seem to use your knowledge output for additional research, product improvements, and so on—not the primary customer that originally requested it or is the normal *customer*.

- Where does this knowledge or reside?
- What is the risk if this knowledge is lost?

How to initiate a knowledge mapping session will be discussed later in this text *conducting a knowledge mapping session.*

Mapping Communities of Practice or Technical Networks Knowledge

Communities of Practice (CoP) and networks are a different use case for knowledge mapping. As communities of practice tend to be composed of like-minded people from multiple organizations or diverse functional groups, the knowledge needed may not exist. Unlike mapping for a functional group, the mapping exercise can also be used to brainstorm the knowledge needed and then determine if it exists and prioritize the need. If the knowledge does not exist, the community or network may choose to create it for the benefits of the members. APQC recommends creating a knowledge map when designing a community of practice at the outset.[*]

Below is an example of how a community could create knowledge:

> Risk assessment (RA) is an area that has many interested groups across a company or division. There are many types of risk assessments, including safety, quality, suppliers, product development, and the programs that support the assessments. There may not be one place to find risk assessment best practices or to share learnings. A community would be interested in taking an inventory of such practices, experts, documents, and learnings and make them more *consumable* or usable by others in the community. In this case they would be interested not only in what knowledge they currently have but also what knowledge they need. The community may host seminars or best practices sharing sessions virtually that could be recorded; this is a new piece of knowledge that can be leverage widely in this community. They also could create a place to list experts, share lessons learned, and best practices. These new databases or lists could house critical knowledge for the practice of *risk assessment.*

When developing a knowledge map for a community or network that does not exist, it is useful to ask the following questions:

- What are the *big* topics for the community? (Could be a brainstorm activity)
- For these *topics* what knowledge is needed?
- For the knowledge needed:
  - Does it exist?
  - If so, where or who will it come from?

---

[*] Based on the APQC Community of Practice Methodology.

- If not, does it need to be created (also, note the urgency of the need)?
- What format is it in?
- Who are the primary customers for this knowledge or content?
- Who were the secondary customers for this knowledge or content?
- Does it need to be validated (verified) before it can be shared?
- What is the risk if this knowledge or content is lost?

Mapping Business Processes Knowledge

When evaluating a business or a technical process it is beneficial to leverage existing process mapping tools and techniques. In the event that process mapping tools are now commonly leveraged, it is possible to utilize similar templates as for mapping functional knowledge. Typically, tools such as a SIPOC diagram (suppliers, inputs, process, outputs, and customers) (Johnston and Dougherty, 2012), work flow diagrams, swim lane exercises, and so on provide a nice framework to develop a knowledge map. For each step in the business process, the existing process flow diagram could be overlaid with the following considerations:

Is there any knowledge or content generated from this step? If so

- Where is it located?
- Who generates it?
- What format is it in? (explicit e.g., reports, memos, and e-mails, or tacit in someone's head)
- Who can access it?
- Who are the primary customers for this knowledge or content?
- Who were the secondary customers for this knowledge or content?
- What is the risk if this knowledge or content is lost?

Collating the responses from these questions across the process provides a rich source describing knowledge created and which of that is the most critical knowledge. It should be noted that explicit knowledge is much more tangible and easier to describe, define and organize, as illustrated by this chapter on knowledge mapping. Many other chapters in this book will present insights on the importance of tacit knowledge and practices to get it to flow.

The following perspective by Christopher Smalley shares some insights to identify tacit knowledge in the flow of work. Chris has spent 34 years in the biopharma industry and has witnessed firsthand the importance of tacit knowledge. These examples and many more like them could be very valuable when mapping knowledge of a function, community or a process (technical or business).

TACIT KNOWLEDGE
BY: CHRISTOPHER SMALLEY

In the regulated biopharmaceutical industry, there is a clear expectation to capture data, information, and knowledge. With that being said, there are several facets to knowledge, just as there are several iterations of data and information that contribute to knowledge. The most used knowledge, but least defined and characterized, is *tacit* knowledge.

Tacit knowledge is knowledge based on human memory and is comprised of that person's experiences, which include learning and education. It defines the difference between someone who purports to be a teacher but is simply reading from a book or the documentation of others, and the learning is identified as being routine. Contrast this with a true teacher who makes the training material come alive with real-life experiences and achieves an apprenticeship-like learning.

Tacit knowledge can also be learned alone. A mother may tell her child innumerable times that the stove is hot and they should not touch it. If the child learned, that would be an explicit knowledge. As parents know, most times it is only when the child touches the stove, experiencing the heat and then the pain, did they learn tacit knowledge. The experience becomes the lesson, not the teaching.

How might an organization use and benefit from tacit knowledge? Production organizations use tacit knowledge when they acknowledge the value and contributions of senior, experienced operators over junior operators or those new to a process. Although all operators have equal access to procedures, recipes, tools, and equipment, the experienced operators are able to identify key decision points and respond to keep processes from foaming, failing, alarming, or in some way deviating from the intended outcome. So, the organization benefits in this way.

Using a car for our example, the gas gauge provides data on how much fuel remains in the car's tank. This is a valuable data, but this data can be used to create information. When coupled with the odometer, the data from the gas gauge can measure the number of miles per gallon that the car is achieving. If that information on fuel economy is trended, it provides knowledge about how the car is performing. If the trend shows that fuel economy is declining, then additional sources of data can be brought into play. One might be to check the air pressure in the tires—another data source. Another might be the driver listening to the sounds that the transmission is making—is it taking longer time before it shifts to a higher gear indicating that there might be a problem with the transmission. All of the sources of data described up to this point are referred to as explicit data—the data are read-off

(Continued)

of an instrument or device and are unambiguous. The last piece of information described was the car driver listening—or equipment operator listening, as it were. In listening for perhaps a higher pitch whine in the transmission, or feeling for the shutter of the shift, the equipment operator is developing tacit information. Tacit knowledge may not be adequately articulated verbally, but it can be as important, or even more important, than explicit knowledge in understanding systems and processes.

So, let us now move into examples of tacit knowledge in pharmaceutical applications. We all have heard the *urban legends* of tacit knowledge, but let us look at two examples:

> *Tablet compression*: A supervisor had years of experience as a table compression operator before being promoted into supervision. On this day, he stopped to speak with a current operator whose equipment was running a 500 mg tablet at 92% of machine rated speed. The supervisor told the operator that he had a bad punch and needed to tear down the machine. If you are familiar with this type of equipment, you would know that tearing down a tablet compression machine, except for PM, is a major undertaking and is not performed without good cause. The operator performed the teardown and found an upper punch that had a crack in the neck. The crack was not yet showing up in the normal monitored parameters such as tablet weight, appearance, hardness, and the like, but was detected by the supervisor's ear.

> *Tablet coating*: The use of natural materials in any manufacturing process will accentuate the importance of tacit knowledge. Shellac is a natural material used for coating, and depending on the volume of tablets that need to be coated, the traditional copper coating pan with manual ladling of the coating solution is used. In this case, the coating specialist was adding shellac to a pan of tablets, adjusting the hot air and exhaust, rotational speed of the pan, and other parameters consistent with the batch record. When it came time to add the last ladle of shellac, the coating specialist threw the shellac on the back of the pan, where it dried onto the pan and not the tablets due to the hot air. When asked, the coating specialist explained that he had to add the amount of shellac called for in the batch record, but if he added that all to the tablets, he would wind up with one large unworkable mass. The coating specialist's tacit experience taught him that applying all of the shellac to the tablet would not result in acceptable tablets.

(Continued)

So, both stories tell of how tacit knowledge contributes to making better pharmaceuticals and helps to explain why an operator trained in the procedures and following the batch record might not be successful in manufacturing a satisfactory batch. *The challenge for the organization is to capture this tacit knowledge, and make it institutional knowledge by updating batch records and procedures by incorporating that tacit knowledge.*

It is incumbent for the industry to encourage and share such expressions of tacit knowledge and leverage knowledge management practices (e.g., knowledge mapping, lessons learned, and communities of practice) to capture and apply such learnings.

Conducting a Knowledge Mapping Activity

Plan and Define

Generating a knowledge map can be as simple as completing a template; however, in the experience of the author, using a standard methodology to plan and define the activity provides additional benefit. It helps set the stage, manage the process and expectations regarding participation as well as defining the required outputs.

Leveraging learning's gained from standardized *Lean Six Sigma* processes, the author developed a process for knowledge mapping. This process includes the following steps: plan, define, analyze, review, recommend, and implement. These steps provide a robust mapping process as well as a mechanism to provide feedback and recommendations (Figure 26.2).

Planning is an important phase of the knowledge mapping exercise, as it sets the stage for the level of engagement required from participants and

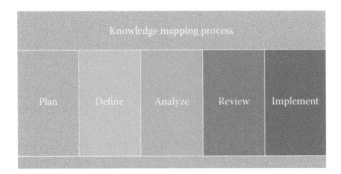

FIGURE 26.2
Knowledge mapping process elements.

TABLE 26.1

Knowledge Mapping—Phased Requirements and Deliverables

	Who	**Timing**[a]	**Output**
Planning	KMapping[b] Facilitator with target audience and management	~1 hour[a]	Agree timing, agenda, attendees, and output
Define	Facilitator with target audience	~1 hour[a] with target audience	Draft knowledge map
Analyze	Facilitator	~1–3 hours[a]	
Review	Target audience	~1 hour[a]	
Recommend	Facilitator	~1 hour[a] with target audience management	KMapping[b] assessment report
Implement	Target audience	TBD depending on observations and resources	Target audience implement changes as agreed

[a] Estimated based on activity.
[b] Knowledge mapping.

outlines the expected deliverables. Table 26.1 outlines a typical knowledge mapping plan.

A key success factor is to ensure that appropriate sponsorship from local and senior management is in place. The next step is to carefully consider the make-up of the participants. Then, with suitable template questions in hand the *define* stage of the knowledge mapping exercise can begin.

There are multiple ways that knowledge mapping exercises can be facilitated; however, the following considerations are suggested for a *functional group*:

- *Facilitator*: An experienced knowledge mapping SME*/facilitator—it is also helpful if the facilitator is familiar with the terminology of the focus area under consideration but not required.

- *Size of focus group*: 10 or less but must be relevant for the scope of team, for example, if it is a large group consisting of 500 people and multiple functional areas, it is best to divide up into subteams—facilitated separately and complied later.

- *Attendees*: Best to include a mix of colleagues at all levels in the organization.

- *Focus group type*: "In person"—meaning meeting may take place with attendees either physically present or virtually via an online meeting tool.

- *Timing*: One hour for initial mapping process, for larger groups, not more than 1.5 hours at a time.

* Subject Matter Expert.

The following considerations are suggested for a *Community of Practice or Network*:

- *Facilitator*: An experienced knowledge mapping SME/facilitator—it is also helpful if the facilitator is familiar with the terminology of the focus area under consideration but not required.
- *Size of focus group*: 10 or less but must be relevant for the scope of the community or network. Often this would be the core team of the CoP or network that is responsible for curation of the community or network.
- *Focus group type*: *In person*—meaning meeting may take place with attendees either physically present or virtually via an online meeting tool.
- To determine the needs of the community or network, it is useful to conduct a brainstorming activity prior to leveraging the templates to determine the major topic areas. This can be accomplished face-to-face or virtually, however it must be planned to ensure success. Once identified, the evaluation of the existence of required knowledge can be captured on the templates. For knowledge that does not exist, prioritization of knowledge generation can be identified via modification of the template columns.
- *Timing*: Up to one hour for initial brainstorming and then an additional hour for capturing the state of the respective knowledge.

Facilitation

When facilitating a knowledge mapping exercise the dialog created during the session is extremely valuable and often highlights the different ways knowledge is currently generated and captured within a group. These knowledge pathways are often not formally documented and may not even be previously recognized by the group. The inconsistency of generating content and knowledge in different formats may or may not be an issue to the organization; however, it may not be known that colleagues are generating content in multiple formats and this may present difficulties in knowledge sharing and access in the future. In the experience of the author, feedback from knowledge mapping sessions routinely contain sentiments such as

- "I had no idea that XXX team used our reports."
- "I keep that material on my hard drive as I am the only one that uses it, I did not know XYZ was interested in it."
- "This was an interesting conversation, we never take the time about how we work, we just do it."

- "It is not even Sally's job to be the go to person for this knowledge, but she seems to be-based on this conversation today."
- "I didn't know that [... .] had that information."

Due to the rich dialog and active learning during the activity, it is recommended that participants join the focus group at the same time and not complete a knowledge mapping template on their own. Two template examples are provided in the appendices at the end of the chapter.

I Have a Map, Now What? Analyze and Review

Once the knowledge map is captured the *analyze* phase can begin. It is good practice to send the captured map upon original completion to the group that participated in the generation. It is helpful to give the participants an additional opportunity to provide information in the event something was missed or captured incorrectly prior to starting the define phase. *Analyze* is the stage when the *know-how* of the facilitator comes to fruition. The facilitator will review the capture from the session (the knowledge map template) and assess the vulnerability of the respective knowledge elements in the context of the customers, findability, and loss aspects of the knowledge. The facilitator can then take the findings and contextualize the risk and suggest a best practice location(s) for storage and accessing of such knowledge if they could be improved. All recommendations must be aligned with the mission of the group, company, and within bounds of regulatory requirements.

Knowledge mapping facilitator skillset:

- Good team facilitation skills
- Understanding of KM capabilities of the organization
- Understanding of content management policies and tools for the organization
- Understanding of records management policies for the organization (GxP records, vs. nonregulated)
- Ability to develop a proposed plan of action resulting from the knowledge mapping exercise
- Ability to engage stakeholders, including management, for the initiation of the knowledge mapping exercise, as well as the readout

Recommend and Implement

In order to action learnings from a knowledge mapping exercise, the learnings and results must be compiled in a manner that is understandable and can be implemented.

A standard template for reporting the knowledge mapping activity and output is recommended. Suggested items include the following:

- Date and list of attendees (including facilitators).
- Timeline for reviewing and developing recommendations.
- Brief statement regarding any insights gleaned from the conversation/knowledge mapping activity in the defined phase.
- Highlight areas that would benefit from modifying the mechanism how knowledge elements are created or stored (e.g., content created is stored typically on hard drives, making it not accessible for others, or several colleagues in the same group create very different knowledge outputs, is that an area that would benefit from standardizing the format?
- Highlight any best practices noted that may be shared with other groups that did not participate in the exercise but could benefit.
- Highlight any KM practices or partners that may be able to assist with implementation.

It is important to note: what has *not* been recommended is the facilitation group takes responsibility for implementing the learnings. It is very important for groups to own the responsibility for collecting and curating their respective knowledge. Good curation behaviors are learned in an organization and in order to sustain best practices for creating, retention, and reuse of knowledge, the people that create knowledge must sense of ownership. With that being said, the knowledge management team or experts in the content management group may be able to assist with the implementation of learnings from knowledge mapping activities.

Summary

Taking the time to thoughtfully evaluate the knowledge created or needed by a functional area, community, a technical, or a business process can greatly improve how others can leverage that knowledge when needed. Remember the example of Sally?

"I really miss Sally but so thankful that we have a map of her 'go to' places for product XYZ. That spreadsheet has really paid off in saving me time! Also I had to call a scientist for some background information the other day regarding an old report, and thankfully the group that produces those reports was also on the map that Sally contributed to...."

Knowledge mapping can be a useful KM practice to add to the knowledge management practitioner's tool kit. Within the pharmaceutical industry and respective organizations, there is a goldmine of knowledge—as a previous colleague used to say: "We have more knowledge than we could buy [via consultants], if we could just find it" (Christopher Smalley, 2007). All businesses could benefit from finding the right knowledge at the right time. For the pharmaceutical industry, efficient and effective leveraging of knowledge further enables the ability to effectively deliver life changing and lifesaving products to patients in a timely manner.

References

Accelerators of Knowledge Management Maturity, APQC, 2016, APQC Knowledge Base.

Ackoff, R. L. (1989). From data to wisdom. *Journal of Applied Systems Analysis, 16*(1), 3–9. Retrieved from http://doi.org/citeulike-article-id:6930744.

Drucker, P. F. (1999). Management challenges for the 21st century. *Harvard Business Review, 86*(3), 74–81. doi:10.1300/J105v24n03_22.

Grey, D. (1999). Knowledge mapping: A practical overview by Denham Grey.

International Conference on Harmonization, *Pharmaceutical Quality System - Q10*, 2008. found at www.ICH.org.

ISPE Guide. (2005). *A risk-based approach to electronic records and signatures.* Tampa, FL: ISPE.

Jafari, M and Akhavan, P. (2009). A Framework for the selection of knowledge mapping techniques. *Journal of Knowledge Management Practice, 10*(1), 1–8. Retrieved from http://webpages.iust.ac.ir/amiri/papers/A_Framework_For_The_Selection_Of_Knowledge_Mapping_Techniques.pdf.

Johnston, M and Dougherty, D. (2012). Developing SIPOC diagrams. *ASQ Six Sigma Forum Magazine, 11*(2), 14.

Wyeth Validation CoP Design Team Workshop, Collegeville, PA 2007.

Appendix I: *Functional Knowledge* Map Template Example

Date _____   Group _____   Attendees: _____

Functional Work Stream	Knowledge Element Created	What Format	Who Generates?	Primary Customer for K[a]	Secondary Customer for K[a]	Where Does It Reside?	What Is the Risk If Lost? (H, M, L)	Gaps Identified	Notes/ Comments

[a] Knowledge.

Appendix II: *Community of Practice* Template Example

Date _____ Group _____ Attendees: _____

Community Topics/	Knowledge Element Needed	Does It Exist?	Who Has It?	What Format?	Primary Customers for K?[a]	Secondary Customer for K?[a]	Need Validation?	What Is the Risk If Lost? (H, M, L)	Gaps Identified	Notes/ Comments

[a] Knowledge.

27

The Practical Application of a User-Facing Taxonomy to Improve Knowledge Sharing and Reuse across the Biopharmaceutical Product Lifecycle: A Case Study

Adam Duckworth, Vincent Capodanno, and Thomas Loughlin

CONTENTS

The word *taxonomy* is likely to evoke one of the three possible responses: fond memories of biology class, unpleasant references to tedious document management systems, or a blank stare. This case study shows how one team at Merck & Co., Inc. have evolved a practical and user-centric approach to taxonomy that is paying big dividends in how product knowledge is managed, shared, and used.

Editorial Team

In 2009, the concept of *taxonomy* may have been familiar to some within Merck's Manufacturing Division (MMD), but its meaning at that time was largely tied to the typical high school biology teachings on the classification of life—*Kingdom, Phylum, Class,* and so on. Seven years later, there is a much broader understanding and appreciation to the extent that it is the focus of project workstreams (or entire projects) and even stands alone as a distinct strategic objective within the business strategy of a major MMD organization whose leader has touted the benefits delivered by taxonomy as "critical to the future success of the organization." This is quite the paradigm shift and is the result of a concerted and robust knowledge management (KM) effort previously described by in *Pharmaceutical Engineering* magazine (Lipa et al. 2013) and by Yegneswaran, Thien, and Lipa earlier in this book. Through an iterative approach that started small with a pilot focused on improving product knowledge flow, taxonomy is increasingly being recognized as a means to unite the workforce around a common language, as well as facilitate knowledge sharing, knowledge reuse, workflows, and expertise location for improved outcomes across a wide range of business activities.

Origin of Taxonomy for Merck Manufacturing Division

Merck & Co., Inc. manufactures, packages, and distributes more than 100 products to approximately 140 markets around the world. The manufacturing division is composed of approximately 18,000 highly skilled scientists, engineers, technicians, supply chain managers, and support persons who are dispersed in more than 50 locations and 20 countries globally. Within MMD is an organization, *global science technology & commercialization* (*GSTC*), where the roots of the present day taxonomy capability first took hold. Using knowledge flow mapping to identify KM improvement opportunities for a prioritized set of business processes, a pilot project focused on improving the flow of product knowledge across the product lifecycle and treating that knowledge as an asset was commissioned with a taxonomy identified as one of the five critical success factors—Figure 27.1 (Lipa et al. 2013).

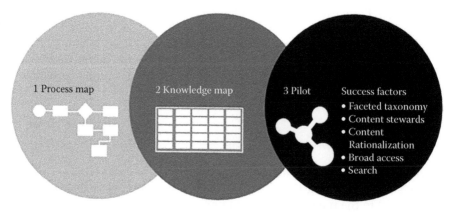

FIGURE 27.1
Pilot project identification and critical success factors.

Taxonomy Terminology

Before MMD's taxonomy journey is discussed, it is important to align on the terminology used in this case study as definitions for many key concepts in the taxonomy field can be highly variable. In addition, to position taxonomy work for success, one must consider the messaging and how it is perceived by the business leaders and users (the customers) so as not to overwhelm or disengage them. The term, *taxonomy*, was introduced at the outset of the MMD knowledge management effort and has been adopted to reflect a variety of distinct concepts among which only a handful of people really need to distinguish. As Heather Hedden has stated with respect to the modern day perception of taxonomy, "The term taxonomy has taken on a broader meaning that encompasses all of the following: specific-subject glossaries, controlled vocabularies, information thesauri, and ontologies" (Hedden 2010). One might consider dictionaries, terms, term sets, facets, concepts, metadata, and semantic architecture all to fall under that broad meaning, as well.

MMD has certainly taken this liberty and has done so purposefully in order to maintain the veil of simplicity and successfully build awareness and sponsorship of taxonomy work across a broader set of business stakeholders who are more interested in the required investment and tangible outcomes than the gory details. In fact, taxonomy is sometimes simply described as MMD's *business language* with taxonomy work being *building and applying an in-depth understanding of our business language to organize and reuse knowledge.*

Facet has also been an important term and can be viewed as a *category* or "one of the fundamental dimensions in which content can be analysed" (Lambe 2007). When introducing the concept of a facet, column headers in a spreadsheet or field names in a SharePoint library are often highlighted as recognizable examples. An analogy to a diamond ring is also commonly used to introduce the concept of a facet—similar to how the facets of a diamond ensure that from whichever direction someone is looking they have a rewarding view; the facets of a taxonomy ensure that each MMD users group will have a meaningful view of the company's product knowledge.

Within each facet is a list of values individually called *terms* and collectively the *term set*. Although this nomenclature has been used, it is only on occasions since the main visibility business users have to terms is when selecting them as metadata to apply to documents. Therefore, *terms* are generally referred to as *tags*, with *tagging* being the act of applying a tag to a document or other content. The use of *tags* and *tagging* has resonated well with the MMD audience and draws parallels to familiar online tools such as Evernote, OneNote, and Flickr, which use similar terminology. The term *metadata* has specifically been avoided and has been found to disengage users.

To elaborate with a simple, modern example, consider a collection of digital photos that might be organized by some criteria such as location, event, and date taken. These categories are the *facets* and this collection of facets represents the *taxonomy* for photo organization. The *location* facet might have a list of values to choose from, the *term set*, and when one of those values is applied to a specific photo, the photo is being *tagged*. Further background on taxonomy terminology, and the taxonomy field in general, can be found in these two popular resources: *Organising Knowledge: Taxonomies, Knowledge and Organizational Effectiveness*, by Patrick Lambe (2007), and *The Accidental Taxonomist*, by Heather Hedden (2010).

The Journey

The successful application of taxonomy is really just about finding that adequate balance between flexibility and standardization of the language of our business. From a purely technical perspective, one might suggest that language differences do not matter very much; technology can be used to manage those differences and ensure that information can be retrieved by people in a variety of ways, such as through the use of synonyms or by defining more complex relationships between terms and phrases. In addition, this perspective holds weight, especially as machine learning technologies and artificial intelligence capabilities mature. But this position only has potential when considering explicit knowledge (i.e., codified information) and breaks down when tacit knowledge and collaboration are considered.

In a world where day-to-day priorities change rapidly, MMD is creating an environment that takes advantage of our people resources and enables them and their knowledge to shift to the company's most pressing needs. In this environment where people from different backgrounds will often be convened quickly for short-term projects, a common language—standardization—can take considerable waste out of these in-person interactions such that we do not waste time mentally translating the nuances of one person's mental model to another's. Minor differences in our language such as *analytical method* versus *analytical procedure* versus *analytical protocol* may or may not have different meanings to different people but at a minimum add up to unnecessary distraction if not a deeper misalignment, and hinder the speed at which these teams can meaningfully focus on the task at hand. This magic balance between flexibility and standardization is indeed the taxonomist's white whale and the fundamental essence of the journey itself, that even if achieved is fleeting; a merger or acquisition, external partnerships, organizational restructuring, and disruptive technologies all continually influence where that line should be drawn.

But though that end goal may be impossible to achieve, it does establish direction—*true north*—and when GSTC recognized that the language of biopharmaceutical product development and commercialization was skewed far toward too much unnecessary *flexibility* (and the associated inefficiencies this brought), support quickly grew for a standard taxonomy as a component of the overall KM capability to establish better balance.

As mentioned earlier, taxonomy was considered one of the five critical success factors of the broader KM product knowledge capability, along with content stewardship, search, broad access, and historic content rationalization. The capability would need to enable knowledge flow for content that could be in use for decades with anywhere from 5,000 to 50,000 pieces of unstructured content being generated for each biopharmaceutical product across its lifecycle; unstructured content included reports and memos, e-mail communications, presentations, images, and similar types of files, some of which were scanned from existing paper records. A controlled taxonomy was identified as an opportunity because the core work processes to develop and manufacture new products has changed little in the past 100 years. A review of chemical process documentation typical of the 1930s, for example, would look very similar to common documents created today, such as process descriptions explaining the product manufacturing recipe. This suggested that a consistent, structured taxonomy describing the knowledge generated over the product lifecycle would be a broadly applicable and enduring asset.

This KM approach has been previously described (Lipa et al. 2013) and for the purposes of this case study, taxonomy is the specific topic of focus. Based on a foundational principle of *think big, start small, but start*, the journey began with a small pilot that was initiated in 2011 and has since been followed by a series of expansion events that introduced additional users and/or knowledge scope and built on lessons learned from the preceding effort. An overview of these step changes is provided in Figure 27.2 and described below.

1. *API technical knowledge pilot*: The pilot with a scope of two active pharmaceutical ingredients (APIs) emerging from the research division and 15 users. This established the first taxonomy design for product technical knowledge—the domain of knowledge pertaining to the science, techniques, and equipment to develop, manufacture, and characterize biopharmaceutical products across the product lifecycle.

2. *API technical knowledge*: The expansion of the taxonomy to all (~50) APIs in the small molecule modality and ~300 core users, including a taxonomy redesign based on lessons learned from the pilot. Although this greatly expanded the user base of the taxonomy, it did not expand the scope of content that was previously covered in the pilot.

Taxonomy capability evolution

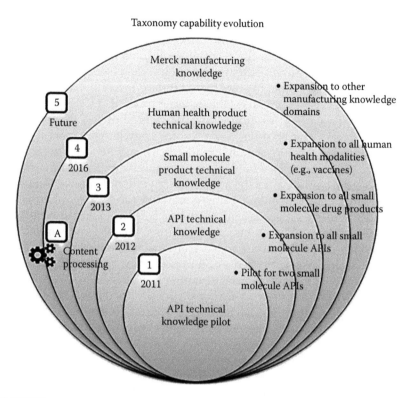

FIGURE 27.2
Taxonomy capability evolution.

3. *Small molecule product technical knowledge*: The expansion of the taxonomy within the small molecule modality from API materials to formulated drug product, adding an additional 300 core users. This expanded the taxonomy both in user base and content scope, introducing knowledge for formulation and packaging as an example.

4. *Human health product technical knowledge*: The expansion of the taxonomy from the small molecule modality to all modalities within the human health business, such as vaccines and monoclonal antibodies. This also expanded the taxonomy in both user base and content scope, introducing knowledge for sterilization as an example.

5. *Merck manufacturing knowledge*: Ongoing refinement and the initial efforts to expand beyond the knowledge domain of product technical knowledge to new knowledge domains such as supply chain.

A. Content Processing: *A process and supporting technology, including natural language processing, to inventory, automatically tag (auto-classify) and consolidate existing product technical knowledge from many disparate repositories to select standard, SharePoint repositories. As will be discussed in* the following sections, the initial focus on MMD's taxonomy work was to develop a taxonomy that could be applied to new content at the time of creation. But this would not address the existing product technical knowledge amassed during the past 15 years for currently-marketed products, which was located across hundreds of repositories, inaccessible to many users, and inconsistently organized. In parallel to the expansion of the taxonomy capability, *content processing* was established in 2014 to address this problem with a key feature of automatically interpreting existing content using natural language processing and tagging it with terms from the new standard taxonomy. This would also set the stage for bulk retagging as part of ongoing sustainment to adapt to changes in the taxonomy as it evolved. As of the writing of this case study, more than 150,000 documents spanning 50+ biopharmaceutical products have been auto-classified and consolidated. This work will not be further explored in this chapter but has been an invaluable application of taxonomy to improve the findability of existing knowledge and position it for reuse. For additional information, see the 2014 KM World presentation, *Knowledge Management Across the Product Lifecycle* (Duckworth and Arnold 2014).

The remainder of the case study highlights the approaches and lessons learned for the first three iterations, from pilot to the expansion across small molecules.

API Technical Knowledge Pilot

Think big, start small, but start. That is a fundamental principle of the MMD knowledge management approach and the API technical knowledge pilot that is evidence of this approach to gain real life experience while making meaningful progress. The intent of the pilot was to experiment with a potential solution that maximizes the opportunity to learn while minimizing disruption to the business. This was also an opportunity to build broader awareness of new concepts, such as taxonomy, and improve collaboration between functional silos. Although initiating sponsorship at the executive level had been secured and a select group of business stakeholders had participated in the original KM strategy development effort, this would be one of the first times a broader audience was engaged with KM and the pilot was recognized as a key activity that would kick start a fundamental change management effort.

Aside from thinking big and starting small, four other key principles helped guide the pilot.

1. *Use standard technologies already available within the company*: This would help ensure the team started small and would serve to prove whether or not more complicated technologies were truly required.

2. *Do not treat this as an IT project*: The leader of the pilot was from within a business unit that would be one of the main customers of the taxonomy, not an IT Project Manager. The individual was provided capacity to focus on the project and supported by a strong business team representing the breadth of business functions who would need to contribute long-term for realization of improved product knowledge flow. IT was considered a supporting function with a representative identified to lead one of the project work-streams focused on technology.

3. *Recognizing the traditional concept of folders is outdated*: As part of an early strategic decision, the limitations of a traditional folder structure and the benefits of a transition to tagging were recognized. The capability would be folderless with organization of content being accomplished through taxonomy and tagging.

4. *Design for extensibility*: The intent of a pilot is to demonstrate success quickly and scale up, so it is critical to design with extensibility as a goal.

Approach: Two late-stage development pharmaceutical compounds early in their lifecycle were identified to take part in the pilot. This served to remove the complexity of the existing knowledge mature products would have had and allowed the team to establish a completely new model for these two compounds just coming out of the research area. The pilot was run as a project with multiple workstreams, such as templates, technology, security, and taxonomy. The approach to design the taxonomy began with the participating business functions aggregating and affinitizing a list of their common documents. After the list quickly grew to 1000 items, a prioritization effort narrowed the list to ~250 of the most important document types required for the critical business processes of technology transfer and regulatory filing. The primary facet of the taxonomy was designed to mirror these document types and structured in a similar fashion to a table of contents as shown in Figure 27.3. It represented a mixture of material types (e.g., raw materials), document types (e.g., list), topics (e.g., methods), and business activities (e.g., validation), which will be highlighted again later when the challenges of the pilot are discussed.

This taxonomy was entirely hierarchical with a depth of four levels. A guiding principle was to limit the number of choices per hierarchical level to a maximum of 10, so that users would not be overwhelmed by choices; each branch of the hierarchy would end in a recognizable topic. In effect, this was a large, but standardized, virtual folder structure that theoretically could yield 10,000 (10^4) options but practically resulted in ~250 available terms at the lowest level in this case. Each terminus of the hierarchy was expected to be specific enough to hold 10–50 documents per drug substance on a specific topic. The filtering ability provided by this facet, combined

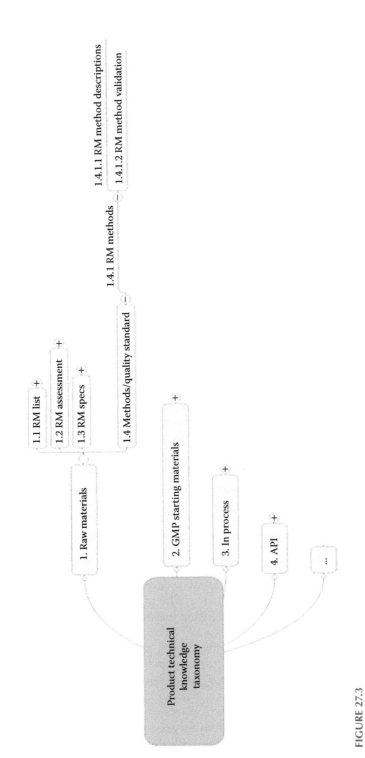

FIGURE 27.3
Initial taxonomy structure.

Name	Analytical Tech Pkg ACPN	.doc
Level 1 taxonomy	1 Raw materials ⌄	
	Select level 1 taxonomy	
Level 2 taxonomy	1.4 Methods and quality standard ⌄	
	Select level 2 taxonomy	
Level 3 taxonomy	1.4 RM methods ⌄	
	Select level 3 taxonomy	
Level 4 taxonomy	1.4.1.1 RM method descriptions ⌄	
	Select level 4 taxonomy	

FIGURE 27.4
Taxonomy as it appeared in SharePoint solution.

with other standard facets are not shown in Figure 27.4, such as *creation date* and *clinical phase*, were expected to provide a level of granularity to enable fast and intuitive retrieval.

SharePoint 2007 was chosen as the content management technology for the pilot as it was the predominant enterprise standard at the time and was not customized in any way for this effort. Working within the constraints of SharePoint required the team to implement the taxonomy as a set of four metadata columns as shown in Figure 27.4. These fields were not dependent on each other, meaning a selection of *1 Raw Materials* for *Level 1* did not restrict the set of terms visible to the tagger at *Level 2*. The metadata fields were also single-select, meaning only one term could be selected at each level.

In parallel to the first two pilot teams using this new taxonomy in support of their work, validation activities were also occurring. A team tasked with standardizing templates across business units mapped the taxonomy terms to the standard templates being developed to assess coverage. Interviews were conducted with dozens of stakeholders to aggregate feedback. In addition, a third party was engaged to assess our progress and provide recommendations. A key activity facilitated by this third party was a card-sorting exercise where ~25 common documents were tagged by ~100 stakeholders; this unfortunately demonstrated a high degree of tagging inconsistency across the group. The pilot was formally concluded with a read out to sponsors highlighting lessons learned and with recommendations for next steps.

Lessons learned: The first taxonomy design proved unwieldy for several reasons, but the pilot did serve its purpose and provided many lessons that were carried forward into the second, larger release and taxonomy redesign.

- Based on feedback from participants, support for the approach using taxonomy rather than folders grew considerably and participants recognized its value along with the value of standardization.

Although the taxonomy resulted in mirroring a very rigid folder structure, it *was* standardized and this served to begin to align functional silos on a common language. More importantly, it built the recognition among stakeholder groups that their work was not as unique from each other's as they thought.

- The taxonomy allowed for only a single selection and similar information could fall under several branches of the taxonomy. This contributed to an inconsistency in tagging that would result in difficulty for users to find information. Although the standardized language was indeed a success, the taxonomy structure would have to be reconsidered.

- The technology was insufficient to provide an adequate user experience. The hierarchical taxonomy had to be implemented across four separate SharePoint columns that were not dependent on each other. Likewise, on the retrieval end, the standard filter and search tools were not sufficient to take advantage of the benefits of the taxonomy.

- The use of standard, pretagged templates was identified as an excellent opportunity to minimize the overhead of tagging documents and to drive consistency across users.

API Technical Knowledge

With the lessons from the pilot and mitigations to address the identified gaps, sponsors agreed to expand the pilot for the two initial development compounds to a full-scale production capability for all small molecule APIs at Merck & Co., Inc. A new taxonomy design would be required to address the shortcomings of the pure hierarchy approach, and improved technology would be needed to make the tagging and search interface more intuitive. The pursuit of standard document templates across business units would continue.

Approach: After further exploration into taxonomy design approaches, a methodology was chosen to redesign the taxonomy through a combination of workshops, focus groups, and card-sorting exercises as outlined in Figure 27.5.

The facets and terms were driven out through a combination of reviewing external examples, a verb–noun activity and ~30 user interviews to understand current practices and pain points. For the verb–noun activity, stakeholders were asked to write down the verb–noun pairs that represented their work (e.g., *release material, scale-up process,* and *validate method*). After reviewing and affinitizing the output, the verbs would later be translated to an initial term set for one facet and the nouns for another.

Based on the output of the workshops and user interviews, the taxonomy was reimagined as a set of generic facets with terms from each facet being combined in a variety of ways to achieve coverage of the broad scope of content generated during the lifecycle of a biopharmaceutical product. This is known as a faceted taxonomy with the combining of terms across facets

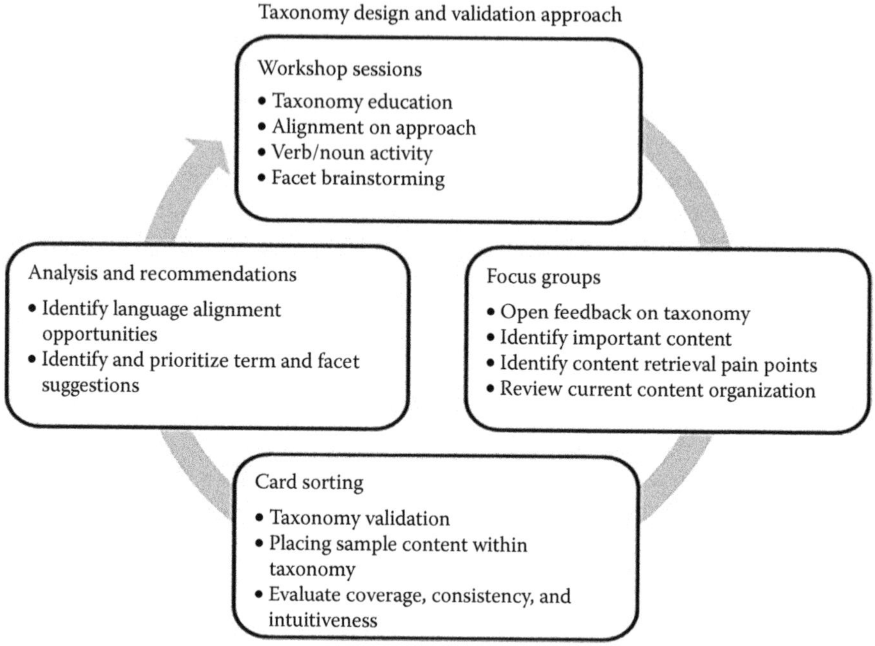

Taxonomy design and validation approach

Workshop sessions
- Taxonomy education
- Alignment on approach
- Verb/noun activity
- Facet brainstorming

Analysis and recommendations
- Identify language alignment opportunities
- Identify and prioritize term and facet suggestions

Focus groups
- Open feedback on taxonomy
- Identify important content
- Identify content retrieval pain points
- Review current content organization

Card sorting
- Taxonomy validation
- Placing sample content within taxonomy
- Evaluate coverage, consistency, and intuitiveness

FIGURE 27.5
Taxonomy design and validation approach.

referred to as *crosscutting*. As described by Patrick Lambe, "facets tend to be much simpler and shallower than tree and hierarchy taxonomies, because each facet only takes one aspect of the content into account and does not need to account for different combinations of attributes. This makes them much easier to navigate and use. In fact, many facets are simply lists" (Lambe 2007). Mathematically, four independent facets with only ten terms each could yield as many tagging combinations (10^4) as the unwieldy hierarchy used in the pilot. Use of facets would greatly improve flexibility while reducing complexity. Four core facets were defined and made mandatory for all content:

- Topic answers the question of what is the author writing about. Term examples include *structure elucidation, occupational exposure*, and *comparability*.
- Document type answers the question of how is the information presented or in what format? Term examples include *report, protocol*, and *proposal*.
- Document purpose answers the question of during which business activity was the information generated or used? Term examples include *screening, release*, and *transfer*.

- Material type answers the question of for which component of a biopharmaceutical product the information was created. Term examples include *API, excipient,* and *raw material.*

Card-sorting exercises and topic-based focus groups (e.g., sourcing, quality assurance, and engineering) were used to refine and validate the underlying term sets. These facets were configured as multiselect, so users could choose one or more terms from within each facet to tag content. The number of terms and depth of hierarchy were kept to a minimum. Practical examples of the cross-cutting ability of these facets are shown in Table 27.1. Note that multiple tags are separated by semicolons (e.g., analytical methods; stability) and parent terms are not shown for facets with hierarchy for the purposes of simplicity.

In addition to these four primary facets, other facets (Figure 27.6) were included to meet specific information retrieval needs. *Subject matter experts (SMEs)* enables authors to tag additional contributors, so future searchers would have a list of contacts they could engage regarding the content. *Keywords/context* was provided so that additional commentary on the content could be added, or keywords to allow product teams to organize content for specific activities.

It was recognized at the time that lack of tagging compliance was a major risk. Regardless of training or enticements, even the *best* content contributors would not exceed a certain effort threshold. The poor work of some contributors could quickly devalue the good work of others. Although the depth and breadth of the facets were kept to a minimum, much focus was still placed on

TABLE 27.1

Examples of Facet Crosscutting

Topic +	Doc Type +	Purpose +	Material	= Content
Analytical methods	Memos/ reports	Transfer	Intermediate	Report that summarizes the analytical results between the transferring and receiving sites for an intermediate as part of the method transfer deliverables
Analytical methods; stability	Protocols	Transfer	API	Protocol used for a transfer of API stability testing between two parties
Analytical measurement/ testing	Certificates	Release	Intermediate	Certificate of analysis showing data for release of an intermediate batch
Hold points/ hold times; stability	Protocols	Processing/ unit operation; execution	Intermediate	Protocol that outlines the specific requirements for the intermediate hold time studies in supply

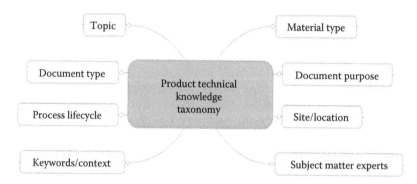

FIGURE 27.6
Additional facets added to product technical knowledge taxonomy.

development of standard document templates that were pretagged to minimize tagging effort and drive consistency.

Along with the redesign of the taxonomy, new technology was implemented to improve how users interact with the taxonomy. Since the pilot, Merck & Co., Inc. had moved from SharePoint 2007 to SharePoint 2010 and the choice was made to enhance out-of-the-box SharePoint 2010 with additional tagging and search tools. This enabled features such as type-ahead suggestions during tagging, tag browsing, and advanced filtering options both directly in the SharePoint libraries and the search interface, much akin to what a user would see in shopping sites like Amazon.com.

Over the course of a year, the KM capability was deployed for all small molecule APIs in development and commercialization, approximately 30 APIs. Several hundred scientists and engineers began using the capability as their primary repository for product technical knowledge, abandoning folders, and adopting a standardized taxonomy for organizing content. Although there was an open channel for user feedback at all times during the deployment, a focused effort to evaluate the impact of the new capability on the business was performed after ~18 months of use; the entire cross-functional team that supported the late stage development and commercialization of a key API was surveyed to determine how the new KM capability enabled their work.

Lessons-learned: Building on the momentum of the pilot and addressing its challenges enabled a successful scale-up to all active development compounds.

- A faceted taxonomy reduced complexity, improved flexibility, and enabled an improved user experience. It reduced the number of terms and allowed for them to remain broad in meaning because they were used in combination across facets. The individual facets with the reduced number of terms also allowed for faster tagging on upload and improved filtering functionality akin to popular shopping sites like Amazon.com, which resonated with users.

- Involving a broad selection of staff in the taxonomy redesign process positioned the taxonomy for success as numerous subject matter experts were invested in it, it was intuitive because it was built by the business and after deployment, and every business had at least one power user who could answer questions, provide guidance, and contribute to ongoing governance.
- Users initially viewed tagging as an extra work and would not see the full benefit of a structured taxonomy until it had been in use for a while. Sponsorship at all levels was critical to ensuring users stayed the course until benefits could be realized. In addition, those benefits must be communicated regularly in a targeted way for each business unit that is participating; success stories have proven to be an excellent way to describe and communicate benefits.
- Even that extra minute or two to tag a document on upload was identified as a barrier to adoption and negatively affects the user experience. Pretagged templates were successful in minimizing this but other mechanisms would have to be explored, such as auto-classification and natural language processing, to further reduce or eliminate this burden.
- The survey results from the API team demonstrated benefit across a wide range of leading indicators (e.g., reduced time to find information) and outcome measures (e.g., reduced cost of manufacturing process development) with some select measures and feedback highlighted in Figure 27.7.

2 out of 3 respondents say we increased speed, improved effectiveness and reduced cost of technology transfer.

"...vital for information retention, especially when the process moves into supply or when on SME has left the company."

"...sped up informed decision-making."

"...more effective information sharing..."

"...allows for new information to be discovered."

92% say it took less time to compile the regulatory filling.

FIGURE 27.7
Selected measures and feedback on product technical knowledge taxonomy.

Small Molecule Technical Knowledge

Shortly after the implementation for all small molecule APIs (Iteration 2) was underway, a parallel project was commissioned to expand the KM capability from the small molecule API functions to drug product functions. Individuals supporting these areas generally do not overlap, so this effort would involve introducing a completely new audience (~300 users) to the capability and the concept of applying taxonomy to manage knowledge assets. Although the transition from the pilot to Iteration 2 represented an expansion of taxonomy to a larger user base, Iteration 3 was unique in that it extended the taxonomy to a new knowledge domain—it was not just more people, it was new, unique content. The taxonomy was designed for extensibility and this would put that intent to the test.

Approach: Considering the intent of extensibility, it was suspected that the taxonomy could be adopted with minor changes to support the additional knowledge domain. However, the business representatives from the new business units were taken through the full suite of design activities (e.g., verb/noun exercises), so they would build the same appreciation for the value of taxonomy and come to the realization themselves that the existing taxonomy could be adapted with minimal changes to accommodate their needs. After the workshops and user interviews, the existing taxonomy was introduced to the new stakeholders and overlaps with the output of their workshops became immediately obvious. This independently demonstrated that the different business areas were similar enough to share a single taxonomy.

Card sorting and subsequent analysis quickly revealed the need for an additional facet specific to the dosage form of the product (e.g., tablet and capsule) and additional terms to the topic facet pertaining to formulation development. These are concepts unique to drug product that would not be necessary in the API domain and represented sensible additions to the taxonomy. There were conversations between API and drug product business areas regarding existing terms, and minor adjustments were made to the existing taxonomy; however, these were indeed minor and generally speaking the existing taxonomy was accepted as is with 80% of the terms directly applicable across both areas with little to no ambiguity or alignment discussion required and 5% were truly new terms due to incorporating a new part of the business with the remaining 15% terms that required some discussion to align on meaning and usage.

Lessons-learned: Applying lessons learned from the large-scale deployment for API and leading with a focus on people enabled a successful expansion of the taxonomy to both new user groups and a new knowledge domain.

- Taking representatives from the new business units through the fundamental activities of taxonomy design positioned those groups to more readily accept the change. Forcing an existing taxonomy standard on those users when they had no direct hand in its development would have bred resistance in many of them at the outset. Although it took longer in the near term, designing their taxonomy independently and then demonstrating the overlap with the existing standard enabled those users to come to the realization on their own, which was a most important step in positioning them to accept and feel ownership over the taxonomy in support of long-term success.

- Operational definitions for terms are critical to expanding the taxonomy to new user groups or knowledge domains. The meaning for most terms identified during the previous efforts had not explicitly been documented, and while this did not greatly hinder progress, it did make evident the importance of documented definitions, especially during discussions on terms that may mean different things to different audiences. For example, scientists that support API development consider raw materials to be the materials that are used to synthesize the API, whereas scientists supporting formulation development consider the API itself to be a raw material. Making these definitions visible in the user interface during content creation or retrieval can drive alignment on the language.

- The relative ease of extending the taxonomy to a new knowledge domain demonstrated the extensibility of the universal facets, topic, document type, and document purpose. Theoretically, these facets could be extensible to any knowledge domain and supplemented with more specific facets where necessary.

- The negotiations between API and drug product business areas established the basis for what has formed the present day taxonomy governance committee, and participants in these discussions were asked to stay on in that longer term role. From these discussions, principles on which to base change decisions emerged that have been since further refined. It is important to listen intently to these early discussions to glean insights into what these principles may be for your specific situation.

Recommendation Summary

In addition to the three iterations of taxonomy development described in the preceding sections, three pilots are in progress for vaccines and biologics modalities, the animal health line of business and a new knowledge domain of issue management (identification and resolution of biopharmaceutical manufacturing issues). As a result of this work, an extensive body of knowledge has been compiled on the practical approach to taxonomy design, implementation, and sustainment, with some key recommendations summarized below.

1. Define the intent of a taxonomy capability for an organization early on and specifically decide if that intent includes aligning user groups on language for the purposes of improved collaboration and flexibility. This is a critical decision point as it determines how much effort must be placed on driving a structured, user-facing taxonomy across the organization versus applying taxonomy in the background to meet use cases for small user groups or one-off activities, such as those often focused on data analysis. If improved collaboration is the intent, then leadership must clearly set the expectation that employees will change the way they speak about their work and also shift from use of the traditional folder structure to taxonomy and tagging. In this case, the initiative must be treated as a transformational change because the intent is to change core behaviors of large groups of people.

2. Place accountability for taxonomy design and implementation with a team that represents a cross section of the impacted business units. The likelihood of success greatly increases if those impacted by a change are vested in it from the beginning. Supplement the team with taxonomy subject matter expertise as needed to facilitate the work.

3. Do not let perfect get in the way of better—start small and iterate quickly. Demonstrate value through success stories or quantitatively if possible. Depending on the use case of the taxonomy, many of the benefits may not be immediate. Seek to structure a pilot where at least some benefit can be realized in the near term and used to extrapolate to the broader business benefit, which might require large amounts of content to be tagged or a certain amount of time to pass before being realized.

4. Use a simple language when discussing taxonomy concepts with the general population and relate it to tools they are likely using outside of work.

5. Invest in technology to minimize the barriers of the transition from a traditional folder structure to tagging. This is a fundamental behavior change for many people. Although sponsorship and success stories are an integral part of making the change successful, ensuring a sufficiently good user experience through the application of modern technologies is equally if not more important to generate pull for the change rather than it being required by management.

6. Assume that the taxonomy will not be perfect. Place as much effort on building an environment to monitor and improve taxonomy usage as on the original taxonomy design. Establish metrics early and ensure your system is designed to capture the required data.

Future

Empowered with a rich body of knowledge on driving business outcomes using taxonomy, MMD is now focusing on leveraging taxonomy to truly become the biopharmaceutical manufacturer of the future. Some of the areas being explored include the following:

Natural language processing: NLP is an approach that enables technology to derive meaning from human language and has already been successfully applied to automatically tag historic product knowledge with terms from the new taxonomy. Opportunities exist to apply NLP during content creation and upload to reduce the burden of tagging on users by automatically suggesting tags and by applying it within a search interface to enable search predictions and advanced filtering. NLP will also enable the automated, ongoing enrichment of existing content to optimize findability, as well as validate our user-defined taxonomy through analysis of existing content. MMD will continue to experiment in this area, introducing machine learning, and other semantic technologies as opportunities arise.

Knowledge domain expansion: Within the knowledge domain of product technical knowledge, MMD is continuing to expand to additional modalities, such as vaccines and biologics, and perhaps other lines of business, such as animal health. But beyond this, other knowledge domains are being explored, including manufacturing platform technology and supply chain. The proven approach of expanding through small iterations will continue to be used while taking advantage of opportunities to apply existing, core facets, such as *material type*, which have broader applicability beyond the current knowledge domain and user base.

Industry standard alignment: Being initiated in 2008 at a time when some of the first industry standards were only emerging, the taxonomy model was designed internally by Merck & Co., Inc. Now standards are gaining traction with various regulatory bodies and consortia driving initiatives to align global regulators, manufacturers, suppliers, and distributors. For example, IDMP (identification of medicinal products) is a developing standard to "address the demand for the unique global identification of medicinal products throughout the lifecycle of a product" (Deloitte 2015). An opportunity exists not only to align with these standards, but apply deep knowledge gained from the activities described in this case study to influence these standards, as well.

IT system integration: The current taxonomy is only in use for content in one particular SharePoint-based system. However, product technical knowledge resides in multiple systems and always will; knowledge workers must still search multiple repositories with each delivering a disconnected user experience. Work is already under way to extend the reach of taxonomy to content in other sources. This will be accomplished through a combination of background retagging via auto-classification and direct updates to the taxonomy models across IT systems with content delivered through an integrated search capability.

Business process integration: Business process management is an emerging capability within the division and there are many integration points with a taxonomy capability. By applying knowledge mapping or information flow modeling approaches to documented processes, both the requirements and business benefit of taxonomy are more easily elicited. In addition, process owners become a key stakeholder in taxonomy as a key enabler to their process.

Governance: Taxonomy is being positioned as an enterprise standard and will require a single point of accountability and associated governance model to ensure that the standard meets the needs of the business and is followed. Since the current taxonomy is in use in only one IT system, the owner of the system has also been the owner of the taxonomy. As the taxonomy expands to other systems and knowledge domains, this is no longer an effective model, and it becomes imperative to establish an appropriate governance model to sustain enterprise taxonomy and define the relationships between that single point of accountability and other key stakeholders such as organizational leaders, business process owners, IT system owners, and master data managers.

Conclusion

Achieving sustainable results from taxonomy work is no easy task. Not only can it be difficult to simply describe the term *taxonomy* but it seems companies fail to recognize some common pitfalls such as treating taxonomy as an IT deliverable or as a one-off project. Hopefully, this case study has highlighted that the practical application of taxonomy can drive real business outcomes and offered key insights that are applicable across industries and knowledge domains. MMD has approached taxonomy as a business capability with the requisite sponsorship, business unit engagement, and focus on continuous improvement, realizing sufficient value to garner continued investment and anticipating much greater value across Merck & Co., Inc.'s strategic priorities as the capability scales.

Acknowledgments

As highlighted in this case study, taxonomy work represents a journey of learning and experimentation with many challenges to overcome along the way. It requires a breadth of expertise that no single individual can possess. And it requires a willingness to accept and learn from failure since there were no known examples to guide us. A journey like that requires quite an astounding team, with which we were abundantly surrounded.

First and foremost, we would like to thank Mike Thien, Jean Wyvratt, and Marty Lipa for their foresight and vision that led to the creation and continued success of the MMD KM program and underlying taxonomy capability. Thanks are also due to Eric Margelefsky, Diana Hou and James Corry for their business leadership, and Joni Valerio, for exceptional project management, during those early years of capability build; we are especially appreciative of Eric's continued support of KM.

We are also indebted to Douglas Arnold for his thought leadership and advice that spans well beyond his IT role, as well as our other IT partners, Yunnie Jenkins and Jason Williams, who worked tirelessly to ensure we had the technology required to be successful. And last but certainly not least, considering this case study repeatedly highlights the importance of focusing on people and behaviors during the course of taxonomy work, Samantha Bruno deserves endless gratitude for her application of change management expertise to drive the adoption of taxonomy across the organization.

References

Deloitte. Identification of medicinal products. *www.deloitte.com.* February 2015. http://www2.deloitte.com/content/dam/Deloitte/ch/Documents/life-sciences-health-care/ch-en-life-sciences-the-challenge-of-idmp.pdf (accessed May 27, 2016).

Duckworth, A., and D. Arnold. Knowledge management across the product lifecycle. *Presented at KM World*, Washington, DC, November 2014.

Hedden, H. *The Accidental Taxonomist*. Medford, NJ: Information Today, 2010.

Lambe, P. *Organising Knowledge: Taxonomies, Knowledge and Organizational Effectiveness*. Oxford: Chandos, 2007.

Lipa, M., S. Bruno, M. Thien, and R. Guenard. A case study of the evolution of KM at Merck. *Pharmaceutical Engineering*, 45(4), 4–104 2013.

28

Knowledge-Based Product and Process Lifecycle Management for Legacy Products

Marco Strohmeier, Christelle Pradines,
Francisca F. Gouveia, and Jose C. Menezes

CONTENTS

Biologic drug substance manufacturing processes of commercial products developed before Quality by Design (QbD) were validated based on general knowledge but with limited understanding of the criticality levels of process parameters (PPs) and quality attributes (QAs). Here Roche describes an approach showing how critical process parameters (CPPs) and critical to quality attributes (CQAs) can be defined from the manufacturing history available for a legacy product (i.e., data- and evidence-based), which combined with formal risk assessments (knowledge-based), may lead to significant improvements of the filed control strategy. Introducing postapproval changes in a legacy filing, originating from increased process understanding and better risk mitigation strategies, is aligned with FDA's regulatory initiative *Established Conditions* (Draft Guidance, May 2015) and the expectations on the forthcoming ICH Q12.

Editorial Team

OVERVIEW   Legacy biological/biotechnology products have a wealth of historical data on process parameters, quality attributes, and manufacturing information of various types that from a knowledge management perspective are largely unused over lifecycle. The subject matter experts (SMEs) involved in development, launching, and in the commercial life of those products have in-depth knowledge (viz., at all 5M levels of an Ishikawa diagram—machines, methods, materials, manpower, measurements) that could be used in a formal knowledge-driven risk identification exercise. The ability to challenge and integrate the SMEs knowledge generated during commercial life is a fundamental requirement to ensure continued stability and capability of the manufacturing process [1–3].

In this chapter we outline general principles and approaches necessary to align process validation activities with the product and process lifecycle concept for biological/biotechnology legacy products that include scientific, data-driven, and risk-based approaches to continuously improve control strategies and the assurance of final product consistency and quality (Figure 28.1). The discussion herein, however, does not imply that Roche supports or uses the current proposal globally.

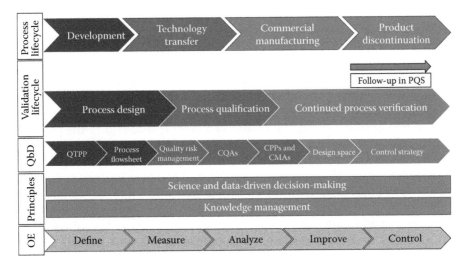

FIGURE 28.1
Elements to be aligned and in place to continuously improve the control strategy and revamp a legacy product validation plan.

Introduction

The practical concept of process validation in drug manufacturing has been widely adopted since its origin in the 1970s. For many years, process validation emphasis was on full compliance with initially established operating procedures, leading to a perception that process validation was essentially a documentation exercise of quality assurance (QbT, Quality by Testing). Good manufacturing practices (GMPs) as defined by the U.S. Food and Drug Administration in 1987 and also in ICH Q7, enforced the notion of a one-off type of activity [4]. According to ICH Q7 "Process Validation as part of GMP is the documented evidence that the process (…) can perform effectively and reproducibly (…) meeting specifications and quality attributes (…)."

However, since "cGMPs for the twenty-first century—a risk-based approach" was launched by FDA in 2002, a debate was started about QbT and specially in regard to a systematic approach to confirm the effectiveness of all process steps and conditions that was needed to ensure that drug products are of the necessary quality, safety, and efficacy [5].

In January 2011, FDA published a revised Process Validation Guidance [1] that evolved from its earlier 2002 initiative. That document sets a shift from a documentation focused to a production end-to-end approach—not only across process flow sheet but also across product lifecycle—integrating process/product development, scale-up activities, and continued process verification (CPV) during routine manufacturing. The impact of process input parameters (alone or in combination) on product quality and process performance (the outputs) should be established from end-to-end of the production process and also throughout product lifecycle. In 2012, ICH adopted this concept in its Q11 guideline and EU changed the volume 4 of GMPs to be included in Annex 15, all aspects of a risk-based and over lifecycle validation procedure as new GMP requirements [2,6].

Before that paradigm shift, process validation for commercial products was performed following a conservative approach. After identification of an appropriate process outline, repeated executions of a process would be performed to demonstrate the ability to produce product of intended quality in three nonconsecutive batches (viz., *golden-batches*). Such one-off validation approach is still accepted by health authorities. However, for those three consecutive batches only sparse data are generated as the validation exercise is performed only once before commercial production. That created significant challenges as to capturing small process variations over longer periods of time as small process variations are inexperienced during development. The lack of risk-based elements and the limitations (amount and type) of data generated in the classical approach, had a significant impact on:

- The understanding of process input and output correlations, and thus subsequent CPP definition based on criticality of their impact on CQA, as required for definition of process robustness indexes.
- The ability to evaluate batch-to-batch consistency at each process step.
- The ability to evaluate the existing control strategy based on data generated during routine production.
- The identification of in-process variability (as indicated by CQAs and key performance indicators—KPIs) that could in turn indicate the need for additional or improved in-process controls (i.e., related to current levels of process observability and controllability).
- The definition of meaningful acceptable ranges for IPCs.
- The definition of key performance indicators based on statistical meaningful data.
- Maintaining process consistency (robustness and reproducibility) during process monitoring as an integral element of the third-process validation stage (cf FDA [1]): continued process verification.
- The evaluation of changes and deviations.

With these limitations, the content of a dossier for a legacy product was generated based on general expert knowledge but with limited anticipation of problems to be experienced over lifecycle. In addition, data generated in the validation exercise followed a conservative *one-factor-at-a-time* development to establish proven acceptable ranges (PAR) for each process parameter. Such data does not map well with the knowledge space as no process parameters (PPs) interactions are taken into account. The resulting PAR typically contain the nominal operating ranges (NORs), but may not entirely overlap with rigorously derived and formally defined design and operating spaces (cf. QbD), respectively. There are instances in which simultaneous changes inside PAR of several interacting critical PPs—a situation not considered in classical development—may lead to undesirable events (i.e., out-of-trend (OOT) or even out-of-specification, (OOS)).

However, legacy products have two advantages over new ones. First, they have massive amounts of highly informative data on final product quality attributes available that can be related to existing chemical, manufacturing, and controls in the CMC filing (i.e., IPCs and PPs). Second, there is a wealth of process and product knowledge accumulated in SMEs throughout manufacturing history.

The current document describes an integrated approach for legacy products, developed without any QbD elements (viz., typically filed before 2005 or ICH Q8 and Q9) that aggregates the data- and knowledge-based aspects above [7,8]. The goal is to use concepts of risk- and knowledge-based process validation, well established over the past decade in regulatory documents, to elevate legacy product validations to the next level, in terms of compliance robustness and operational performance. Under the framework of PQS (cf. ICH Q10) [9], that incorporates quality risk management (QRM) and science-based justifications—cf. ICH Q11 [6], the main elements of this approach are supported by:

- An initial risk assessment to identify
  - Gaps within existing validation studies and in process controls
  - Knowledge gaps in the dossier
  - Identification of meaningful in process control acceptable ranges
- Through
  - Process input and output correlation analysis
  - Criticality assessment of PPs
  - A comprehensive description of known root-causes for variations in PPs and QAs.

When using QRM principles as described in ICH Q9 [8] the identified gaps and associated activities can be prioritized according to the risk level, ensuring transparency to senior management within the context of their

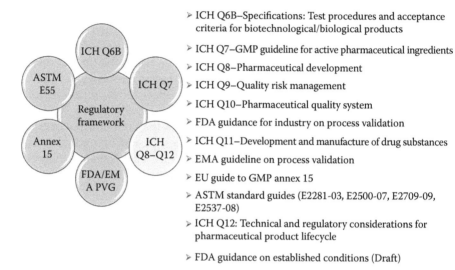

> ICH Q6B–Specifications: Test procedures and acceptance criteria for biotechnological/biological products

> ICH Q7–GMP guideline for active pharmaceutical ingredients

> ICH Q8–Pharmaceutical development

> ICH Q9–Quality risk management

> ICH Q10–Pharmaceutical quality system

> FDA guidance for industry on process validation

> ICH Q11–Development and manufacture of drug substances

> EMA guideline on process validation

> EU guide to GMP annex 15

> ASTM standard guides (E2281-03, E2500-07, E2709-09, E2537-08)

> ICH Q12: Technical and regulatory considerations for pharmaceutical product lifecycle

> FDA guidance on established conditions (Draft)

FIGURE 28.2
Regulatory framework of legacy products—an evolving perspective of process validation.

responsibility for aligning the business strategy with QRM principles (cf. ICH Q10, [9]). One of the most important elements of QRM is the risk assessment exercise, linking inputs and outputs to identify critical relations, and implement effective strategies for process supervision, deviation management, and change management.

The current framework integrates the principles covered by the International Conference on Harmonization (ICH) guidelines for industry [4,6–11], EU Guide to GMP Annex 15 [2], ASTM E55 Committee standard guides (namely, E2281-03, E2500-07 and E2709-09, E2537-08) [12–15], and FDA and EMA guidelines on CPV [1,2], and can be summarized under the concept of risk-based process validation applied to legacy products (Figure 28.2), a general concept providing a broader framework for QbD realization [11,16].

Many companies find it easier to postpone improvements to facilities, processes, and analytics, or simply refrain from planning for advancements at all in order to avoid the intricate nature of implementing such changes, especially for products registered in multiple countries [17]. However, a knowledge and risk-based approach to process validation can significantly support the value-adding focus and contribute to enhanced control and understanding of the manufacturing process. A discussion on how to align process validation activities with the process and product lifecycle concept is provided therein.

The Approach

A stepwise approach (Figure 28.3) is used to improve understanding of the manufacturing process of legacy [biological/biotechnology] products. The approach includes the following elements/steps:

- Identification of relevant outputs (i.e., critical quality attributes—CQAs—and key performance indicators—KPIs for each process step influenced by that specific unit operation.
- Knowledge-based identification of all process parameters and material attributes for each unit operation that might influence relevant outputs.
- Determining the functional relationship through data-driven approaches (e.g., correlation analysis and multivariate methods) between material attributes and process parameters to process relevant outputs.
- CPP definition based on the joint assessment of their variability and criticality through a functional relationship with a relevant output using various sources of knowledge (e.g., scientific knowledge derived from platform technologies).
- Identification of gaps and/or improvement opportunities in the overall control strategy of the legacy product (i.e., quality controls, manufacturing controls and/or validation protocols).

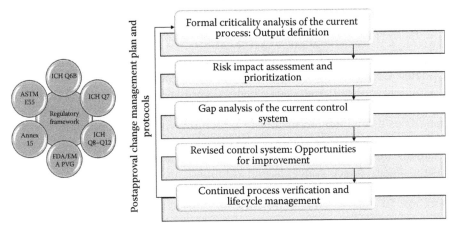

FIGURE 28.3
A lifecycle approach to process validation of a legacy product.

This stepwise approach:

- Combines evidence-based assessments with SMEs prior knowledge to bring increased understanding on observed variability that may have a potential impact on quality attributes.

- Reveals mismatches between knowledge and evidence components, potentially helping reconcile perceptions with reality, and can be used for prioritization of actions to improve the current control strategy.

- Supports implementation of an efficient lifecycle management (LCM) plan for continued process verification (CPV) ensuring a robust validated state over the entire remaining commercial lifecycle.

- Ensures completion of the process validation exercise and mitigation of significant risks identified in the legacy filing (viz., potential or already verified).

- Enables a complete postapproval change management procedure that is both comprehensive and supported by a formal criticality assessment of potential risks and evidence-based backing—that is, contains strong science- and knowledge-based components.

A short overview of existing published alternatives to the presented approach is provided below for this very recent area.

A discussion on parameter and attribute selection based on prior knowledge and data to support CPV programs is done by Boyer et al. [18]. The authors detail on how to use process capability indexes (Cpks) and process performance index (Ppk) based on parameters and attributes, and they provide a discussion about when these indexes should be included in the CPV plan. In addition, a comprehensive roadmap on how to implement and update a CPV plan is presented for new and legacy products.

Gouveia et al. [19] discussed a general approach to include QbD elements into legacy filings following FDA's regulatory initiative on *established conditions* as enabler, according to which improvement opportunities are not restricted to the available CMC history and to the current knowledge available from SMEs. The authors propose that scaled-down experiments using design-space concepts and qualified models should also be used with the existing lifecycle aspect in addition to other elements of the QbD toolbox. In their approach the criticality assessment is supported by univariate and multivariate analysis of variance (ANOVA and MANOVA, respectively) end-to-end, but no process capability indexes are defined to support ranking of criticality, which is done through formal risk assessments. Agarwal and Hayduk [20] use an approach somewhat comparable to the one presented here.

Assessing CQAs is relatively straightforward, but assessing the criticality of PPs can be much more challenging. Identification of CPPs and also of critical interactions between, otherwise noncritical process parameters, is essential to determine the critical relationships between input and output parameters and improve the control strategy. Bozzone [21] describes the use of the Z score—namely, a measure of distance from the limit of specification—and its use over lifecycle. It is perhaps a better metrics that could also be used to evaluate input parameters when the size or the distribution of the data raise concerns, since Cpk measures distances from the average and not distances to the limits of specification as the Z score. De Long et al. [22], defined an R_p index defined as the proportion of the allowable range used by the process (R_p is between 0 and 1, with R_p close to 0 indicating high process performance). The authors point out that Cpk and Ppk can be misleading when the data are not normal.

The main objective of the proposed approaches is to understand what is already known about critical interactions between input and output parameters and which elements should be further included in the validation plan to be able to consistently meet the established goals under the defined operating conditions.

Procedure Overview

As indicated, the procedure:

- Provides a science and risk-based approach for identifying potential risks not anticipated or described in the legacy filing (i.e., it will be incremental to the currently filed dossier)

and is bounded by

- The available data and information on the current process control system, being of sufficient quality, to enable meaningful statistical analysis required by the evidence-based component.

The two components used in the identification of potential gaps in the current control strategy are complementary:

- The knowledge-based component will explore the empirical and prior-knowledge available on SMEs, whom in turn will provide estimates of the likely impact that changes in process parameters (inside PAR or NOR ranges), may have on-process relevant outputs for each unit operation; this assessment will be scored using a Severity ranking.
- The evidence-based component, on the other hand, is an estimate on the quality of the existing control strategy—that may or may not be incomplete; this assessment involve the process observability

score based on the availability of data to describe the input/output relationship (Certainty ranking), and the process controllability score (Occurrence ranking) through capability indexes (e.g., Cpk, Ppk or other appropriate metrics are accepted) calculation for input parameters.

The variability observed on input parameters from each unit operation can be computed, and descriptive statistics for the distributions over time of each of those properties, therefore, can be obtained. This will enable ranking of variability sources in terms of their impact (linkage and criticality) to the final-product quality specifications.

The current framework has been created with the purpose of conducting an end-to-end criticality assessment of existing commercial processes influenced by a complex matrix of input/output parameters across several unit operations (UOs). Upstream (USP) and downstream (DSP) processing involves multiple steps and each step introduces potential critical relationships between input and output parameters that must be evaluated and controlled to ensure successful production.

A general flow of a chinese hamster ovary – based (CHO) cell culture process for production of a recombinant protein will be further used for demonstration purposes and to illustrate the principles and tools deployed in the current document (Figure 28.4).

A vial from the working cell bank (WCB) is thawed in culture media. The cell culture is then generated and expanded through successive transfers in shake flasks. When the cell density reaches the target value, the culture is used to inoculate intermediate bioreactors to generate enough cell mass to inoculate the production bioreactor (i.e., main fermentation bioreactor). Both intermediate and main production bioreactors are monitored and controlled in a similar way to ensure production requirements. At the end of the production process, cells are harvested and clarified using centrifugation and depth-filtration. As indicated by the process layout, considerable variations can be propagated over the flow sheet including among others, variability in media components and in the feeding strategy, temperature and pH shifts, variability of gas exchange parameters, and shear stresses [23].

Downstream processing combines a sequence of operations to purify the harvested material. Purification steps are in general chromatography based and require sensitive and sophisticated equipment that can be an object of considerable variations over the lifecycle of commercial processes. As such, performance of each chromatographic step should be timely evaluated against process outputs in order to identify potential gaps in the current control system.

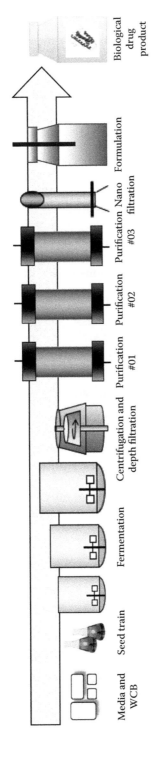

FIGURE 28.4
General representation of the manufacturing flow sheet for a biological/biotechnology drug product.

Risk and Knowledge Management Components

Under the current framework [1–3,24], legacy products are expected to experience a *new existence* in terms of manufacturability. It is often the case that the original filing contains PAR and NOR ranges for many of those parameters, but not for all. There are multiple causes for that, namely the type of validation procedure, lack of a lifecycle perspective in the original filing as discussed earlier, and also limitations on equipment and technology capabilities at time of development. Moreover, considering the currently accepted science-based framework for quality as a manufacturing science (aka QbD), a formal criticality analysis on which of those PPs and QAs are really critical and impactful to the final product quality (drug substance or drug product) may have never been made or completely documented. Consequently, when approaching such dossier, the technical team usually faces a significant challenge in demonstrating that the relationship between CPPs and CQAs derived from earlier studies in development is fully understood and used in commercial manufacturing.

The systematic approach described earlier (based on prior-knowledge harvest or recollection, using formal risk assessments) is therefore required to identify and prioritize areas for improvement and innovation (e.g., equipment changes, shift control set-points within NOR/PAR).

Formal Criticality Analysis of the Current Process

Output parameters reflect the performance of a given unit operation and indicate whether the process gave the desired outcome for each intermediate critical quality attribute. As such, output parameters reflect the step contribution in terms of performance (e.g., yield, impurity removal factor) or influence in the properties of the final product (e.g., purity and isoform composition).

Under the current framework, an output filtering exercise is performed by SMEs to identify relevant parameters for each UO according to the following criteria:

- Existing knowledge (SMEs or corporate technology-platform knowledge).
- Introduced/not introduced by the UO.
- Data availability (e.g., validation studies).
- Detection limit before the UO.
- Covered [correlated] by other output parameter.

The UO relevant outputs and the final process outputs are used to provide evidence that the process is robust and consistently meets specifications.

Risk Impact Assessment and Filtering

Knowledge and Risk Assessment Methodology

The knowledge and risk assessment methodology forms part of the five-stage roadmap (Figure 28.3) that make up the approach.

As general requirements for each unit operation (UO):

- All available inputs from the different data sources (i.e., online data, trend and offline process data, analytical measurements, and experimental data) are collected and listed for a number of production batches considered to be representative of process performance over time (e.g., 30 end-product, lots manufactured within 3 production years).
- The necessary descriptors to estimate the *observability* [Certainty] and *controllability* [Occurrence] scores are available and provided in an appropriate template (see Annex).

In the risk analysis stage, the team should discuss and rank the likely impact [Severity] that deviations in process inputs might have in process relevant outputs for each UO according to predetermined criteria (Table 28.1). The observability component [Certainty] should be evaluated based on the scientific demonstration of a given relationship between the input and output parameter. Whenever possible, a quantitative threshold should be established for score ranking (e.g., degree of correlation, parameter importance in a multivariate model). The controllability component [Occurrence] is intended to demonstrate the process capability to operate within specified limits to control the input, providing evidence if additional control measures on input are required. In all situations the PAR/NOR are not specified in the batch record or in the dossier, the team should evaluate if those limits could be specified based on existing knowledge or historical data.

Scoring Matrixes and Prioritization of Follow-Up Actions

CPP identification should be determined by multiplying the S score by the C score in order to evaluate if the input parameter is noncritical, critical, or potentially critical from a process control perspective, as depicted in Table 28.2. The filtering exercise will provide a first indication of the current validation status as well as an indication of potential misalignments between the knowledge and evidence components—specifically, the observability score—characterizing the actual manufacturing process. It should be highlighted that a Certainty score of 10 will immediately classify an input parameter as potentially critical (pCPP) because the implementation of a revised control strategy cannot be based on uncertainty. This directly triggers additional experiments and investigations to close this gap in knowledge.

TABLE 28.1

Established Criteria for Severity (S), Certainty (C), and Occurrence (O) Ranking

Score	Severity (S)	Score	Certainty (C)	Score	Occurrence (O)
2	Variation in process input across the acceptable range (PAR/filed) alone, or if affected by an interaction, causes no measurable/detectable variation in process output	2	Supported by data (e.g., manufacturing data and validation studies)	2	Cpk > = 1,2* (or other statistically significant metric)
6	Variation in process input across the acceptable range (PAR/filed) alone, or if affected by an interaction, causes variation in process output within expected output range	6	Based on experience/common knowledge/correlation found in historical data (correlation or significant variation over production history)	6	0.8 < Cpk < = 1.2* (or other statistically significant metric)
10	Variation in process input across the normal operating range (NOR) alone, or if affected by an interaction, causes variation in process output within expected output range	10	Nothing is known (validation data not available or not possible to perform data analysis (e.g., no significant parameter variation over process history)	10	Cpk < 0,8* (or other statistically significant metric) or Cpk cannot be determined or there is no defined PAR/NOR

* Calculated based on historical data and compared to PAR (depending on what is provided in MMR and dossier).

TABLE 28.2

SC Score Matrix for CPP Identification

		Input Parameter Classification (Non-CPP/pCPP/CPP)		
	10	CPP	CPP	pCPP
Severity (S)	6	CPP	CPP	pCPP
	2	Non-CPP	Non-CPP	pCPP
		2	6	10
		Certainty (C)		

TABLE 28.3

SCO Score Matrix to Evaluate the Current Validation Status and Control Strategy

		SCO Score		
	100	200	600	1000
	60	120	360	600
SC Score	36	72	216	360
	20	40	120	200
	12	24	72	120
	4	8	24	40
		2	6	10
		Occurrence (O)		

The current process control strategy can be evaluated by combining the SC score with the *controllability* component [Occurrence] (Table 28.3), providing a reliable indication of the current status and a procedure to prioritize neces-sary follow-up actions (i.e., SCO score) in agreement with:

- Low SCO score [8–40]: well controlled parameter; it is not expected to have impact in output parameter.
- Medium SCO score [72–120]: process is capable to control CPP within PAR; activities to improve process control might be required if there is at-scale evidence of an optimal operating window within PAR/NOR: (1) designed studies to improve knowledge-based com-ponent and (2) at-scale verification if improvements have high likelihood.
- High SCO score [200–1000]: process capability to control CPP is questionable; activities to improve the control system are required: (1) process design studies to revise NOR/PAR followed by at-scale verification studies and (2) in-process controls (IPCs) update (e.g., redundant measurements) due to high variability.

Gap Analysis of the Current Control System

Once the initial CPP identification has been completed, there remains the critical part of follow-up actions. The formal knowledge and risk assessment results in (1) a documented characterization of the product/process by a mul-tidisciplinary team and (2) a prioritization of parameters or critical areas/UOs that need to be addressed. Based on the SCO score, the team should agree on a cut-off value (e.g., Pareto analysis) for prioritizing action items depending on time and resources availability showing manufacturing excellence based on smart combination of the business and quality aspects. The criteria to establish the cut-off value should be recorded in the assessment report. All follow-up actions are to be addressed by the responsible person and should be documented in separate technical or validation reports (according to the PQS structure).

Revised Control Strategy: Opportunities for Improvement

According to ICH Q11 [6] a control strategy consists in a planned set of controls derived from *current* product and process understanding that assures process performance and product quality. Change control programs are considered essential elements of PQSs as stated in EU GMP Annex 15 [3]: "A formal system by which qualified representatives of appropriate disciplines review proposed or actual changes that might affect the validated status of facilities, systems, equipment or processes. The intent is to determine the need for action that would ensure and document that the system is maintained in a validated state."

As part of the control system revision, the global criticality assessment can potentially trigger:

- Inclusion of additional output parameters (e.g., UO relevant outputs).
- Revision of output parameters criticality.
- Inclusion of additional input parameters (e.g., derived from existing measurements).
- Resetting operating ranges for existing input parameters or defining PAR/NOR for new ones.
- Revised performance indicators target ranges.
- Changes in starting materials specifications.

All these elements should be considered in the validation master plan in agreement with the company's internal procedures.

Continued Process Verification and Lifecycle Management

The present document describes a structured science-based approach to align process validation approaches with process and product lifecycle management activities of legacy products. After considering all relevant scientific and technical aspects involved at each critical unit operation, a systems engineering perspective is used to integrate the different components. A whole process analysis connecting raw materials, unit operations, and end-product properties is adopted comprehensively, in parallel to a time-wise integration of data, information and knowledge acquired throughout process development, industrialization, and commercialization. Continuous improvement can only happen if these two perspectives are present and combined under proper data-, information-, and knowledge-management systems. Under the current framework, it is expected to (1) comply with the requirement that all manufactured product lots (batches) should be as good as any of the batches previously required for validation, even after manufacturing changes have been introduced and (2) manage any deviation backed by deep process understanding and scientific knowledge. In this context, the resulting quality risk management report, a living document, will become the knowledge base enabling continuous improvement.

Documentation Management

Results of the herein described assessments for each UO should be described in a comprehensive report that should include the following:

- Scope and date.
- List of team members (name and functional role).
- Raw data used in the assessment (input and output parameters).
- References to documentation used to support the knowledge-based component (e.g., reference to validation studies, literature, and internal data sources).
- Calculations to estimate the evidence-based component (observability and controllability scores).
- Formal CPP assessment exercise (reference to template) and its outputs (i.e., SC scores for CPP identification and SCO scores for assessing the validation status).
- Cut-off SCO criteria to prioritize follow-up actions.
- Report-out and action list (with indication of the responsible to further investigate the items assigned and write subsequent technical reports).

Conclusion

The current proposal aims to address a regulatory expectation, relevant to establishing a process validation lifecycle concept for legacy products, fully aligned with the Pharmaceutical Quality System (PQS, cf. ICH Q10) within each company and following fundamental science-based and knowledge management principles.

It enables a comprehensive post-approval lifecycle management approach that is both comprehensive, supported by a formal criticality assessment of potential risks and evidence-based backing—that is, contains strong science- and knowledge-based components.

The example provided in the Annex section demonstrates how the approach can be applied to evaluate and improve the control system of a legacy biologic product.

Glossary of Terms

Inputs Operating parameters with validated set points or specified ranges that define process execution.

Outputs Parameters that reflect the outcome of a given process step and provide a measurement of process performance or product quality.

QTPP A prospective summary of the quality characteristics of a drug product that ideally will be achieved to ensure the desired quality, taking into account safety and efficacy of the drug product.

(p)CQA A physical, chemical, or microbiological property or characteristic that should be within an appropriate limit, range, or distribution to ensure the desired product quality ([p] = potential). Considers the relevant mechanisms of action.

(p)CPP, (p)CMA A process parameter or material whose variability has an impact on a critical quality attribute and therefore should be monitored or controlled to ensure that the process produces the desired quality ([p] = potential).

KPIs Performance measurement to evaluate the robustness of an operation. Its assessment should lead to the identification of improvement opportunities and is therefore linked to performance improvement initiatives.

Capability indexes (e.g., CpK) Statistical metrics to evaluate if process input and output parameters are in a state of statistical control.

Severity Knowledge-based ranking of the expected impact that changes in input parameters may have on output parameters for each unit operation of the manufacturing process.

Certainty Observability ranking supported by data to describe the relationship between the input and output parameter (e.g., correlation analysis, multivariate data analysis and experimental data).

Occurrence Controllability ranking demonstrating the ability to control an input parameter.

SC Score Numeric assessment that quantifies the criticality of an input parameter.

SCO Score Numeric assessment of the severity of impact and ability to control an input parameter.

Design Space The multidimensional combination and interaction of input variables (e.g., material attributes) and process parameters that have been demonstrated to provide assurance of quality.

Control System A planned set of controls (quality control, manufacturing control, and validation protocols) derived from current product and process understanding that ensures process performance and product quality.

IPC In-process controls.

QRM A systematic process of organizing information to support decision making based on identification of hazards and evaluation of risks management associated with those hazards.

PAR Proven acceptable ranges.

NOR Nominal operating ranges.

CMC chemistry, manufacturing, and control.

CPV Continued process verification.

LCM Lifecycle management.

Annex: A Practical Demonstration of the Knowledge and Data-Driven Approach

A primary goal of process validation is to demonstrate that each step or unit operation are in a state of control and that all variation sources are identified, monitored toward relevant process outputs and in a state of control. The following example is provided for illustrative purposes and exemplifies how the current approach should be applied to a purification step (Purification #01) of a legacy recombinant protein (Figure 28.5).

Stage 1—Identification of Process Relevant Outputs

By utilizing extensive SME and platform knowledge a complete assessment was conducted in order to identify process step relevant outputs influenced by Purification #01 operations. The assessment was conducted taking into consideration the likely impact that deviations in process inputs within the PAR/NOR ranges might have in overall process outputs.

Product Specific Impurities and CQAs

The product specific impurities and CQAs indicated in Table 28.4 were considered in Purification #01 assessment, to be potentially affected by input parameters variability.

Process Specific Impurities

Typically, process specific impurities are introduced by cell disruption and raw materials used throughout the production flow. The purification steps must assure that all critical process or product related impurities are controlled/reduced to be within acceptable levels in final bulk. If a criticality assessment based on available data is not possible, further studies to

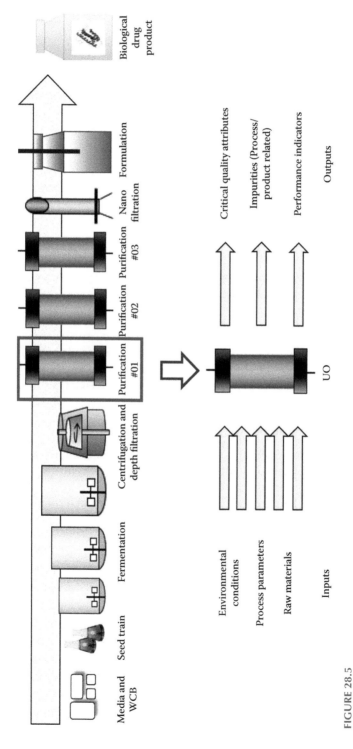

FIGURE 28.5
A stepwise approach for quality risk assessment of legacy products.

TABLE 28.4

Identification of Product Specific CQAs Potentially Affected by Purification #01

Category	CQA Description	Relevant for Process Step CPP Identification?
Product variants	Size-related variants	Yes
	Charge-related variants (Isoforms)	Yes
	Glycosylated variants	Yes
	Structural variants	Yes
Adventitious agents	Viral purity	No
	Bioburden	No
	Endotoxins	No
DS composition	Appearance	No
	pH	No
	Protein content	Yes
	Sialic acid	Yes
	Purity	Yes
DS Strength	Potency	No
	Bioassay	No

TABLE 28.5

Identification of Product Related Impurities Potentially Affected by Purification #01

Impurity	Point of Introduction	Relevant for Process Step CPP Identification?
DNS (DNA) (pg/IU)	USP/Cell Culture	Yes
CHO Protein (ppm)	USP/Cell Culture	Yes
Antifoam (µg/mL)	USP/Fermentation	No
Methotrexate (µg/mL)	USP/Fermentation	No
Solvent A (ppm)	DSP/Purification #01	Yes
Solvent B (ppm)	DSP/Purification #02	No

demonstrate removal of identified critical impurities would be required. For those impurities already identified (Table 28.5) the validation status is assessed following the knowledge/data-driven risk assessment methodology previously described..

Performance Indicators

Performance outputs are defined as indicators of how well the unit operation performed but are not directly linked with product quality. An assessment of key performance indicators (KPIs) based on historical data, process validation data, and SME knowledge is performed to derive a complete output list relevant to monitor the performance and consistency of each unit operation (Table 28.6).

TABLE 28.6

Identification of Performance Indicators Potentially Affected by Purification #01

Performance Indicator	Process Step	Relevant for Process Step CPP Identification?
Viability at transfer (%)	USP/Seed Train	No
Doubling time (h)	USP/Seed Train	No
IVC (cells/mL day)	USP/Fermentation	No
Viability (%)	USP/Fermentation	No
Yield at harvest (g/L)	USP/Fermentation	No
Max VCD (cells/mL)	USP/Fermentation	No
Culture duration (days)	USP/Fermentation	No
Specific productivity (pg/cell day)	USP/Fermentation	No
Specific oxygen uptake rate (mmol/cell h)	USP/Fermentation	No
Max pCO_2 (mmHg)	USP/Fermentation	No
Step yield (%)	DSP/Purification #01	Yes
Step yield (%)	DSP/Purification #02	No
Step yield (%)	DSP/Purification #03	Yes

Stage 2—Risk Impact Assessment and Prioritization

Identification of Process Parameters and Material Attributes

For each unit operation, all available process and material parameters are listed and collected for a number of production batches considered to be representative of process performance over time (Table 28.7). Subsequently, the data are analyzed by uni- and multivariate statistical methods in order to investigate if meaningful correlations with outputs parameters can be established. In the current assessment, production data spanning three production years were considered (90 production batches).

Knowledge and Data-Driven Components Estimation

In order to illustrate the risk impact assessment for a particular input parameter, the protein pooling procedure will be further detailed.

In the risk analysis step the team should discuss and rank the likely impact of the start/end protein collection trigger in process relevant outputs—Severity score. The data-driven component (Certainty and Occurrence) are simultaneously evaluated and intended to demonstrate a given relationship between pooling parameters and each process step relevant outputs. This way, the data-driven component leverages the extensive SME's knowledge and arguments with process-specific data to provide insights if additional control measurements are required to enhance the overall control system.

Figure 28.6 describes the approach used to rank the Certainty component. From the online UV elution profiles the absorbance values at which the

TABLE 28.7

Collection of Input Parameters for Purification #01. Initial Scientific Evaluation Supported by SME Knowledge and Validation Studies

Input Category	Process Parameters	PAR/NOR (Dossier)	Scientific Rationale
Equipment	Temperature	5 ± 4 (°C)	Evaluation of temperature effect in specific outputs not fully documented
	Column volume	30–40 L	Description and harmonization of column packing available
	Column diameter	42–46 cm	
	Compression factor	ND	
	-//-		
Materials	Resin lot-to-lot variability	ND	Assessed by functional testing to ensure proper performance within *load density* limits
	Resin capacity	1 g protein/L resin	
	Buffer stock solution molarity	5 mM	Potentially critical effect on product quality/process performance
	(...)		
Process	Flow rate	15–31 cm/h	Potentially critical effect on product quality/process performance
	pH equilibration buffer	6.9 ± 0.2	Potentially critical effect on product quality/process performance
	Volume wash buffer	1.5–3.0 CV	Potentially critical effect on product quality/process performance
	pH wash buffer	6.9 ± 0.2	Potentially critical effect on product quality/process performance
	pH Elution buffer	6.9 ± 0.2	Potentially critical effect on product quality/process performance
	Protein pooling (Star/ End UV Signal)	ND	Evaluation of pooling criteria effect in specific outputs not evaluated
	Protein pooling (End UV Signal)	ND	Moderate impact in step yield expected
	(...)		

Note: ND—Not determined

protein collection was initiated and ended were retrieved for all production batches considered in the assessment (Figure 28.6a-b).

A correlation analysis was performed against all relevant process outputs according to the following formula:

$$\text{Correlation}(X, Y) = \frac{\Sigma(x - \bar{x})(y - \bar{y})}{\sqrt{\Sigma(x - \bar{x})^2 \, \Sigma(y - \bar{y})^2}}$$

Where $\bar{X}$ and $\bar{Y}$ are input and output sample means, respectively. The correlation coefficients (Figure 28.6c) were used to rank the relationship between the pooling criteria and all relevant process outputs together with multivariate

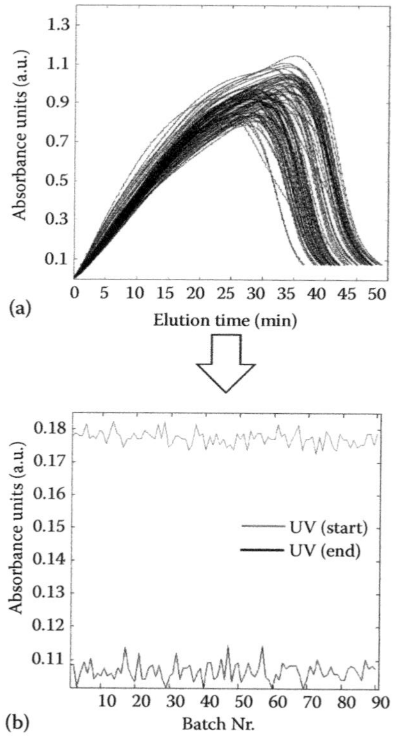

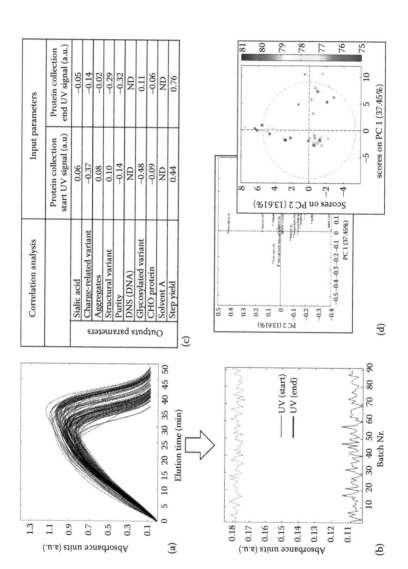

Correlation analysis		Input parameters	
		Protein collection start UV signal (a.u)	Protein collection end UV signal (a.u.)
Outputs parameters	Sialic acid	0.06	−0.05
	Charge-related variant	−0.37	−0.14
	Aggregates	0.08	−0.02
	Structural variant	0.10	−0.29
	Purity	−0.14	−0.32
	DNS (DNA)	ND	ND
	Glycosylated variant	−0.48	0.11
	CHO protein	−0.09	−0.06
	Solvent A	ND	ND
	Step yield	0.44	0.76

(c)

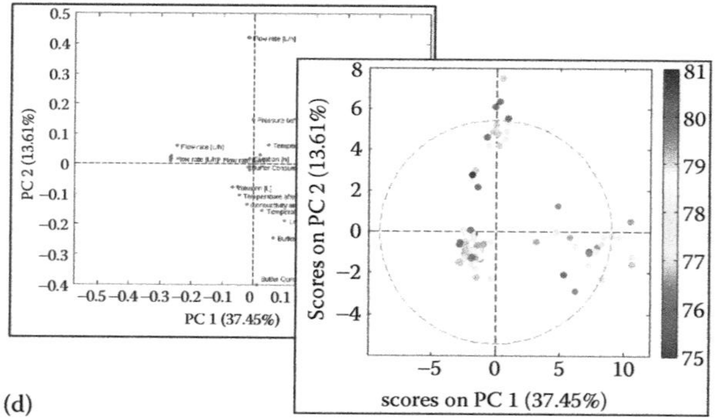

FIGURE 28.6
Estimation of the *observability* component for protein pooling: (a) online elution UV-based profiles from purification #01 step, (b) protein collection start/end absorbance units retrieved from online measurements, (c) correlation analysis with relevant output parameters, and (d) principal component analysis (PCA) of Purification #01 input parameters (score plot color scheme refers to step yield % over production history).

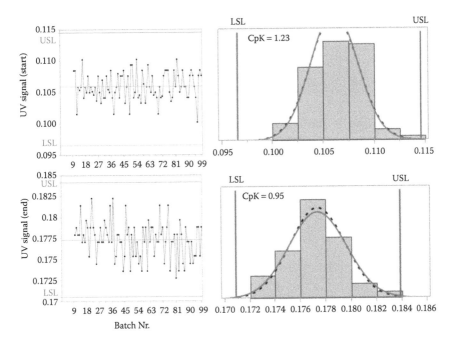

FIGURE 28.7
Estimation of the *controllability* component for protein pooling (Fractionation start/end).

data analysis evidences (Figure 28.6d) to identify potential interaction effects within input parameters potentially critical to specific outputs. The consistency of the protein pooling procedure was evaluated through process capability index (CpK) calculation (Figure 28.7). Since no PAR/NOR are defined in the dossier, LCL and UCL were estimated as

$$LCL = \bar{X} - k\sigma;\ UCL = \bar{X} + k\sigma$$

where:
  $\bar{X}$ is the mean of individual measurements
  k is the number of standard deviations
  σ is the process standard deviation

With all elements in place it is thus possible to rank Severity, Certainty and Occurrence scores toward CPP identification, gap analysis of the current control system, and prioritization of follow-up actions, if required (Table 28.8).

Stage 3—Gap Analysis and Follow Up Actions for Purification #01

The knowledge and risk-assessment should be documented and used to support a designated revision of the control system while investigations and further studies are taking place. Results of the studies should be summarized in

TABLE 28.8

CPP Identification for Process Relevant Outputs Based on SC Scores. Risk Ranking and Filtering of SCO Scores for Gap Analysis of the Current Control System and Prioritization of Follow-Up Actions

Process Phase/Step	Input Parameter	Output Parameter	Severity (S)	Knowledge Source	Certainty (C)	Knowledge Source	Occurrence (O)	Knowledge Source	SC Score	SCO Score
Purification #01 (Elution)	Protein collection start	Size-related variants	6	SME/Platform knowledge	2	Internal Project E2E	2	Internal Project E2E	CPP	24
		Charge-related variants (Isoforms)	10	SME/Platform knowledge	2	Multivariate Process Analysis	2	Multivariate Process Analysis	CPP	40
		Glycosylated variants	10	SME/Internal Project	2	Validation Report	2	CpK = 1.23	CPP	40
		Structural variants	6	SME/Internal Project	2	Solvent A clearance demonstrated by spiking studies	2		CPP	24
		Sialic acid	6	SME/Internal Project	2		2		CPP	24
		Purity	2	SME/Platform knowledge	2		2		Non-CPP	8
		DNS (DNA)	2	SME/Platform knowledge	2		2		Non-CPP	8
		CHO protein	2	SME/Platform knowledge	2		2		Non-CPP	8
		Solvent A	2	Validation report	2		2		Non-CPP	8
		Step yield	6	SME/Platform knowledge	2		2		CPP	24

(Continued)

TABLE 28.8 (*Continued*)

CPP Identification for Process Relevant Outputs Based on SC Scores. Risk Ranking and Filtering of SCO Scores for Gap Analysis of the Current Control System and Prioritization of Follow-Up Actions

Process Phase/Step	Input Parameter	Output Parameter	Severity (S)	Knowledge Source	Certainty (C)	Knowledge Source	Occurrence (O)	Knowledge Source	SC Score	SCO Score
Purification #01 (Elution)	Protein collection end	Size-related variants	6	SME/Platform knowledge	2	Internal Project E2E	6	Internal Project E2E	CPP	72
		Charge-related variants (Isoforms)	10	SME/Platform knowledge	2	Multivariate Process Analysis	6	Multivariate Process Analysis CpK = 0.95	CPP	120
		Glycosylated variants	10	SME/Internal Project	2	Validation Report	6		CPP	120
		Structural variants	6	SME/Internal Project	2	Solvent A clearance demonstrated by spiking studies	6		CPP	72
		Sialic acid	6	SME/Internal Project	2		6		CPP	72
		Purity	6	SME/Platform knowledge	2		6		CPP	72
		DNS (DNA)	2	SME/Platform knowledge	2		6		Non-CPP	24
		CHO protein	2	SME/Platform knowledge	2		6		Non-CPP	24
		Solvent A	2	Validation report	2		6		Non-CPP	24
		Step yield	10	SME/Platform knowledge	6		6		CPP	360

TABLE 28.9

Risk Evaluation and Follow-Up Actions for Protein Pooling Criteria Optimization (Purification #01)

Process Phase/Step	Input Parameter	Output Parameter	Severity (S)	Certainty (C)	Occurrence (O)	SC Score	SCO Score	Risk Evaluation	Risk Mitigation Strategy/ Recommended Actions
Purification #01 (Elution)	Protein collection end	Step yield	10	6	6	CPP	● 360	Platform and process knowledge confirm moderate influence of protein pooling criteria in quality attributes. Due to the steepness of the end part of the elution peak variations in the end-point protein collection significantly impact protein yield	Improvement of fraction collection trigger is recommended. Further investigate baseline checks and auto-zero procedures to minimize variability in protein collection
	Protein collection end	Charge-related variants (Isoforms)	10	2	6	CPP	120		
	Protein collection end	Glycosylated variants	10	2	6	CPP	120		
	Protein collection end	Size-related variants	6	2	6	CPP	72		
	Protein collection end	Structural variants	6	2	6	CPP	72		
	Protein collection end	Sialic acid	6	2	6	CPP	72		
	Protein collection end	Purity	6	2	6	CPP	72		

(Continued)

TABLE 28.9 (Continued)

Risk Evaluation and Follow-Up Actions for Protein Pooling Criteria Optimization (Purification #01)

Process Phase/Step	Input Parameter	Output Parameter	Severity (S)	Certainty (C)	Occurrence (O)	SC Score	SCO Score	Risk Evaluation	Risk Mitigation Strategy/Recommended Actions
	Protein collection start	Charge-related variants (Isoforms)	10	2	2	CPP	40	Input parameter not expected to have an effect in process outputs	No further actions required
	Protein collection start	Glycosylated variants	10	2	2	CPP	40		
	Protein collection start	Size-related variants	6	2	2	CPP	24		
	Protein collection start	Structural variants	6	2	2	CPP	24		
	Protein collection start	Sialic acid	6	2	2	CPP	24		
	Protein collection start	Step yield	6	2	2	CPP	24		
	Protein collection end	DNS (DNA)	2	2	6	Non-CPP	24		
	Protein collection end	CHO protein	2	2	6	Non-CPP	24		
	Protein collection end	Solvent A	2	2	6	Non-CPP	24		

(Continued)

TABLE 28.9 (*Continued*)

Risk Evaluation and Follow-Up Actions for Protein Pooling Criteria Optimization (Purification #01)

Process Phase/Step	Input Parameter	Output Parameter	Severity (S)	Certainty (C)	Occurrence (O)	SC Score	SCO Score	Risk Evaluation	Risk Mitigation Strategy/ Recommended Actions
	Protein collection end	Purity	2	2	2	Non-CPP	8		
	Protein collection end	DNS (DNA)	2	2	2	Non-CPP	8		
	Protein collection start	CHO protein	2	2	2	Non-CPP	8		
	Protein collection start	Solvent A	2	2	2	Non-CPP	8		

follow-up reports, including a brief outline on the risk evaluation and mitigation actions implemented (Table 28.9).

Stage 4—Revised Control System—Opportunities for Improvement

A specific mitigation action derived from a high priority SCO score on Table 28.9, with indication of CPP to be manipulated, associated ranges, means of control, and expected impacted CQAs and GMP plan for action, aligned with the company PQS.

The current approach contributes to efficient risk management by targeting process robustness while providing the tools for continuous monitoring and ongoing risk assessment.

Stage 5—Continued Process Verification and Lifecycle Management

The capabilities to collect all CPPs, CQAs, and control actions taken according to the revised control strategy; aggregating all that information and monitoring the correlations found during the legacy criticality assessment for each new lot produced; revisiting and updating the risk management and ranking; computing performance statistics (e.g., Cpk) and deciding on improvement opportunities within the current revised control strategy and company PQS.

Bridging scientific knowledge and process derived information will increase the effectiveness of the validation program, providing the necessary evidence that the process is fully understood, well controlled, and performing in a consistent manner. The resulting documents should represent a body of work to support nonconformance investigations, postapproval changes, and/or to address questions arising during inspections or regulatory reviews.

References

1. FDA Guidance for Industry—Process Validation: General Principles and Practices, January 2011.
2. EMA Guideline on Process Validation for Finished Products—Information and data to be provided in regulatory submissions, February 2014.
3. EudraLex, Vol. 4—EU Guidelines for Good Manufacturing Practice for Medicinal Products for Human and Veterinary Use. Annex 15: Qualification and Validation, 2015.
4. ICH Guideline Q7, Good Manufacturing Practice Guide for Active Principle Ingredients, November 2000.
5. Pharmaceutical cGMPs for the 21st Century—A Risk-Based Approach, FDA, 2002.
6. ICH Guideline Q11, Development and Manufacture of Drug Substances, May 2012.

7. ICH Guideline Q8(R2), Pharmaceutical Development. Int. Conf. Harmonization, November 2009.
8. ICH Guideline Q9, Quality Risk Management. Int. Conf. Harmonization, June 2006.
9. ICH Guideline Q10, Pharmaceutical Quality System, Int. Conf. Harmonization, April 2009.
10. ICH Guideline Q6B, Specifications: Test Procedures and Acceptance Criteria for Biotechnology/Biologic Products, March 1999.
11. ICH Guideline Q12, Technical and Regulatory Considerations for Pharmaceutical Product Lifecycle (concept paper), September 2014.
12. ASTM E2500-07: Guide for Specification, Design and Verification of Pharmaceutical and Biopharmaceutical Manufacturing Systems and Equipment, 2007.
13. ASTM E2281-03: Standard Practice for Process Measurement and Capability Indices, 2008
14. ASTM E2709-09: Standard Practice for Demonstrating Capability to Comply with a Lot Acceptance procedure, 2014.
15. ASTM E2537-08: Standard Guide for the Application of Continuous Quality Verification to Pharmaceutical and Biopharmaceutical Manufacturing, 2008.
16. FDA Guidance for Industry—Established Conditions: Reportable CMC Changes for Approved Drug and Biologic Products (draft guidance), May 2015.
17. Seymour, M., Ramnarine, E., Baker, D., Vinther, A. (2015). High at the End of the Tunnel for PAC (Post-Approval Change) Complexity. PDA Letter Nov/Dec 48–49.
18. Boyer, M., Gampfer, J., Zamamiri, A., Payne, R. (2016). A roadmap for the implementation of continued process verification. *PDA J. Pharm. Sci. Tech.*, 70, 282–292.
19. Gouveia, F.F., Campos, M.B., Felizardo, P.M. (2016). Quasi-QbD Frameworks for Commercial Products: Established-Conditions. IFPAC Annual Meeting Washington DC, Jan 24–27, USA.
20. Agarwal, A., Hayduk, E. (2016). Retrospective implementation of quality by design for legacy commercialized enzyme replacement therapies. Cell Culture Engineering XV–Palm Springs (CA), May 8 to 13, USA.
21. Bozzone, S. (2016). Continuous Use of Development Data in Validation Plans and the Process Lifecycle. PDA Annual Meeting San Antonio (TX) March 14–16, USA.
22. De Long, M., Blaettler, D., Goerke, A.R. (2016). Development of Process Performance as a Quality Metric: Looking Back to Indicate Future Risk. PDA Annual Meeting San Antonio (TX) March 14–16, USA.
23. Gronemeyer, P., Ditz, R., Strube, J. (2014). Trends in upstream and downstream process development for antibody manufacturing. *Bioengineering*, 1, 188–212. doi:10.3390/bioengineering1040188.
24. WHO Technical Report Series No. 996, 2016, Annex 3 WHO good manufacturing practices for biological products.

Conclusion

Nuala Calnan, Martin J. Lipa, Paige E. Kane, and Jose C. Menezes

"We shall not cease from exploration
And the end of all our exploring
Will be to arrive where we started
And know the place for the first time."

—T.S. Eliot, Four Quartets

This lifecycle approach to knowledge excellence in the biopharmaceutical industry began by setting out the reasons why knowledge management (KM) is good for business. Chapter 1 affirmed that there is a renewed effort within the industry to lower the cost of goods and reduce inventory levels by enhancing the efficiency, agility, and velocity across the complex and fragmented supply chain. Faced with the challenges of increased mergers and acquisitions, outsourcing activities, localization of manufacturing in the emerging markets, and rapidly changing requirements in product demand the authors assert that it is now imperative for the industry to transform. A key element of this transformation is the recognition that critical knowledge about product performance and patient risk is created at each node in the supply chain right across the lifecycle of the product, starting as early as predevelopment. This largely untapped resource of organizational knowledge, and more importantly each organization's ability to flow that knowledge to where and when it is needed, presents an enormous opportunity for each business to unlock.

The 28 chapters herein have explored the importance and the role of managing the free flow of knowledge, in a variety of complex environments, to overcome these challenges in a manner that ultimately benefits the patients and stakeholders of the company, including employees and shareholders.

Therefore, as the industry accelerates toward the third decade of twenty-first century on this knowledge excellence journey, we can revisit the *House of Knowledge Excellence* and recognize the contributions of the different authors and their organizations. The case studies provided in Chapters 13, 14, and 18 through 28 have clearly highlighted different aspects of the current state of the journey. The four pillars of people, processes, technology, and governance serve their purpose, almost like an Ishikawa diagram, to organize the different elements that underpin this journey. Consolidating the messages from industry practitioners, thought leaders, and academics, the current state of knowledge management, and use, has been described.

It may be worth reflecting on the journey so far in order to envisage some of the key influences poised to alter the course.

The biopharmaceutical industry embarked on its unprecedented drive for modernization at the dawn of the twenty-first century and started largely with a technology-driven effort to achieve Quality by Design (QbD) by bringing better science into manufacturing, aimed at achieving and sustaining more predictable quality from early development in R&D, through process development and into commercial manufacturing. However, it soon became clear through these initial efforts that there were fundamental challenges regarding how the industry was leveraging its knowledge and how incomplete the understanding was about core aspects of the industry's products and processes. Indeed, risk management was often engaged as a way to harvest prior knowledge (and tacit knowledge) in order to support decisions related to poorly understood systems. Then in 2008, with the publication of ICH Q10, knowledge management entered the industry lexicon.

Those two realities—uncertainty in decision-making and poor retention and usage of prior knowledge over the lifecycle—were both directly addressed in the regulatory framework that emerged from ICH with the publication of several key guidance documents over the past decade, that is, ICH Q8 through Q11 (see Figure A.1). However, that effort was focused almost entirely on quality-related aspects and linking the process to the product, whereas the safety and efficacy aspects (as experienced by a very heterogeneous patient population) were assumed to be assured by reliably matching a prespecified quality target product profile (QTPP).

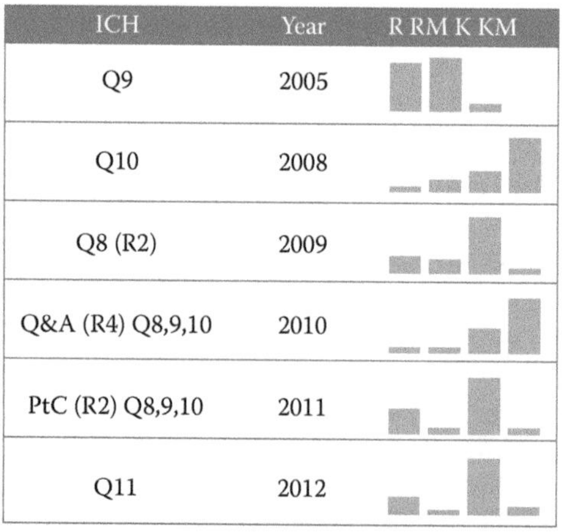

ICH	Year	R RM K KM
Q9	2005	
Q10	2008	
Q8 (R2)	2009	
Q&A (R4) Q8,9,10	2010	
PtC (R2) Q8,9,10	2011	
Q11	2012	

FIGURE A.1
ICH citations for risk, risk management, knowledge and knowledge management. Normalized instances by document to allow comparisons.

Figure A.1 depicts 872 instances of risk (R), risk management (RM), knowledge (K), and knowledge management (KM), in the most important ICH quality guidelines, normalized by document to allow comparisons. The regulatory drive toward promoting KM use by the industry over this period is clear.

By the time of publication of ICH Q11 *development and manufacture of drug substances* in 2012, leaders in the industry had matured in their use of the new quality-as-a-manufacturing-science instruments, and with that grew their expectations for the new regulatory initiatives purportedly available. But then a new trend emerged, again led by FDA, that questioned the foundations on which the whole quality paradigm was based, that is, the integrity of the data itself! What followed was a global regulatory focus on data quality and its integrity, which questioned how safe it might be for organizations to base decisions on such data, and what levels of trust might be necessary to assure these decisions were sound and made in the best interest of the patient. Today, the data integrity expectations extend to the whole organizational culture responsible for generating and managing the data resources on which quality is assured for both current marketed products and new highly sophisticated therapies.

So, what started as a technology-driven quality initiative has now emerged as a much broader transformation, which is questioning the very roots of what it means to be a "learning organization," one that possesses a culture of quality and thinks holistically about their processes, their products, and the patients they serve.

This is just the beginning of the knowledge excellence journey, what follows are just some of the potential waypoints of the road ahead, as the KM field accelerates exponentially in the coming years.

ICH Q12 Lifecycle Management

As new considerations related to lifecycle management and the real adoption of QbD emerge, one possible trend will drive capabilities to deliver enhanced levels of integration of past, present, and future knowledge about a product. This can be seen from FDA's launch of their initiative for companies to revisit their "quality-commitments" for their commercial marketed products in the "established conditions" draft guidance, issued in May 2015. In Chapter 28, Roche showed how CPPs and CQAs can be defined from the manufacturing history available for a legacy product (i.e., data- and evidence-based), and then how they can be combined with formal knowledge-based risk assessments to derive significant improvements in the field control strategy. There is therefore a regulatory expectation that companies can retrospectively consider the use of key QbD elements (e.g., QRM) to address and justify improvements in their current legacy control strategies, supporting post-approval changes using risk- and knowledge-based approaches. Managing science-based development and manufacturing, with QbD elements, over the entire product lifecycle will require a strong effort from the companies in knowledge management.

QRM and KM

QRM is to KM what PAT is to QbD! Although presently the industry may lack better ways to manage knowledge through sophisticated tools and platforms, risk-management tools can provide very effective capabilities to capture, retain, and support knowledge-driven lifecycle activities. One can preserve the context and rationale of different changes or process optimization by aggregating multiple formal risk-assessments performed over the lifecycle. For example, the continuous improvement rationale for a particular product and process can be leveraged across multiple instances of products using similar processes that a company might have. Indeed, QRM can even be a true enabler of establishing the desired quality culture by supporting the use of knowledge-based risk assessment activities to help transform an organization from a quality-by-testing or compliance-driven organization into a "learning organization."

Novel Therapies and Accelerated Submissions

Knowledge management will be key in supporting complex and accelerated new product filings. The account in Chapter 21 by Merck & Co., Inc. reveals this coming trend, as have been seen from other companies such as Vertex, who are producing drug products for small patient populations with very agile and state-of-the-art continuous manufacturing platforms.

Agile Filings

Better filings, which are easier to manage over the lifecycle, possessing greater clarity in terms of quality commitments, and presented in a more transparent format to facilitate ease of communication with regulatory authorities, can be achieved through KM. This includes how the drug master files (or e-CTDs) are put together, in a rigorous but "Lean" way, in terms of both the type and quality of the information provided to support the company's quality commitments.

Quality Metrics

The ongoing initiative by FDA in the area of quality metrics, intended to distinguish the risk profile of capable and compliant companies from those unable to consistently meet quality standards, has triggered much recent debate (Villax, 2013, 2017). The need to reliably report key quality data, from across the supply chain, will drive knowledge management capabilities and technology developments. Indeed, the concurrent initiative on quality culture (ISPE, 2015 and 2016) is likely to have a significant impact on how the industry will transform from the pre-QbD into the post-Q12 years.

Excellence

Through KM, companies can achieve Class-A operational excellence not only for an individual product but for the portfolio and technology platforms they are operated in. Such efforts will be a key business aspect for organizations striving for excellence. The megatrend behind Industry 4.0 and the role that data-generating devices and real-time information management will have is very impressive within the current KM context. The cumulative amount of patents related to Industry 4.0 topics has grown 12-fold in the last five years, with ca. 5100 patents granted in 2015 (Williams, 2016). New manufacturing paradigms connecting the physical and the augmented-reality worlds with better decision-making capabilities using big-data approaches will bring with it better ways of extracting insights using analytics. This is a promising and fertile field combining those new patents with the so-called internet-of-things. The digitalization of industry, and in particular of pharma and biotech, is progressing at an unprecedented pace. The most advanced economies are actually promoting the acceleration through their digitization. The usage of digital interactions between users (patients) and suppliers (pharma companies) as well as the level of digitalization/smart automation in manufacturing in the biopharma sector has been recently discussed in a McKinsey Co. report (Bughin et al. 2016). A true army, both in size and specialization, of knowledge- and intellectually intensive workers is now available, and they need better tools and platforms to develop their capabilities and potential impact to the fullest. This will certainly be a major area for companies to support.

Post-Approval Market Surveillance

Post-approval surveillance presents a real challenge to the four-pillar framework. Capturing all the information related to quality, safety, and efficacy of medicines on the market and transferring that knowledge back into the companies will clearly be a difficult undertaking. Furthermore, it represents an area where a lot of integrated groundwork will have to happen beforehand. Using that knowledge to influence any necessary optimization programs and retaining that knowledge for future products within the same technology platform will drive innovation in knowledge management. This was highlighted in Chapter 7, where findings from a recent knowledge management (KM) study by the Irish medicines regulator was undertaken to create new knowledge from the available data about sources of product quality defects and recalls, to examine potential solutions and prevention strategies for these market-based failures. This is just a start of where market surveillance will go. A visionary thought would be to really enable effective leveraging of prior knowledge for more new product development from similar marketed products; namely, being able to base formal elements of the filing on evidence from comparable products in terms of their targets and/or functioning modes.

Partnerships and CoPs

A well-networked industry that cooperates and shares best practices and experiences through communities of practice (CoP) with a wide variety of organizations such as ISPE, PDA, and other public fora as well as with the regulators through pilot programs, joint initiatives, and bidirectional consultations, will thrive. The coming years will see more and more new drugs coming to market out of collaborative efforts, requiring knowledge to flow internally and across institutional boundaries. Some new products will really require a much better integration than those in the past, such as advanced cell therapies. There are therapies and technologies that require the close proximity of private and public entities, not only during development but also actually throughout the entire lifecycle. This will drive the need for KM by effective and agile management, to ensure the most in terms of patient outcomes.

Patient at the Center

Drawing knowledge from the marketplace, such as from electronic patient records systems, patient outcomes, and through patient communities, was addressed in Chapter 5. Showing how unleashing the power in the information that patients have access to nowadays, before and after medical diagnosis, and during their treatment, improves the outcomes for all. This is the new reality for the medical and patient communities, but the industry seems, as yet, to be absent from that space.

Closing

The current times are extraordinary. As the industry transforms to rapidly adjust to a new world, a new reality will emerge, one that supports the future of knowledge and knowledge management making true the promise of "learning organizations."

References

Bughin, J., Hazan, E., Manyika, J., Woetzel, J. (2016). Digital Europe: Pushing the Frontier, Capturing the Benefits, June 2016 (www.mckinsey.com/business-functions/digital-mckinsey/our-insights/digital-europe-realizing-the-continents-potential) (accessed March 13, 2017).

Williams, Z.D. (2016). https://iot-analytics.com/industrial-technology-trends-industry-40-patents-12x (accessed June 30, 2016).

ISPE. (2015). Quality Metrics Initiative: Wave 1 Report.

ISPE. (2016). Quality Metrics Initiative: Wave 2 Report.

Villax, G. (2013). "The Dean's List" www.hovione.com/sites/default/files/assets/files/fda-deans-list-guy-villax.pdf (FDA needs to step it up. (2013). *Chem. Eng. News*, 91(16), 3).

Villax, G. (2017). Do patients want capsules full of compliance, or full of quality? Why quality culture wins every time. QRM Summit 2017, May 10–11, Lisbon, (Portugal).

Index

Note: Page numbers followed by f and t refer to figures and tables, respectively.

9 780367 875725